2020 IEEE Silicon Nanoelectronics Workshop (SNW 2020)

Honolulu, Hawaii, USA
13 – 14 June 2020

IEEE Catalog Number: CFP20SNW-POD
ISBN: 978-1-7281-9736-4

**Copyright © 2020 by the Institute of Electrical and Electronics Engineers, Inc.
All Rights Reserved**

Copyright and Reprint Permissions: Abstracting is permitted with credit to the source. Libraries are permitted to photocopy beyond the limit of U.S. copyright law for private use of patrons those articles in this volume that carry a code at the bottom of the first page, provided the per-copy fee indicated in the code is paid through Copyright Clearance Center, 222 Rosewood Drive, Danvers, MA 01923.

For other copying, reprint or republication permission, write to IEEE Copyrights Manager, IEEE Service Center, 445 Hoes Lane, Piscataway, NJ 08854. All rights reserved.

*** This is a print representation of what appears in the IEEE Digital Library. Some format issues inherent in the e-media version may also appear in this print version.

IEEE Catalog Number: CFP20SNW-POD
ISBN (Print-On-Demand):   978-1-7281-9736-4
ISBN (Online):   978-1-7281-9735-7
ISSN: 2161-4636

Additional Copies of This Publication Are Available From:

Curran Associates, Inc
57 Morehouse Lane
Red Hook, NY 12571 USA
Phone: (845) 758-0400
Fax: (845) 758-2633
E-mail: curran@proceedings.com
Web: www.proceedings.com

TABLE OF CONTENTS

HAFNIUM OXIDE AS AN ENABLER FOR COMPETITIVE FERROELECTRIC DEVICES 1
Thomas Mikolajick, Halid Mulaosmanovic, Patrick Lomenzo, Michael Hoffmann, Stefan Slesazeck, Uwe Schroeder

STATE-OF-THE-ART IN SILICON QUANTUM COMPUTER DEVELOPMENT 3
Kohei M. Itoh

PERFORMANCE ASSESSMENT OF BEOL-INTEGRATED HFO$_2$-BASED FERROELECTRIC
CAPACITORS FOR FERAM MEMORY ARRAYS 5
L. Grenouillet, T. Francois, J. Coignus, N. Vaxelaire, C. Carabasse, F. Triozon, C. Richter, U. Schroeder, E. Nowak

RECORD FAST POLARIZATION SWITCHING OBSERVED IN FERROELECTRIC
HAFNIUM OXIDE CROSSBAR ARRAYS 7
Xiao Lyu, Mengwei Si, Pragya R. Shrestha, Jason P. Campbell, Kin P. Cheung, Peide D. Ye

ON THE PHYSICAL ORIGINS OF TIME-DEPENDENT STEEP SS IN FEFET 9
Xiuyan Li, Yulong Dong, Jingquan Liu

FERROELECTRIC TUNNEL JUNCTION OPTIMIZATION BY PLASMA-ENHANCED
ATOMIC LAYER DEPOSITION 11
Jae Hur, Yuan-Chun Luo, Panni Wang, Nujhat Tasneem, Asif Islam Khan, Shimeng Yu

PROCESS AND STRUCTURE CONSIDERATIONS FOR THE POST FINFET ERA 13
Chun-Jung Su, Po-Jung Sung, Kuo-Hsing Kao, Yao-Jen Lee, Wen-Fa Wu, Wen-Kuan Yeh

INTEGRATION OF ALD HIGH-K DIPOLE LAYERS INTO CMOS SOI NANOWIRE FETS
FOR BI-DIRECTIONAL THRESHOLD VOLTAGE ENGINEERING 15
Wonil Chung, Dongqi Zheng, Wei-E Wang, Mark Rodder, Peide D. Ye

VERTICAL INAS/INGAASSB/GASB NANOWIRE TUNNEL FETS ON SI WITH DRAIN
FIELD-PLATE AND EOT = 1 NM ACHIEVING S$_{MIN}$ = 32 MV/DEC AND G$_M$/I$_D$ = 100 V^{-1} 17
Abinaya Krishnaraja, Johannes Svensson, Lars-Erik Wernersson

PROBABILISTIC COMPUTING BASED ON SPINTRONICS TECHNOLOGY 21
Shunsuke Fukami, William A. Borders, Ahmed Z. Pervaiz, Kerem Y. Camsari, Supriyo Datta, Hideo Ohno

A SELF-ALIGN GATE-LAST RESISTIVE GATE SWITCHING FINFET NONVOLATILE
MEMORY FEASIBLE FOR EMBEDDED APPLICATIONS 23
W. Y. Yang, E. R. Hsieh, C. H. Cheng, Steve S. Chung

DESIGN AND ANALYSIS OF CORE-GATE SHELL-CHANEL 1T DRAM 25
Md. Hasan Raza Ansari, Jae Yoon Lee, Seongjae Cho, Byung-Gook Park

FLASH MEMORY BASED COMPUTING-IN-MEMORY TO SOLVE TIME-DEPENDENT
PARTIAL DIFFERENTIAL EQUATIONS 27
Yang Feng, Xuepeng Zhan, Jiezhi Chen

A NOVEL HIGH-DENSITY AND LOW-POWER TERNARY CONTENT ADDRESSABLE
MEMORY DESIGN BASED ON 3D NAND FLASH 29
H. Z. Yang, P. Huang, R. Z. Han, Y. C. Xiang, Y. Feng, B. Gao, J. Z. Chen, L. F. Liu, X. Y. Liu, J. F. Kang

SILICON QUANTUM DOT DEVICES FOR SPIN-BASED QUANTUM COMPUTING 31
  Tetsuo Kodera

INTEGRATED CIRCUITS COMPOSED OF NANOWIRE AND SINGLE-ELECTRON
TRANSISTORS OPERATING AT ROOM TEMPERATURE.. 33
  *Tomoko Mizutani, Kiyoshi Takeuchi, Takuya Saraya, Masaharu Kobayashi, Toshiro
  Hiramoto*

A STUDY OF SINGLE-ELECTRON TUNNELING FUNCTIONALITIES IN HIGHLY-DOPED
SILICON-ON-INSULATOR JUNCTIONLESS TRANSISTORS 35
  *T. Teja Jupalli, G. Prabhudesai, M. Hasan, A. Debnath, P. Jeevan Kumar, M. Tabe, D.
  Moraru*

DOUBLE-GATE SINGLE-ELECTRON DEVICES FORMED BY SINGLE-LAYERED FE
NANODOT ARRAY 37
  *Takayuki Gyakushi, Yuki Asai, Beommo Byun, Ikuma Amano, Atsushi Tsurumaki-Fukuchi,
  Masashi Arita, Yasuo Takahashi*

A NOVEL HIGH-DENSITY AND LOW-POWER TERNARY CONTENT ADDRESSABLE
MEMORY DESIGN BASED ON 3D NAND FLASH.. 39
  *H. Z. Yang, P. Huang, R. Z. Han, Y. C. Xiang, Y. Feng, B. Gao, J. Z. Chen, L. F. Liu, X. Y.
  Liu, J. F. Kang*

CHARGE-ASSISTED RECOVERY AND DEGRADATION IN CHARGE-TRAPPING 3D
NAND FLASH MEMORY, EXPERIMENTAL EVIDENCES AND THEORETICAL
PERSPECTIVES.. 41
  Xiaolei Ma, Rui Cao, Fei Wang, Xuepeng Zhan, Jiezhi Chen

APPROXIMATE 3D-TLC NAND FLASH WRITE WITH INITIAL ERROR INJECTION FOR
APPLICATION-LEVEL RELIABILITY IMPROVEMENT OF MACHINE LEARNING-BASED
COMPUTING 43
  Shun Suzuki, Hiroki Aihara, Keita Mizushina, Shin Yamaguchi, Ken Takeuchi

A NOVEL QUASI-SLC(QSLC) PROGRAM/ERASE SCHEME IN ULTRA-DENSIFIED
CHARGE-TRAPPING 3D NAND FLASH MEMORY TO ENHANCE SYSTEM LEVEL
PERFORMANCE 45
  Yuanpeng Li, Qianwen Wang, Xuepeng Zhan, Menghua Jia, Rui Cao, Jiezhi Chen

3840X RELIABILITY ENHANCED ROBUST NAND FLASH OPTIMIZED TO STORE
WEIGHT DATA FOR OBJECT DETECTION AND SEMANTIC SEGMENTATION OF SELF-
DRIVING CAR AT HIGH TEMPERATURE.. 47
  Keita Mizushina, Shun Suzuki, Hiroki Aihara, Ken Takeuchi

SPATIAL COLOR-PERCEIVED DATA CONTROL OF 3D-TLC NAND FLASH FOR IMAGE
DECTECTION... 49
  Chihiro Matsui, Shun Suzuki, Ken Takeuchi

PERFORMANCE ENHANCEMENT OF BF_2^+ IMPLANTED POLY-SI JUNCTIONLESS
TRANSISTORS BY BORON SEGREGATION AND FLUORINE EFFECT 51
  Min-Ju Ahn, Takuya Saraya, Masaharu Kobayashi, Toshiro Hiramoto

CRYSTALLINITY EFFECT ON RELIABILITY OF SIDEWALL DAMASCENED NANOWIRE
POLY-SI GAA FETS 53
  Chuan-Hui Shen, Wei-Yen Chen, Chun-Chih Chung, Yu-En Huang, Tien-Sheng Chao

SUPERIOR SUBTHRESHOLD SLOPE OF GATE-ALL-AROUND (GAA) P-TYPE POLY-SI JUNCTIONLESS NANOWIRE TRANSISTORS WITH HIGHLY SUPPRESSED GRAIN BOUNDARY DEFECTS 55

Min-Ju Ahn, Takuya Saraya, Masaharu Kobayashi, Toshiro Hiramoto

CHARACTERISTICS OF DUAL-GATED POLY-SI JUNCTIONLESS NANOWIRE TRANSISTORS WITH ASYMMETRICAL SOURCE/DRAIN OFFSETS 57

You-Tai Chang, Ruei-Jen Wu, Kang-Ping Peng, Chun-Jung Su, Pei-Wen Li, Horng-Chih Lin

GE AND GESN-BASED NANO-ELECTRONIC AND PHOTONIC DEVICES 59

Xiao Gong, Ying Wu, Dian Lei, Shengqiang Xu, Kaizhen Han

POLARIZATION DEPENDENCE OF INCIDENT ANGLE SENSITIVITY IN SOI PHOTODIODE WITH 2D HOLE ARRAY GRATING 63

Anitharaj Nagarajan, Shusuke Hara, Hiroaki Satoh, Aruna Priya Panchanathan, Hiroshi Inokawa

PHOTON-NUMBER STATISTICS OBSERVED BY SOI MOSFET SINGLE-PHOTON DETECTOR WITH REAL-TIME SIGNAL PROCESSING 65

Revathi Manivannan, Hiroaki Satoh, Hiroshi Inokawa

SCALED TRANSISTORS WITH 2D MATERIALS FROM THE 300MM FAB 67

I. Asselberghs, T. Schram, Q. Smets, B. Groven, S. Brems, A. Phommahaxay, D. Cott, E. Dupuy, D. Radisic, J.-F. De Marneffe, A. Thiam, W. Li, K. Devriendt, A. Gaur, T. Maurice, D. Lin, P. Morin, I. P. Radu

HALF-MESHED AND FULLY-MESHED SUSPENDED GRAPHENE FOR TRANSPORT GAP ENGINEERING 69

Fayong Liu, Manoharan Muruganathan, Shinichi Ogawa, Yukinori Morita, Marek Schmidt, Hiroshi Mizuta

A COMPUTATIONAL PERFORMANCE EVALUATION OF NEGATIVE-CAPACITANCE MOSFETS BASED ON ULTRA-THIN BODY SILICON AND MONOLAYER MOS$_2$ 71

Sheng Luo, Xiaoyi Zhang, Gengchiau Liang

DECONVOLUTION OF HOT CARRIER AND COLD CARRIER INJECTION IN ZNO TFTS 73

P. Bolshakov, R. A. Rodriguez-Davila, M. Quevedo-Lopez, C. D. Young

REALIZATION OF DIVERSE SPIKE-TIMING-DEPENDENT PLASTICITY WITH NANOSECOND TIMESCALE BASED ON METAL OXIDE RESISTIVE SWITCHING MEMORY 75

Ruiyi Li, Peng Huang, Yulin Feng, Zheng Zhou, Xiangyu Wang, Wensheng Shen, Xiangxiang Ding, Lifeng Liu, Xiaoyan Liu, Jinfeng Kang

UNSUPERVISED LEARNING ARCHITECTURE BASED ON SPIKE-TIMING-DEPENDENT PLASTICITY USING FLASH MEMORY SYNAPTIC DEVICES 77

Won-Mook Kang, Soochang Lee, Jangsaeng Kim, Byung-Gook Park, Jong-Ho Lee

EFFECTS OF THERMAL ANNEALING ON TA$_2$O$_5$ BASED CMOS COMPATIBLE RRAM 79

Somsubhra Chakrabarti, Jia Min Ang, Jia Rui Thong, Kunqi Hou, Mun Yin Chee, Putu Andhita Dananjaya, Desmond Loy Jia Jun, Yong Chiang Ee, Wen Siang Lew

HIGHLY SCALABLE 4F^2 CELL TRANSISTOR FOR FUTURE DRAM TECHNOLOGY 81

Kyung Kyu Min, Sungmin Hwang, Jong-Ho Lee, Byung-Gook Park

WEIGHTED SYNAPTIC BEHAVIORS OF HFON BASED RRAM DEVICE BY A NOVEL WAVEFORM MODULATION METHOD 83
Yuechi Ma, Ruiyi Li, Ao Yu, Zehao Wang, Xiangxiang Ding, Yulin Feng, Lifeng Liu

FERROELECTRIC FEW LAYER BLACK PHOSPHORUS FIELD-EFFECT TRANSISTORS FOR SRAM APPLICATION 85
Cheng-Hsien Yang, Yun-Fang Chung, Kuan-Ting Chen, Shu-Tong Chang

BULK-LIMITED EFFECT IN GRADUAL CONDUCTANCE SWITCHING BEHAVIOUR OF HFO$_X$-BASED MEMRISTIVE DEVICES FOR ANALOG SYNAPTIC DEVICE APPLICATIONS 87
Putu Andhita Dananjaya, Desmond Loy Jia Jun, Mun Yin Chee, Samuel Chow Chen Wai, Jia Min Ang, Kunqi Hou, Jia Rui Thong, Somsubhra Chakrabarti, Yong Chiang Ee, Wen Siang Lew

FLOATING-GATE TRANSISTOR AT CRYOGENIC TEMPERATURE: CHARACTERIZATION AND MODELLING OF TUNNELLING AND HOT ELECTRONS INJECTION 89
Michele Castriotta, Enrico Prati, Giorgio Ferrari

IMPACT OF STOPPING VOLTAGE AND HOPPING CONDUCTION ON THE OXYGEN VACANCY CONCENTRATION OF MULTI-LEVEL HFO$_2$-BASED RESISTIVE SWITCHING DEVICES 91
Desmond Jia Jun Loy, Putu Andhita Dananjaya, Somsubhra Chakrabarti, Kuan Hong Tan, Samuel Chen Wai Chow, Mun Yin Chee, Jia Rui Thong, Kunqi Hou, Jia Min Ang, Gerard Joseph Lim, Yong Chiang Ee, Eng Huat Toh, Wen Siang Lew

RU CONDUCTING FILAMENT BASED CROSS-POINT RESISTIVE SWITCHING MEMORY FOR FUTURE LOW POWER OPERATION 93
Siddheswar Maikap, Asim Senapati

INFLUENCE OF GATE TO DRAIN UNDERLAP ON NEGATIVE DIFFERENCIAL RESISTANCE IN FERROELECTRIC FET 95
Kitae Lee, Sihyun Kim, Byung-Gook Park

STUDY ON ETCH SLOPE IN FIN AND SOURCE/DRAIN ETCH PROCESS OF VERTICALLY-STACKED NANOSHEET GATE-ALL-AROUND MOSFET 99
Sihyun Kim, Kitae Lee, Byung-Gook Park

RELIABILITY ANALYSIS OF P-TYPE SOI FINFETS WITH MULTIPLE SIGE CHANNELS ON THE DEGRADATION OF NBTI 101
Tzuting Cho, Renrong Liang, Guofang Yu, Jun Xu

IMPACTS OF BIAXIAL TENSILE STRAIN IN DOUBLE-GATE TUNNELING FIELD-EFFECT-TRANSISTOR (DG-TFET) WITH A MONOLAYER WSE$_2$ CHANNEL 103
Qianwen Wang, Pengpeng Sang, Yuan Li, Jiezhi Chen

TOWARDS NOVEL CHANNEL DOPING PROFILES IN SHORT CHANNEL BULK MOSFETS FOR OFF-STATE CURRENT REDUCTION AND SUPERIOR CHANNEL ELECTROSTATICS 105
Harshit Kansal, Aditya Sankar Medury

A MOBILITY STRESS RESPONSE MODEL OF FINFET: SILICON VS GERMANIUM 107
Cheng-Hsien Yang, Yun-Fang Chung, Kuan-Ting Chen, Shu-Tong Chang

COULOMB-BLOCKADE CHARGE-TRANSPORT MECHANISM IN BAND-TO-BAND TUNNELING IN HEAVILY-DOPED LOW-DIMENSIONAL SILICON ESAKI DIODES 109
G. Prabhudesai, K. Yamaguchi, M. Tabe, D. Moraru

EFFECTS OF CO-DOPING ON THE TRANSPORT CHARACTERISTICS OF NANOSCALE N-TYPE SILICON-ON-INSULATOR TRANSISTORS 111
C. Pandy, A. Debnath, K. Yamaguchi, T. Teja Jupalli, G. Prabhudesai, Ramakrishnan V. N., Y. Neo, H. Mimura, D. Moraru

HIGH SPEED NANOANTENNA THERMOPILES FOR LONG-WAVE INFRARED DETECTION 113
Gergo P. Szakmany, Gary H. Bernstein, Edward C. Kinzel, Alexei O. Orlov, Wolfgang Porod

ADIABATIC CAPACITIVE LOGIC USING VOLTAGE-CONTROLLED VARIABLE CAPACITORS 115
Rene Celis-Cordova, Alexei O Orlov, Gregory L Snider

RESISTIVE APPROACH FOR EXTRACTION OF BIAS-DEPENDENT PARASITIC RESISTANCE, MOBILITY AND VIRTUAL GATE LENGTH IN GAN HEMT 117
Pragyey Kumar Kaushik, Sankalp Kumar Singh, Ankur Gupta, Ananjan Basu, Edward Yi Chang

UNDERSTANDING TRANSIENT RESPONSES OF SILICON NANOWIRE PHOTOCONDUCTORS 119
Yaping Dan, Jiajing He, Chulin Huang

OPTIMAL APPROACH TO SCALING OF THE NEMS FOR LOW STAND-BY CMOS APPLICATIONS 121
Sumit Saha, Sanjog Joshi, Tejas Naik, Mayank Goel, V. Ramgopal Rao, Maryam Shojaei Baghini

ELECTROLUMINESCENCE OF ER:O-DOPED NANO PN DIODE IN SILICON-ON-INSULATOR AND ITS CURRENT-VOLTAGE CHARACTERISTICS AT ROOM TEMPERATURE 123
Takafumi Fujimoto, Keinan Gi, Stefano Bigoni, Michele Celebrano, Marco Finazzi, Giorgio Ferrari, Takahiro Shinada, Enrico Prati, Takashi Tanii

PERFORMANCE INVESTIGATION OF UNIVERSAL GATES AND RING OSCILLATOR USING DOPING-FREE BIPOLAR JUNCTION TRANSISTOR 125
Abhishek Sahu, Abhishek Kumar, Shree Prakash Tiwari

TOWARDS A CHIP-SCALE MILLIMETER-WAVE SPECTRUM/SIGNAL ANALYZER USING SPIN-WAVE DIFFRACTION AND INTERFERENCE 127
H. Aquino, D. Connelly, A. Papp, A. Orlov, J. Chisum, G. H. Bernstein, W. Porod

Author Index

Hafnium oxide as an enabler for competitive ferroelectric devices

Thomas Mikolajick[1,2],

Halid Mulaosmanovic[1], Patrick Lomenzo[1], Michael Hoffmann[1], Stefan Slesazeck[1], and Uwe Schroeder[1]

[1]NaMLab gGmbH, Germany, [2]Institute of Semiconductors and Microsystems, TU Dresden, Germany,
Email: Thomas.mikolajick@namlab.com; thomas.mikolajick@tu-dresden.de

Abstract — Ferroelectric materials offer the promise to realize low power memory devices and show negative capacitance operation that could lead to novel electronic devices. Although intense research on realizing different memory device concepts based on three different readout schemes have been subject to intense research, the commercial success is limited to low density ferroelectric random access memories based on a direct capacitor readout. The complexity of integrating ferroelectric materials into CMOS processes has limited successful implementations. Ferroelectricity in hafnium oxide related material systems could overcome this limitations for memories and at the same time enable new devices based on negative capacitance.

Keywords: ferroelectric, memory, FeRAM, FeFET, FTJ

I. INTRODUCTION

Ferroelectrics are materials that have two stable polarization states that can be switched by applying an electric field. As such, ferroelectrics are ideal candidates to realize devices with two or more non-volatile states. However, typical ferroelectric crystals are quite complicated and impose severe restrictions with respect to their compatibility with standard CMOS fabrication. As a result, ferroelectrics are successfully used in speciality memory products since more than a quarter of a century, but are limited to niches were the specific advantage of a low power write together with non-volatility is a must. The unexpected discovery of ferroelectricity in doped hafnium oxide first reported in 2011 [1] revived the hopes for scalable and therefore competitive ferroelectric devices.

II. FERROELECTRIC DEVICES

When constructing a memory device using a ferroelectric, two aspects need to be considered: The readout mechanism and the ferroelectric material.

A. Mechanisms for readout of polarization

While setting the polarization in a ferroelectric is straight forward by applying an electric field in the correct direction that is larger than the coercive field, the readout is not trivial. In principle three different approaches are pursued (se fig. 1):

A <u>Direct read of the current that is flowing during switching of the device.</u> Note that the current is often used to charge a capacitor and the capacitor voltage is subsequently read.

B <u>Coupling the ferroelectric to a field effect transistor (FET) to change the VT of the device based on the polarization state</u> and read out the current of the FET.

C <u>Modulation of the tunnelling barrier in a thin layer of ferroelectric or a double layer of ferroelectric and thin dielectric</u> and read out of the current that is flowing through the structure.

The three options are illustrated in fig. 1. While these readout mechanisms result in a bistable (or even multilevel) device, in the last decade there has also been intense research in stabilizing a close to unpolarised state in a ferroelectric to realize a state of negative capacitance.

B. Ferroelectric Materials

Perovskites like BTO and PZT layered perovskites like SBT and also organic ferroelectrics like PVDF:TRFE have been investigated to realize ferroelectric devices since the 1950s [2]. Only PZT and SBT based 1T/1C FeRAM have made it to the market and achieved limited success in niche applications. However, the complicated structure including the weakly bound oxygen makes these materials hard to integrate and have limited their scaling success [3]. Polymer ferroelectrics are not a real alternative due to the limited thermal stability. However, more than half a century of research into such devices have helped to understand the limitations and opportunities [4]. The discovery of ferroelectricity in HfO_2 based materials has therefore quickly attracted the attention from both industrial and academic research since it immediately solved the integration issue of traditional ferroelectrics. However, a new material always comes with challenges of its own that need to be solved. These include the rather high fields that need to be applied during operation of this material leading to limited endurance and other possible reliability issues [5][6]. Fig. 2 compares the simplified crystal structures of the relevant ferroelectric materials-

C. Devices using ferroelectric HfO_2

The similarity of the material stack to hi-k/metal Gate technology immedialtely spurred the interest to integrate feroelecric hafnium oxide into FeFETs [7][8]. For the classicel 1T-1C FeRAM the reliability still needs to be improved, but recently, promising results have been shown [9][10]. For ferroelectric tunnel junctions it still seems to be challenging to get the material into the tunneling range with low enough background leakage. The two-layer FTJ has the potential to overcome this issue [11][12]. Finally, the existance of the S-curve leading to negative capacitance was recently experimentally verified [13] but the posibility to achieve a stabilized negative capcitance for steep slope devices is still a matter of intense scientific discussion.

III. SUMMARY AND OUTLOOK

The big potential of ferroelectric materials to enable new device functionalities could only partially be exploited using perovskite and related ferroelectric materials. Fluorite structure ferroelectrics based on HfO_2 show great potential to overcome the limitations and could enable competitive devices.

REFERENCES

[1] T. Boeske at al., Appl. Phys. Let 99, 10, pp. 102903-102903-3 (2011)
[2] S. Slahuddin and S. Datta, Nnao letters 8, 2,pp. 405-410 (2008)
[2] T. Mikolajick et al., IEEE-TED 67, 4, pp. 1434 – 1443 (2020)
[3] C. U. Pinnow and T. Mikolajick, JES 151, 6, pp. K13-K19 (2004)
[4] C. S. Hwuang and T. Mikolajick, in Advances in Non-Volatile Memory and Storage Technology, Woodhead Publishing, pp. 393-441 (2019)
[5] U. Schroeder et al., ESSDERC, pp. 364-368 (2016)
[6] P. Buragohain, ACS Appl. Mater. Interfaces 11, 38, pp. 35115-35121 (2019)
[7] T. S. Böscke et al., IEDM, pp. 24.5.1-24.5.4 (2011)
[8] J. Müller et al, Symposium on VLSI Technology, pp. 25-26 (2012)
[9] T. Francois et al., IEDM, pp. 15.7.1-15.7.4 (2019)
[10] J. Okuno et al., Symposium on VLSI Technology (2020)
[11] S. Fujii et al., IEEE Symposium on VLSI Technology, pp. 1-2 (2016)
[12] B. Max et al., Proceedings of ESSDERC, pp. 142-145 (2018)
[13] M: Hoffmann et al., Nature 565, 7740, pp. 464-467 (2019)

2020 IEEE Silicon Nanoelectronics Workshop

978-1-7281-9736-4/20 $31.00 © 2020 IEEE

1.1

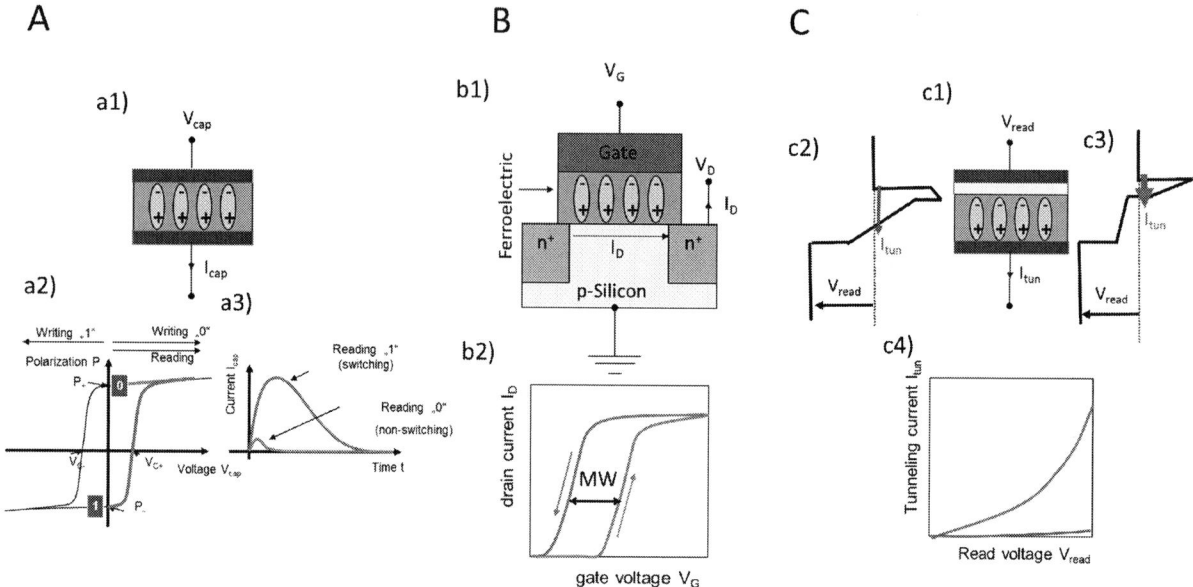

Fig. 1. Possibilities to read out the polarization state of a ferroelectric. A) Direct reading of the switched charge. a1) Ferroelectric capacitor in a given polarization state a2) Ferroelectric hysteresis showing the different pulses applied to the ferroelectric capacitor during writing and reading. a3) Current response of the ferroelectric capacitor after a switching and a non-switching readout event. B) Readout by coupling the ferroelectric to an FET channel. B1) ferroelectric FET (FeFET) with ferroelectric as the gate dielectric b2) Transfer characteristics for the two possible polarization directions. C) Two layer ferroelectric tunnel Junction (FTJ). c1) schematic of two layer FTJ. c2)c3) band diagrams for the two possible polarization states. C4) IV curves for the two possible polarization states.

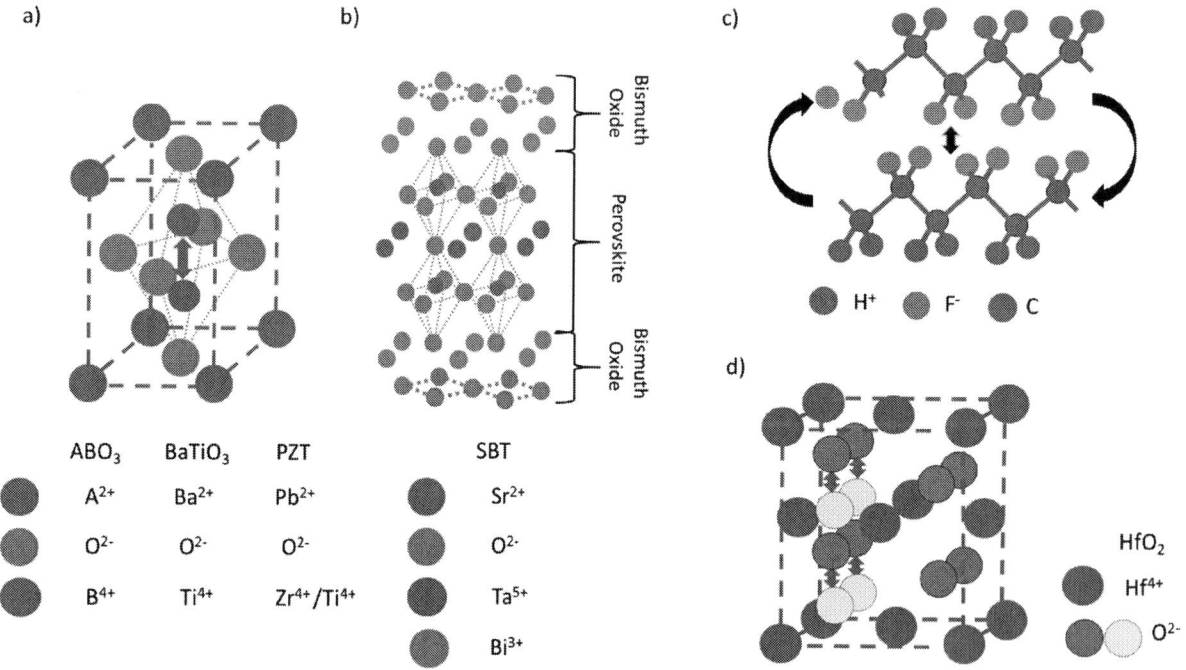

Fig. 2. Schematic crystal structures of the most important ferroelectric materials. a) Perovskite with the general structure ABO_3 and the specific examples BTO and PZT. b) Layered perovskite using the example of SBT c) Polymer ferroelectric PVDF d) Fluorite structure ferroelectric based on orthorhombic HfO_2.

2020 IEEE Silicon Nanoelectronics Workshop

978-1-7281-9736-4/20 $31.00 © 2020 IEEE

State-of-the-Art in Silicon Quantum Computer Development

Kohei M. Itoh

School of Fundamental Science and Technology and Quantum Computing Center, Keio University, Japan
Email: kitoh@appi.keio.ac.jp

Abstract — **State-of-the-art in silicon quantum computer development is introduced.**
Keywords: Semiconductor, Silicon, Quantum Computer

I. OVERVIEW

A. Quantum Computing

Quantum computing will enable some of the mission impossible by the extension of today's computing framework. Expected areas of applications include optimization, machine learning, security, data search, etc. Examples of quantum computer software research in the finance [1] and chemical industry [2] sectors are introduced.

B. Why silicon quantum computing (See Ref. 3 for an excellent review)

A quantum computer will not be almighty. It is going to be a very specialized machine that will be supplementing tasks that will be mostly performed by the classic computers such as CMOS electronics and AI machines. If so, it is natural to integrate the classical and quantum computers together in one silicon chip using the same process steps. In this way, the size of each silicon quantum bit (qubit) can be scaled down to the order of a few tens of nm for the large integration. Moreover, unlike the case of superconducting qubits that require an operating temperature less than 0.1 K, silicon spin qubits can be operated at much higher temperature ~5K [4-6], i. e., no dilution refrigerator is needed. The chip and electronics space for the normal helium refrigerator is much bigger and cryogenic CMOS circuits can be operated on the same chip at such "high" temperature together with qubits.

C. Variation in silicon qubits

The most advanced silicon qubits are categorized into the following four groups.

i) Gate-defined MOS spin qubit [7, 8]
A MOS gate structure fabricated on a silicon wafer electrically attracts only one electron under the gate and utilize the two levels of the single electron spin as a qubit.

ii) Gate-defined Si/SiGe spin qubit [9, 10]
A single electron spin confined by the gate bias in the layer of strained Si grown on the lattice mismatched SiGe virtual substrate is employed as a qubit.

iii) Gate-defined Si/SiGe singlet-triplet qubit [11-13]
Two or three electrons confined in a pair of gate-defined Si/SiGe quantum dots are employed as a qubit

iv) MOSFET structure converted to spin qubit [4, 14, 15]
Advanced MOSFET structure is cooled down to confine single electrons under the certain positions of cthe hannel region to employ their spins as qubits

v) Donor qubit [16-18]
A single phosphorus donor is placed near the surface by ion-implantation or scanning tunneling microscope manipulation followed by Si epi-growth. An electron, i. e., single spin bound to the phosphorus at <5K are employed as a qubit.

D. Substrates

Naturally available silicon (^{nat}Si) is composed of 92.2% ^{28}Si, 4.7% ^{29}Si, and 3.1% ^{30}Si. Among the three, only ^{29}Si possesses nuclear spin and its fluctuation is shown to be one of the major sources of qubit decoherence (loss of information) [See Ref. 19 for a review on the isotope engineering for Si quantum information processing]. Therefore, high-quality 300 mm Si epi-layer depleted of ^{29}Si nuclear spins are being developed [20, 21]. For the same reason, isotopically purified ^{28}Si strain layers grown on SiGe have been proven useful [10, 14, 22].

E. Interconnect

Transferring of quantum information from one qubit to distant ones turns out to be a big challenge. While a small number (~20) of quantum bits can be made into 1D or 2D arrays to allow sequential nearest-neighbour information swapping to transfer information from one qubit to the other, a longer distance requires transferring of the electron spin qubit information to a different kind of a mobile qubit such a microwave photon [23, 24]. The microwave photon qubit travels through a waveguide fabricated on the chip.

F. Architecture and integration with classical CMOS circuits

A variety of circuit layouts of qubits, readout elements (qubits or single electron transistors), qubit manipulators for logic operation (microwave antennas), gate electrodes, waveguides, and interconnects to classical Si CMOS elements has been proposed [1, 25-27].

G. Conclusion and outlook

Rapid advancements are being made towards realization of practical silicon quantum computers. Based on the current state-of-the-art, outlook for the next decade will be given.

ACKNOWLEDGMENTS

This work was supported by Spintronics Research Network in Japan

REFERENCES

[1] Y. Suzuki, S. Uno, R. Raymond, T. Tanaka, T. Onodera, and N. Yamamoto, "Amplitude Estimation without Phase Estimation," *Quantum Information Processing* **19**, 75 (2020).
[2] Q. Gao, H. Nakamura, T. P. Gujarati, G. O. Jones, J. E. Rice, S. P. Wood, M. Pistoia, J. M. Garcia, and N. Yamamoto,

"Computational Investigations of the Lithium Superoxide Dimer Rearrangement on Noisy Quantum Devices," arXiv:1906.10675

[3] L.M.K. Vandersypen, and M.A. Eriksson, "Quantum Computing with Semiconductor Spins," *Physics Today* **72**, 38 (2019)

[4] K. Ono, T. Mori, and S. Moriyama, "High-temperature operation of a silicon qubit," *Scientific Reports* **9**, 469 (2019).

[5] L. Petit, J.M. Boter, H.G.J. Eenink, G. Droulers, M.L.V. Tagliaferri, R. Li, D.P. Franke, K.J. Singh, J.S. Clarke, R.N. Schouten, V.V. Dobrovitski, L.M.K. Vandersypen, and M. Veldhorst, "Spin Lifetime and Charge Noise in Hot Silicon Quantum Dot Qubits," *Physical Review Letters* **121**, 076801 (2018).

[6] C. H. Yang, R. C. C. Leon, J. C. C. Hwang, A. Saraiva, T. Tanttu, W. Huang, J. C. Lemyre, K. W. Chan, K. Y. Tan, F. E. Hudson, K. M. Itoh, A. Morello, M. Pioro-Ladriere, A. Laucht, A. S. Dzurak, "Operation of a silicon quantum processor unit cell above one kelvin," *Nature* **7803**, 350 (2020)

[7] M. Veldhorst, C. H. Yang, J. C. C. Hwang, W. Huang, J. P. Dehollain, J. T. Muhonen, S. Simmons, A. Laucht, F. E. Hudson, K. M. Itoh, A. Morello, and A. S. Dzurak, "A two-qubit logic gate in silicon," *Nature* **526**, 410 (2015).

[8] W. Huang, C. H. Yang, K. W. Chan, T. Tanttu, B. Hensen, R. C. C. Leon, M. A. Fogarty, J. C. C. Hwang, F. E. Hudson, K. M. Itoh, A. Morello, A. Laucht, and A. S. Dzurak, "Fidelity benchmarks for two-qubit gates in silicon," *Nature* **569**, 532 (2019).

[9] A. Pateras, J. Park, Y. Ahn, J.A. Tilka, M.V. Holt, C. Reichl, W. Wegscheider, T.A. Baart, J.P. Dehollain, U. Mukhopadhyay, L.M.K. Vandersypen, and P.G. Evans, "A programmable two-qubit quantum processor in silicon," *Nature* **555**, 633 (2018).

[10] J. Yoneda, K. Takeda, T. Otsuka, T. Nakajima, M. R. Delbecq, G. Allison, T. Honda, T. Kodera, S. Oda, Y. Hoshi, N. Usami, K. M. Itoh, and S. Tarucha,"A quantum-dot spin qubit with coherence limited by charge noise and fidelity higher than 99.9%," *Nature Nanotechnology* **13**, 102 (2018).

[11] K. Takeda, A. Noiri, J. Yoneda, T. Nakajima, and S. Tarucha, "Resonantly Driven Singlet-Triplet Spin Qubit in Silicon," *Physical Review Letters* **124**, 117701 (2020)

[12] R. W. Andrews, C. Jones, M. D. Reed, A. M. Jones, S. D. Ha, M. P. Jura, J. Kerckhoff, M. Levendorf,S. Meenehan, S. T. Merkel, A. Smith, B. Sun, A. J. Weinstein, M. T. Rakher, T. D. Ladd, and M. G.Borselli, "Quantifying error and leakage in an encoded Si/SiGe triple-dot qubit," *Nature Nanotechnology* **14**, 747 (2019).

[13] D. Kim, Z. Shi, C. B. Simmons, D. R. Ward, J. R. Prance, T. S. Koh, J. K. Gamble, D. E. Savage, M. G. Lagally, M. Friesen, S. N. Coppersmith, M. A. Eriksson, "Quantum control and process tomography of a semiconductor quantum dot hybrid qubit," *Nature* **511**, 70 (2014).

[14] M. Urdampilleta, D. J. Niegemann, E. Chanrion, B. Jadot, C. Spence, P.-A. Mortemousque, C. Bauerle,L. Hutin, B. Bertrand, S. Barraud, R. Maurand, M. Sanquer, X. Jehl, S. De Franceschi, M. Vinet, and T. Meunier, "Gate-based high fidelity spin readout in a CMOS device," *Nature Nanotechnology* **14**, 737 (2019).

[15] R. Maurand, X. Jehl, D. Kotekar-Patil, A. Corna, H. Bohuslavskyi, R. Lavieville, L. Hutin, S. Barraud,M. Vinet, M. Sanquer, and S. De Franceschi, "A CMOS silicon spin qubit," *Nature Communications* **7**, 13575 (2016).

[16] S. B. Tenberg, S. Asaad, M. T. Madzik, M. A. Johnson, I, B. Joecker, A. Laucht, F. E. Hudson, K. M. Itoh, A. M. Jakob, B. C. Johnson, D. N. Jamieson, J. C. McCallum, A. S. Dzurak, R. Joynt, and A. Morello, "Electron spin relaxation of single phosphorus donors in metal-oxide-semiconductor nanoscale devices," *Physical Review B* **99**, 205306 (2019).

[17] Y. He, S. K. Gorman, D. Keith, L. Kranz, J. G. Keizer, and M. Y. Simmons, "A two-qubit gate between phosphorus donor electrons in silicon," *Nature* **571**, 371 (2019).

[18] J. T. Muhonen, A. Laucht, S. Simmons, J. P. Dehollain, R. Kalra, F. E. Hudson, S. Freer, K. M. Itoh, D. N. Jamieson, J. C. McCallum, A. S. Dzurak, and A. Morello, "Quantifying the quantum gate fidelity of single-atom spin qubits in silicon by randomized benchmarking," *Journal of Physics Condensed Matter* **27**, 154205 (2015).

[19] K. M. Itoh and H. Watanabe, "Isotope engineering of silicon and diamond for quantum computing and sensing applications," *MRS Communications* **4**, 143 (2014).

[20] D. Sabbagh, N. Thomas, J. Torres, R. Pillarisetty, P. Amin, H.C. George, K. Singh, A. Budrevich, M. Robinson, D. Merrill, L. Ross, J. Roberts, L. Lampert, L. Massa, S. Amitonov, J. Boter, G. Droulers, H.G.J. Eenink, M. van Hezel, D. Donelson, M. Veldhorst, L.M.K. Vandersypen, J.S. Clarke, and G. Scappucci, Quantum transport properties of industrial ^{28}Si/^{28}SiO$_2$," *Physical Review Applied* **12**, 014013 (2019).

[21] V. Mazzocchi, P. G. Sennikov, A. D. Bulanov, M. F. Churbanov, B. Bertrand, L. Hutin, J. P. Barnes,M. N. Drozdov, J. M. Hartmann, and M. Sanquer, "99.992% Si-28 CVD-grown epilayer on 300 mm substrates for large scale integration of silicon spin qubits," *Journal of Crystal Growth* **509**, 1 (2019).

[22] A. J. Sigillito, J. C. Loy, D. M. Zajac, M. J. Gullans, L. F. Edge, and J. R. Petta, "Site-Selective Quantum Control in an Isotopically Enriched Si-28/Si0.7Ge0.3 Quadruple Quantum Dot," *Physical Review Applied* **11**, 061006 (2019).

[23] X. Mi, M. Benito, S. Putz, D. M. Zajac, J. M. Taylor, G. Burkard, and J. R. Petta, "A coherent spin-photon interface in silicon," *Nature* **555**, 599 (2018).

[24] N. Samkharadze, G. Zheng, N. Kalhor, D. Brousse, A. Sammak, U.C. Mendes, A. Blais, G. Scappucci, and L.M.K. Vandersypen, "Strong spin-photon coupling in silicon," *Science* **359**, 1123 (2018).

[25] R. Li, L. Petit, D.P. Franke, J.P. Dehollain, J. Helsen, M. Steudtner, N.K. Thomas, Z.R. Yoscovits, K.J. Singh, S. Wehner, L.M.K. Vandersypen, J.S. Clarke, and M. Veldhorst, "A Crossbar Network for Silicon Quantum Dot Qubits," *Science Advances* **4**, e3960 (2018).

[26] L. M. K. Vandersypen, H. Bluhm, J. S. Clarke, A. S. Dsurak, R. Ishihara, A. Morello, D. J. Reilly, L. R. Schreiber, and M. Veldhorst, "Interfacing spin qubits in quantum dots and donors – hot, dense and coherent," *npj Quantum Information* **3**, 34 (2017).

[27] M. Veldhorst, H. G. J. Eenink, C. H. Yang, A. S. Dzurak, "Silicon CMOS architecture for a spin-based quantum computer," *Nature Communications* **8**, 1766 (2017).

Performance assessment of BEOL-integrated HfO$_2$-based ferroelectric capacitors for FeRAM memory arrays

L. Grenouillet[1], T. Francois[1], J. Coignus[1], N. Vaxelaire[1], C. Carabasse[1], F. Triozon[1], C. Richter[2], U. Schroeder[2], E. Nowak[1]

[1]CEA, LETI, Univ. Grenoble-Alpes, 38000 Grenoble, France - email: laurent.grenouillet@cea.fr
[2]NaMLab gGmbH, Noethnitzer Str. 64a, Dresden, 01187, Germany

Abstract— **We report on the scalability of ferroelectric TiN/HfO$_2$-based/TiN capacitors by integrating them in the Back-End-Of-Line of 130nm CMOS technology. Excellent performance is reported on those scaled bitcells, such as remanent polarization $2 \cdot P_R > 40\mu C/cm^2$, endurance $> 10^{11}$ cycles, switching speeds $< 100ns$, operating voltages $< 4V$, and data retention at 125°C. This demonstration suggests that ultra-low power ($< 10fJ/bit$) FeRAM memories, so far using PZT material that limits integration to 130nm node, could be attractive at 130nm node and beyond thanks to CMOS compatible ferroelectric HfO$_2$.**

I. INTRODUCTION

We are today in a data-hungry society, with a data traffic that reached more than a zettabyte (10^{21} bytes) in 2017. The energy-related consumption is tremendous and significantly growing every year: in 2030, the total electricity demand of information and communication technology (ICT) could reach 20% of the worldwide electricity demand [1]. Therefore, the energy efficiency has to be drastically improved to be able to offset the growth of the data demand.

At the technology level, non-volatile memories are key components to reduce the power required by data-centric electronic systems. Among the diversity of non-volatile memories, ferroelectric memories are attractive from an "energy/bit" standpoint. Their principle of operation relies on the application of an electric field to switch the electric polarization that stores the information. Thus a negligible current flows through the FeRAM device, leading to ultra-low a programming power of ~ 10fJ/bit, which is at least 3 orders of magnitude lower than Flash memories or resistive memories like MRAM, PCM, or RRAM.

II. FERROELECTRIC HfO$_2$ AS AN ENABLER FOR NEXT GENERATION MEMORIES

Although FeRAM products already exist today [2] and demonstrate excellent performance [3,4] they suffer from poor CMOS compatibility due to the use of lead-based perovskite materials like PZT. Moreover, the low coercive field of these ferroelectric materials requires ~100 nm thick layers, limiting the scaling at 130nm node.

The paradigm on ferroelectric memories changed a decade ago with the discovery that ~10nm thick HfO$_2$ films, when properly doped and annealed, crystallize in an orthorhombic non centro-symmetric phase, which is ferroelectric [5]. The excellent CMOS compatibility of this material combined with its atomic layer deposition capability opens the way to non-volatile memory devices [6] such as FeRAM [7,8], FeFET [9] or DRAM with great potential for scaling.

III. BEOL INTEGRATED HfO$_2$-BASED FERROELECTRIC CAPACITORS

Despite the growing number of publications in the field [10], today most of the ferroelectric HfO$_2$ capacitor studies come from academia, with no constraints on the capacitor diameter (typically $> 10\mu m$) nor the crystallization temperatures (usually $> 600°C$). Therefore, we focused our efforts on demonstrating scaling and Back-End-Of-Line (BEOL) compatibility. Here we report on the integration of Hf$_{0.5}$Zr$_{0.5}$O$_2$ (HZO) films since they have the lowest crystallization temperature among HfO$_2$-based ferroelectric candidates [7]. However we very recently obtained similar results with silicon-doped HfO$_2$ [11].

TiN/10nm-HZO/TiN capacitors with diameters ranging from 600nm down to 300nm were integrated between M4 and M5 in the BEOL of 130nm node CMOS technology (Fig.1&2). The highest thermal budget used in this integration, 450°C for 80s, is sufficient to crystallize HZO material in its orthorhombic ferroelectric phase (Fig.3&4) and has no detrimental impact on front-end transistor I_D-V_G characteristics (Fig.5). Remanent polarization values $2 \cdot P_R$ as large as $50\mu C/cm^2$ are reported after wake-up on 550nm diameter scaled capacitors (Fig.6), similarly to what is usually reported on 100μm diameter structures [7]. Switching efficiency maps [12] performed on a single 600nm diameter ($0.27\mu m^2$) TiN/HZO/TiN capacitor demonstrates operating speeds down to 30ns at 3V (Fig.7) and confirms the switching voltage/speed tradeoff [13].

In addition, area scaling statistically decreases the number of defects, leading to record endurance values ($>10^{11}$ cycles) (Fig.8). Despite the very low displacement current signal in the nA range, single 1T-1C bitcells with capacitor area of $0.12\mu m^2$ were successfully measured (Fig.9) for the first time with PUND technique. Fig.9 also shows that the polarization charge scales with the surface between 550nm and 400nm diameter capacitors and that the introduction of the 1T transistor selector does not modify the 1C behavior. Promising data retention up to 125°C is also measured (not shown here).

Very recent demonstrations confirm the potential of BEOL-integrated HZO-based capacitors integrated in 1T-1C FeRAM memory arrays [14]. Also, the potential to go 3D for the capacitor has already been successfully reported [8,15] to increase the charge per footprint if the charge to sense is too small in a 2D configuration. Regarding BEOL compatibility of Zirconium free HfO$_2$-based material whose crystallization temperature exceeds 500°C, it is expected from recent results that ultrafast nanosecond laser anneal may be an interesting option to crystallize the HfO$_2$ ferroelectric films at high temperatures while preserving the rest of the CMOS layers [11,16].

IV. CONCLUSION AND PERSPECTIVES

Ferroelectric HfO$_2$-based capacitors integrated in the BEOL of 130nm CMOS demonstrate excellent performance down to 300nm diameter, with remanent polarization $2 \cdot P_R > 40\mu C/cm^2$, endurance $> 10^{11}$ cycles, switching speeds $< 100ns$ and operating voltages $< 4V$. The reported results are promising to replace the conventional PZT-based FeRAM and scale beyond 130nm node. The possibility to integrate CMOS friendly ferroelectric HfO$_2$ in 3D should help further scale the footprint of these devices.

2020 IEEE Silicon Nanoelectronics Workshop

ACKNOWLEDGMENTS

We acknowledge the support from the European Union's Horizon 2020 research and innovation program under Grant Agreement No. 780302 (3εFERRO project).

REFERENCES

[1] N. Jones, Nature 561, 163-166, 2018
[2] https://www.cypress.com/products/f-ram-nonvolatile-ferroelectric-ram
[3] J. Rodriguez *et al.*, IEEE Transactions on Device and Materials Reliability, vol. **4**, pp. 436-449 (2004).
[4] J. Rodriguez *et al.*, IEEE proceedings of International Memory Workshop (IMW), pp. 157-160 (2016)
[5] T. Böscke *et al.*, Appl. Phys. Lett. 99, 102903 (2011)
[6] T. Mikolajick *et al.*, IEEE IEDM Techn. Digest., pp. 15.5.1-15.5.4 (2019)

[7] T. François *et al.*, IEEE IEDM Techn. Digest., pp. 15.7.1-15.7.4 (2019)
[8] Y.D. Lin *et al.*, IEEE IEDM Techn. Digest., pp. 15.3.1-15.3.4 (2019)
[9] M. Trentzsch et al., IEEE IEDM Techn. Digest., pp. 11.5.1-11.5.4 (2016)
[10] M.H. Park *et al.*, MRS Communications **8**, pp. 795–808 (2018)
[11] L. Grenouillet *et al.*, IEEE Symposium on VLSI Technology, accepted (2020)
[12] T. François *et al.*, IEEE proceedings of International Memory Workshop (IMW), pp. 97-100 (2019)
[13] X. Lyu *et al.*, IEEE IEDM Techn. Digest., pp. 15.2.1-15.2.4 (2019)
[14] J. Okuno *et al.*, IEEE Symposium on VLSI Technology, accepted (2020)
[15] P. Polakowski *et al.*, IEEE proceedings of International Memory Workshop (IMW) (2014)
[16] T. Tabata, Appl. Phys. Express **13** 015509 (2020)

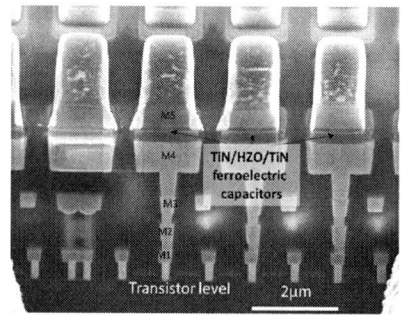

Fig.1: SEM cross section showing the transistor level, the different metal lines, and 300nm diameter TiN/HZO/TiN ferroelectric capacitors integrated between M4 and M5.

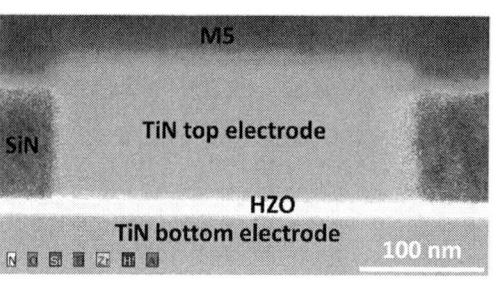

Fig.2: STEM-EDX cross section image of a 300nm diameter BEOL-integrated TiN/HZO/TiN ferroelectric capacitor revealing the different materials. CMP-planarized TiN bottom electrode ensures excellent surface roughness and well-defined interface with HZO.

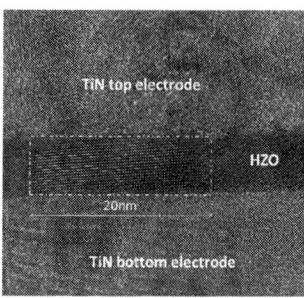

Fig.3: HRTEM image detail of a 300nm diameter BEOL-integrated TiN/HZO/TiN ferroelectric capacitor with 20nm crystallite size (in-plane).

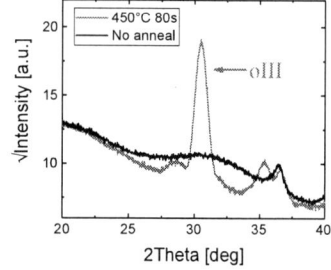

Fig.4: GIXRD spectra on TiN/HZO/TiN stacks with and without 450°C 80s anneal performed after TiN top electrode deposition, highlighting the appearance of orthorhombic oIII-phase.

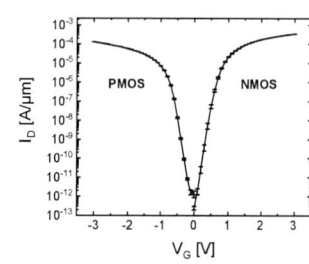

Fig.5: I_D-V_G characteristics of CMOS transistors after 1T-1C HZO BEOL integration, showing neither V_T dispersion nor I_{ON} reduction (mean and sigma, 55 dies tested at wafer scale).

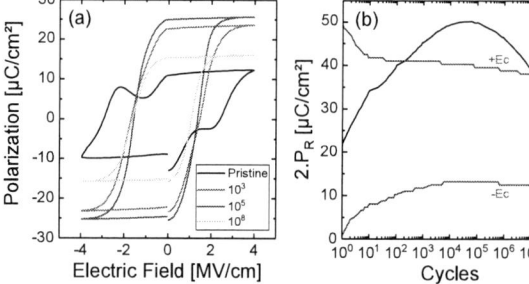

Fig.6: (a) PUND-corrected Electrical Polarization vs Electric Field plot measured on 339 identical 550nm diameter BEOL-integrated TiN/HZO/TiN ferroelectric capacitors routed in parallel (total area 80µm²) at various cycles and (b) corresponding $2.P_R$ and E_C during triangular cycling (4V/100kHz).

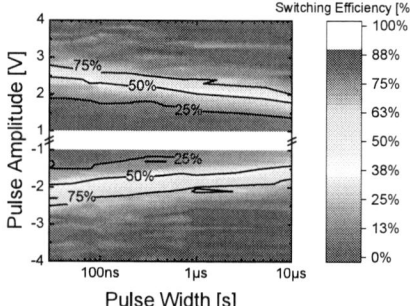

Fig.7: Switching efficiency map on a 600nm diameter BEOL-integrated single TiN/HZO/TiN ferroelectric capacitor.

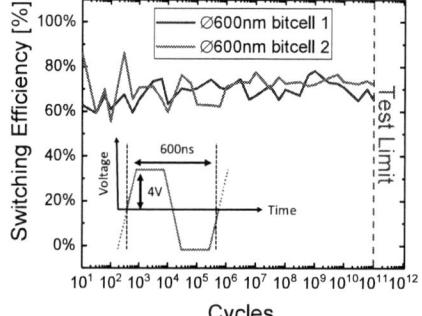

Fig.8: Ferroelectric switching efficiency measured on two 600nm diameter BEOL-integrated single TiN/HZO/TiN ferroelectric capacitors. No breakdown is observed before 10^{11} cycles.

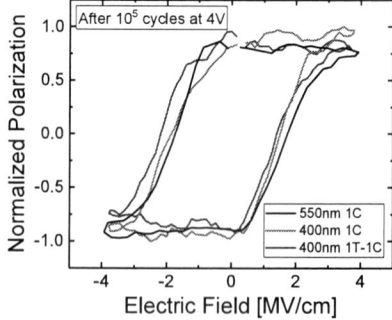

Fig. 9: Normalized polarization measured on TiN/HZO/TiN BEOL-integrated single 1C (550nm and 400nm diameter) and 1T-1C (400nm diameter) through an open transistor (V_G=5V).

2020 IEEE Silicon Nanoelectronics Workshop

978-1-7281-9736-4/20 $31.00 © 2020 IEEE

Record Fast Polarization Switching Observed in Ferroelectric Hafnium Oxide Crossbar Arrays

2.2

Xiao Lyu[1], Mengwei Si[1], Pragya R. Shrestha[2,3], Jason P. Campbell[3], Kin P. Cheung[3] and Peide D. Ye[1,*]

[1]School of Electrical and Computer Engineering, Purdue University, West Lafayette, USA. *Email: yep@purdue.edu
[2]Theiss Research, La Jolla, USA. [3]National Institute of Standards and Technology, Gaithersburg, USA.

Abstract

The polarization switching speed of ferroelectric (FE) hafnium zirconium oxide (HZO) is studied with the device size down to sub-μm in lateral dimension. Ultrafast measurement of transient switching current on metal-ferroelectric-metal (MFM) device with a crossbar array or a single crossbar structure is performed to analyze the switching dynamics. A record fast polarization switching of 360 ps is achieved for 15 nm thick HZO with 0.1 μm² crossbar array device structure. The observed record switching speed is found to be limited by domain wall propagation speed in the nucleation limited switching process. It is further verified after significant reduction of RC delay of the devices and the implementation of crossbar array structure.

Introduction

Ferroelectric hafnium oxide [1] has been widely studied as a promising CMOS-compatible material in commercial ferroelectric device applications. The remarkable endurance and retention performance [2] make FE HfO₂ suitable for non-volatile ferroelectric random-access memory (FeRAM) and ferroelectric FET (FeFET) devices. However, fast operational speed [3-11] is crucial for FE HfO₂-based memory to replace the current commercial memory products. Previous works [3-5] demonstrated that switching speed of FE HfO₂ can reach sub-ns regime in a single MFM device with an area of ~6 μm². Direct fast speed measurement on a much smaller size MFM capacitor is challenging due to small capacitance and high access resistance of metal contacts.

In this work, we fabricated crossbar array devices with 100-200 nm in size along with single crossbar devices with similar total areas for comparison. We performed transient current measurements and analysis with an ultrafast pulse measurement setup. Polarization switching dynamics in single and array crossbar devices is studied and quantitatively characterized. By applying nucleation limited switching (NLS) model [12] to the net switching current, a record fast switching time of 360 ps is obtained from an array of 0.1 μm² crossbar device. It confirms experimentally that the domain wall propagation is the limiting factor for the measured hundreds of ps polarization switch speed instead of the RC delay.

Experiments

The devices were fabricated on an insulating sapphire substrate to exclude the impact of parasitic capacitance from the substrate in fast pulse measurements. The fabrication process is described in Fig. 1. Sputtered Au/W metal stack was used as bottom electrode to achieve small series resistance. Anisotropic wet etch of gold was performed after the dry etch of tungsten to prevent possible short circuit between the top and bottom electrodes. The bottom Ti/Au contact pads were formed by e-beam evaporation for a better contact resistance. The growth of 15 nm thick HZO was completed by atomic layer deposition (ALD) at 200 °C [3]. A 20 nm thick tungsten layer was sputtered after HZO deposition, followed by a rapid thermal annealing (RTA) in nitrogen at 500 °C for 1 min. The top Ti/Au electrodes were fabricated by a lift-off process. Dry etching of top W layer was performed using Ti/Au as the hard mask to realize device isolation.

In order to precisely measure the polarization current flowing through a very small size capacitor, crossbar array structure is used to collectively measure a large number of individual small capacitors at the same time. Crossbar array devices consist of ten 1 μm-wide metal electrode stripes at the bottom and twenty 100 or 200 nm-wide metal stripes on the top, as seen in Fig. 2(a). Single crossbar devices with large

sizes but same total areas, as shown in Fig. 2(b) and Fig. (3), are designed as control devices for comparison. Fig. 4 shows the cross-section schematic and layer details of the MFM structure. All devices show strong ferroelectricity in polarization versus electric field (P-E) as illustrated in Fig. 5. Positive-up-negative-down pulse sequences, as shown in Fig. 6, are generated by an arbitrary pulse generator to perform direct measurement of transient current. The circuit diagram of the ultrafast pulse measurement setup and established methodology can be found in [3]. Signal reflection is efficiently suppressed by applying impedance matched probes, 50 Ohm terminations and the pick-off tee.

Results and Discussion

Fig. 7 shows a representative PUND transient current measurement. Both pulses share the same input voltage waveform with a 200 ps rise time and the corresponding transient current responses of the switching (first) pulse and non-switching (second) pulse are measured. The transient current I_{pulse1}, I_{pulse2} and the net switching current I_{FE} are plotted in Fig. 8 for a crossbar array device with 15 nm thick HZO as a representative case. The switching current is extracted by the subtraction of I_{pulse1} and I_{pulse2}. The time response of ferroelectric polarization switching charge can be determined by integrating the corresponding net switching current as seen in Fig.9. The FE polarization switching follows the NLS model, which can be described by $P = P_S(1 - \exp(-(\frac{t}{t_0})^2))$ in thin film. The t_0 in NLS model is a characteristic switching time constant [10-11] with a record fast 360 ps on 0.1 μm² and 640 ps on 0.2μm² array devices with 15 nm HZO.

These newly fabricated crossbar devices have lower series resistance compared to unoptimized devices in [3]. Fig. 10 shows the comparison of (a) switching speed and (b) RC delay between this work and ref [3]. The optimized RC delay is 2 magnitudes smaller and falls in a few ps regime, while there are only minor differences in terms of switching speed for devices with comparable area. Therefore, the measured switching speed by NLS model is clarified as FE material property instead of being overwhelmed by device RC delay.

Fig. 11 presents the switching speed of crossbar array and single crossbar devices with comparable total area. It is clear that smaller size MFM with array structure boosts switching speed, while keeping the total area large enough for accurate current measurements. The switching speed in smaller array devices is faster while the parasitic effect is similar for both structures. This is because the switch speed is determined by domain wall propagation speed in FE HZO.

Conclusion

Sub-μm crossbar array FE HZO MFM devices were fabricated and the polarization switching of these devices is studied. Record fast polarization switch speed of 360 ps is obtained. The work unveils that domain wall propagation speed in HZO is the limiting factor for switch speed and more aggressively scaled devices will offer much faster switch speed. The work is supported by SRC JUMP ASCENT Center.

References

[1] J. Muller et al., Nano Lett., p.4318, 2012. [2] K. Ni et al., IEEE TED, p. 2461, 2018. [3] X. Lyu et al., IEDM, p. 342, 2019. [4] W. Chung et al., VLSI, p. T89, 2018. [5] M. Si et al., APL, p. 072107, 2019. [6] E. Yurchuk et al., IEEE TED, p. 3699, 2014. [7] J. matthew et al., IEDM, p. 139, 2017. [8] H. K. Yoo et al., IEDM, p. 481, 2017. [9] S. Dunkel et al., IEDM, p. 485, 2017. [10] C. Alessandri et al., IEEE EDL, p. 1780, 2018. [11] K. Karda et al., IEEE EDL, p. 801, 2016. [12] J. Y. Jo et al., PRL, p. 267602, 2007.

2020 IEEE Silicon Nanoelectronics Workshop

2.2

☐ Bottom Au/W Sputtering
☐ Bottom Electrode Etching
☐ Bottom Ti/Au Pad Deposition
☐ ALD HZO
☐ Top W Sputtering
☐ RTA 500 °C in N₂ for 1 min
☐ Top Ti/Au Deposition
☐ Top W Etching

Fig. 1. Device fabrication process flow for W/HZO/W crossbar arrays and single crossbars.

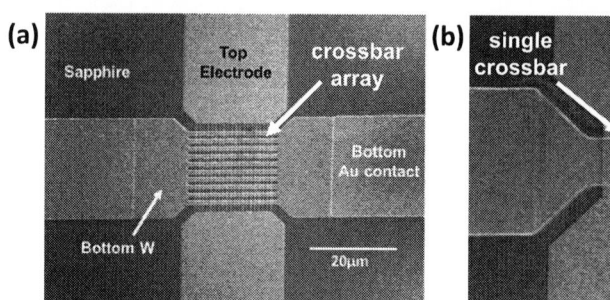

Fig. 2. SEM image of (a) a crossbar array device and (b) a single crossbar device. Insulating sapphire substrate is used for device fabrication considering on ultrafast measurements.

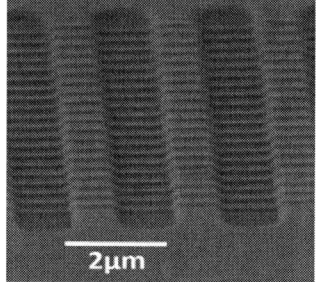

Fig. 3. SEM image of device structure details of a crossbar array device.

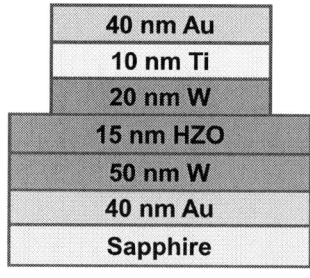

Fig.4. Cross sectional schematic diagram of W/HZO/W device structure.

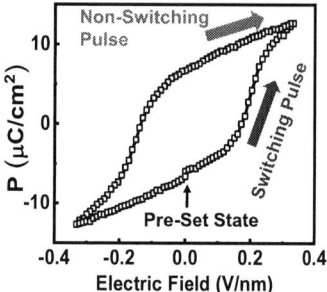

Fig. 5. P-E characteristics of a representative single crossbar MFM device with 15 nm thick HZO.

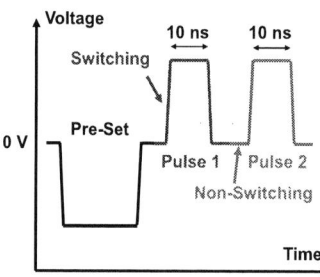

Fig. 6. PUND pulse sequence for transient ferroelectric polarization switching current measurement.

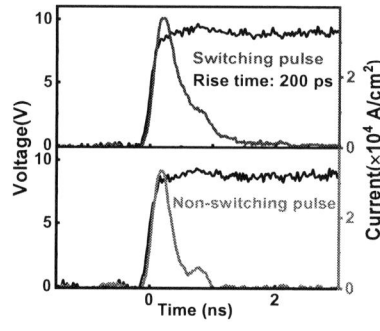

Fig. 7. Ultrafast pulse input with a 200 ps rise time and the corresponding transient current response of the switching (upper) and non-switching (lower) pulse.

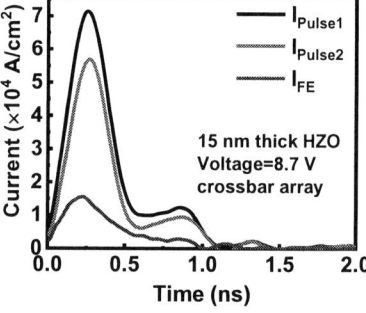

Fig. 8. Measured transient currents and the extracted switching current of a 15 nm thick HZO crossbar array device.

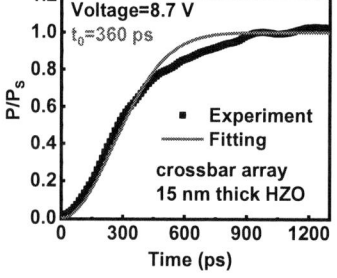

Fig. 9. Normalized transient switched polarization charge density from both experiment and fitting by NLS model of a 15 nm thick HZO crossbar array device.

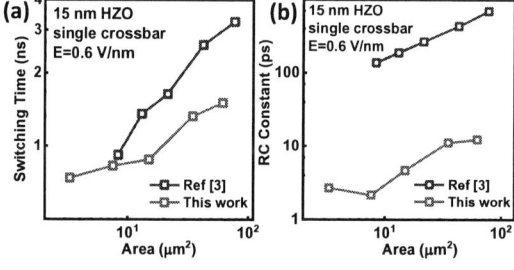

Fig. 10. Comparison of (a) switching time constant and (b) RC delay constant between the single crossbar devices with fabrication process improvement in this work and Ref. 3. The MFM area is similar.

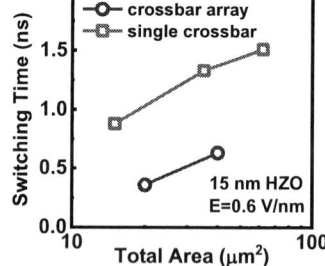

Fig. 11. Switching speed difference of single crossbar and crossbar array devices with similar total area.

978-1-7281-9736-4/20 $31.00 © 2020 IEEE

2020 IEEE Silicon Nanoelectronics Workshop

On the physical origins of time-dependent steep SS in FeFET

Xiuyan Li, Yulong Dong and Jingquan Liu

Department of Micro/Nano Electronics, Shanghai Jiao Tong University, Shanghai 200240, China

Email: xiuyanli@sjtu.edu.cn

Abstract —The time effects on steep SS and its hysteresis in FeFET are experimentally studied and analytically modelled by paying attention to the gate stack leakage. Our results suggest that both polarization switching time and depolarization life time should be considered to analyze steep SS FeFET.

I. INTRODUCTION

Steep SS FeFETs has been intensively discussed from a viewpoint of negative capacitance (NC) effects in FE/PE gate stacks [1]. The reported small SS values, however, showed a strong frequency/time dependence and a correlation with a hysteresis, which are not expectable from the original concept of NC [2,3]. Several models of transient NC effects in AC analysis have been reported to provide some evidences on this [4, 5]. But, steep SS effects are mostly discussed in DC mode. Very recently, we proposed an electrostatic model, with which steep SS in DC mode are explainable [6]. Therefore, the objective of this work is to clarify the time dependence of steep SS by paying attention to the leakage effects in addition to the domain flipping kinetics, and to shed light on the engineering of SS value and hysteresis of FeFET.

II. RESULTS AND DISCUSSION

A. Time effect on steep SS in FeFET in DC mode

The time effect on steep SS in FeFET structure was firstly investigated experimentally in DC mode by connecting a PZT-FE capacitor to a simple n-MOSFET externally (**Fig. 1(a)**). The gate bias (V_G) was swept in a stepwise manner and the stress time (t_s) at each step was varied from 0.1s to 2s. **Fig. 1(b)** shows the I_D-V_G characteristics of FE/MOSFET system with different t_s. It seems that the SS steepness degrades, while its hysteresis shrinks, with increase of t_s. **Fig. 1(c)** summarizes the minimum SS and hysteresis width at a given I_D as a function of t_s. It is interesting that both values show a negative correlation in change of t_s. When t_s is up to 2s, SS approaches 60mV/dec with almost no hysteresis.

B. Modeling of time-dependent steep SS in DC mode

We have understood the steep SS of FeFET in DC mode by analysing the internal potential (V_{int}) gain in an ideal FE/PE stack with single domain in FE layer based on electrostatics [6]. As shown in **Fig. 2**, a V_{int} gain is achievable in FE domain flipping at coercive voltage, corresponding to a depolarization filed formation on FE layer. Note that both amplitude and hysteresis of V_{int} gain are controlled by PE capacitance (C_P) in this case. Thus, a near hysteresis-free V_{int} gain along with maximum amplitude is possible with a suitable C_P.

In the actual FE/PE system in DC mode, the domain flipping kinetics can be ignored but the leakage effects in both FE and PE layers should be taken into account. The equivalent circuit in this case is shown and formulated in **Fig. 3**. The solution of $\delta V_{int}/\delta V$ suggests that the V_{int} gain decreases with t_s exponentially when t_s is comparable to a time constant determined by the capacitances and resistances of FE and PE layers. Namely, a finite "depolarization life time" (τ_{DP}), is induced by leakage effects. The τ_{DP} of the FE/MOSFET system in our experiments was estimated to be 5s. Thus, the SS-t_s curve in Fig. 1(b) is well fitted by the formula.

Concerning the SS hysteresis, note that it is determined not only by the value and hysteresis of V_{int} gain but also by $\partial V_{int}/\partial V$. Here, two cases with PE and FE leakage dominant are considered. Interestingly, the initial hysteresis can be reduced in t_s increase along with SS steepness decay in both cases based on our model (**Fig. 4**). We have confirmed a much larger leakage in FE than in MOS in experiments. The results are also consistent with the expectation from the model. It is worth noting that the correlation between SS steepness and hysteresis affected by t_s is quite different from that by C_P in present review.

C. Time-dependent steep SS in FeFET in general view

In a general view including both DC and AC modes, the impacts of domain flipping kinetics and charge communication with power supply, as pointed out in literatures [3-5], should be considered together with leakage effects to understand the time dependent steep SS in FeFET. By employing a polarization switching time (τ_{sw}) based on KAI model [3], the time dependent V_{int} gain is schematically shown in **Fig. 5**. It demonstrates that there is a frequency/time window to obtain steep SS if the depolarization life time is much longer than the polarization switching time. This may explain the time resolved V_{int} gain and steep SS reported so far [2,3].

III. CONCLUSION

Steep SS in FeFET degrades with time exponentially along with hysteresis shrink in DC measurement. Our results suggest that the leakage effect in FE/PE stack in addition to domain switching dynamics and capacitance matching should be carefully considered for engineering the steep SS and its hysteresis in FeFET.

ACKNOWLEDGEMENT: This work was supported by National Natural Science Foundation of China (61904103, 91964110). We would thank Prof. A. Toriumi for discussion and suggestions.

REFERENCES: [1] S. Salahuddin *et al.*, Nano Lett. **8**, 405 (2008). [2] P. Sharma *et al*, EDL **39**, 272-275 (2018). [3] H. Wang *et al*, IEDM (2018). [4] B. Obradovic *et al*, TED **65**, 5157-5164 (2018). [5] C. Jin *et al*, JEDS 7, 368 (2019). [6] X. Li and Akira Toriumi, Nat. Commun. **11**, 1895 (2020).

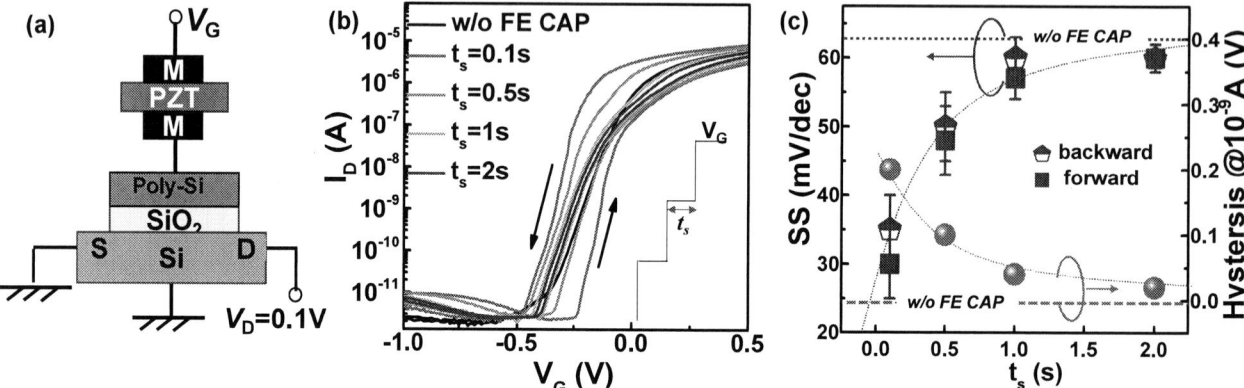

Fig. 1 (a) Schematics of experimental investigation of SS in FeFET. **(b)** I_D-V_G characteristics in FE/MOSFET system with four kinds of t_s. t_s is defined in inset schematics. SS looks shaper while hysteresis is larger with a smaller t_s. **(c)** SS values (minimum) and hysteresis width as a function of t_s. Hysteresis width is defined as V_G difference between forward and backward sweeping when $I_D=10^{-9}$A. SS increases with t_s along with hysteresis shrinking.

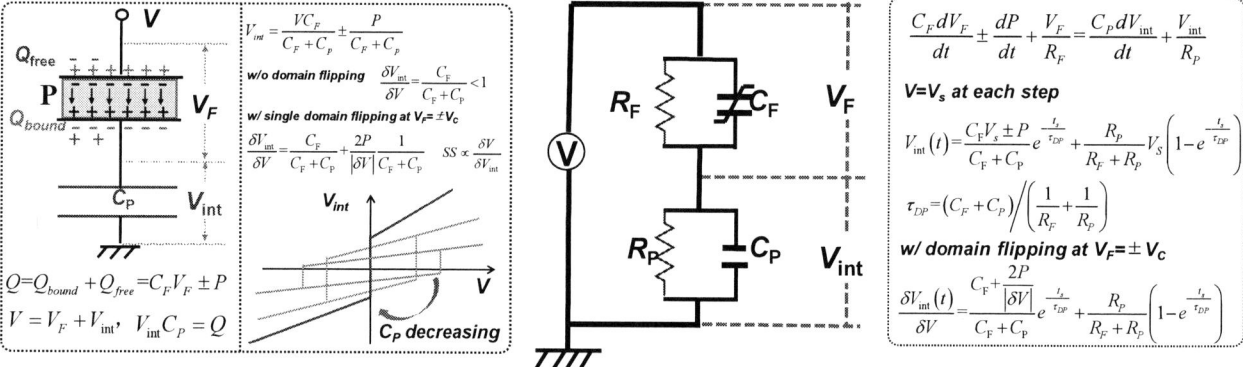

Fig. 2 Schematics and electrostatic analysis of an ideal ferroelectric/paraelectric (FE/PE) stack with single domain. At $V_F=\pm V_C$, V_{int} jump along with V_F drop occur, corresponding to $\delta V_{int}/\delta V > 1$. The amplitude and hysteresis of V_{int} gain are controlled by C_P, and a near hysteresis-free V_{int} gain is possible with a suitable C_P.

Fig. 3 Equivalent circuit of FE/PE system by considering leakage effects in both PE and PE layers, and its formulation by combing Kirchhoff's law with electrostatics. In DC mode, V increases in a stepwise manner as in the experiment and it is assumed to be a constant in each step t. The amplitude of V_{int} gain degrades with t_s exponentially with a fixed C_P.

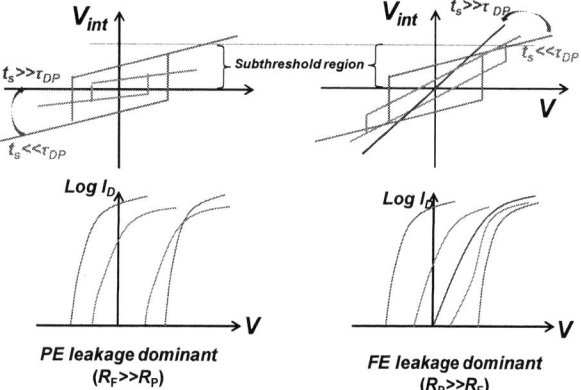

Fig. 4 Schematics to understand the negative correlation between SS value and hysteresis. In the case with PE leakage dominant, both the hysteresis of V_{int} gain and V_{int}-V slope decrease with time. So, SS hysteresis decrease with time. In the case with FE leakage dominant, although hysteresis of V_{int} gain increases, V_{int}-V slope also increase. So, the hysteresis for the whole subthreshold region decreases.

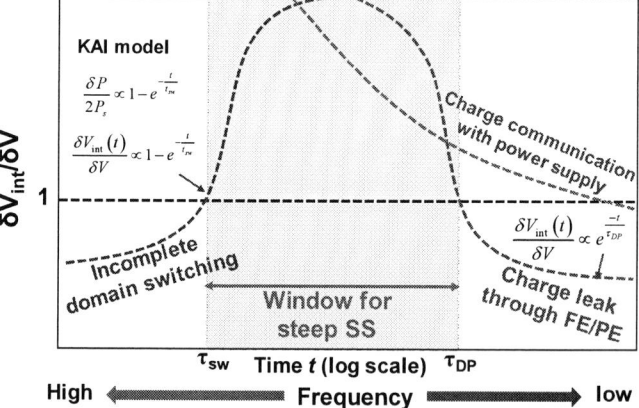

Fig. 5 Time/frequency dependence of internal potential gain in the general view. Both domain switching time and depolarization life time should be considered. Steep SS could not be obtained due to incomplete domain switching in high frequency region, and due to leakage effect in low frequency region. Charge communication with power supply also has an effect. Thus, there is a frequency window for obtaining steep SS in the case of $\tau_{DP} \gg \tau_{sw}$.

2020 IEEE Silicon Nanoelectronics Workshop

Ferroelectric Tunnel Junction Optimization by Plasma-Enhanced Atomic Layer Deposition

Jae Hur, Yuan-Chun Luo, Panni Wang, Nujhat Tasneem, Asif Islam Khan and Shimeng Yu

School of Electrical and Computing Engineering, Georgia Institute of Technology, Atlanta, GA 30332, USA

Email: asif.khan@ece.gatech.edu and shimeng.yu@ece.gatech.edu

Abstract —Ferroelectric tunnel junction (FTJ) based on alloyed HfO_2 and ZrO_2 is emerging as a promising two-terminal device candidate for the crossbar array for high density memory and compute-in-memory. FTJ is able to operate under non-destructive read mechanism as opposed to the ferroelectric capacitor. Herein, we report an optimized fabrication process that boosts the on-state current while suppressing the off-state current leading to an improved performance in $Hf_{0.5}Zr_{0.5}O_2$ (HZO) based FTJs. The plasma-enhanced atomic layer deposited (PEALD) of HZO and the incorporation of an interlayer Al_2O_3 are keys to improve the HZO-based FTJ in terms of the on/off ratio and cycling endurance.

I. INTRODUCTION

Ferroelectric Tunnel Junction (FTJ) is a two-terminal device that is suitable for crossbar array [1], [2]. Since the discovery of ferroelectricity in alloyed HfO_2 and ZrO_2 thin films with post-deposition annealing for crystallization, researchers have investigated process optimizations of the material composition and annealing temperature to improve the quality of HZO based ferroelectrics [3]–[5]. In this work, $Hf_{0.5}Zr_{0.5}O_2$ (HZO) based FTJs with two different fabrication techniques (thermal atomic layer deposition, THALD and plasma-enhanced atomic layer deposition, PEALD) are systematically compared. It was found that the on/off ratio and cycling endurance characteristics showed much improved performance using PEALD for the inter-layer (IL) and HZO. It is the first time that an FTJ device fabricated with PEALD is demonstrated with its improved performance compared to prior works using THALD.

II. RESULTS AND DISCUSSION

Fig. 1(a) shows the basic FTJ structure composed of separate layers of TiN, Al_2O_3, and HZO for top/bottom electrode, IL, and ferroelectric barrier respectively. The fabrication process flow is shown in **Fig. 1(b)**. **Fig. 1(c)** and **(d)** demonstrate the energy band diagram of the FTJ when it is in the off-state and on-state. Here, two types of FTJs were fabricated, one with Al_2O_3 and HZO layers by THALD and another one using PEALD. **Fig. 2(a)** and **(b)** show the typical polarization-voltage (P-V) characteristics of the entire FTJ stack (after wake-up process) with sweeping voltage from ±3.5 V to ±7.0 V. The remnant polarization (P_r) and saturation polarization (P_s) are comparable between the two FTJs. The leakage is, however, noticeable for the THALD FTJ in the P-V loop when compared to the PEALD FTJ as voltage increases. This can be correlated with the electrical breakdown properties of the THALD and PEALD HZO-only thin films sandwiched between top and bottom

electrodes of TiN, as in **Fig. 3(a)**. The breakdown electric-field (E_{BD}) is larger and also the off-state leakage is lower for the PEALD HZO when compared to THALD HZO. **Fig. 3(b)** displays the P-V endurance characteristics of the THALD and PEALD HZO capacitor with electric field stress (E_{pulse}) of ±2.5 MV/cm and using triangular pulse with pulse width (t_{pulse}) of 1 ms. It can be observed that the cycling wake-up process is more prominent for the THALD HZO while the PEALD HZO showed almost no wake-up cycles and stayed rather stable. Moreover, the THALD HZO shows severe fatigue effect right after it experiences the wake-up cycles of 1000. For erase (to off-state) and program sequence (to on-state) of the FTJ devices, 6 V and -4 V, with pulse widths of 50 μs and 10 μs respectively, have been applied to the top Al. The read voltage of the FTJs was chosen to be 1.5 V. In **Fig. 4(a)**, the on/off characteristics of the THALD and PEALD FTJs are shown. The on/off ratio is higher for PEALD FTJ through the endurance cycling. In **Fig. 4(b)**, the I_{ON} increases significantly after experiencing approximately 100s of cycles in THALD FTJ from the beginning, which is correspondent to wake-up characteristics of the P-V endurance curve in **Fig. 3(b)**. In **Fig. 4(c)**, the off-current (I_{OFF}) behavior is shown and it is larger for THALD FTJ owing to its more leakage as demonstrated in **Fig. 2(a)** and **Fig. 3(a)**. Thus, THALD FTJ results in lower on/off ratio along with rather lower I_{ON}. All of these FTJ performance can be understood via the current-voltage (I-V) characteristics of the FTJ in **Fig. 5**. In **Fig. 5(a)** and **(b)**, from 10 to 100 cycles indicated as A, A', B and B', the wake-up effect can be observed for both FTJs at the negative and positive voltages. The peak current value is both larger in PEALD FTJ, which can explain the higher on/off ratio in **Fig. 4(a)**. Indicated with C, from 100 to 2000 cycles, the fatigue effect in THALD FTJ attributes to the lower I_{ON} in **Fig. 5(a)**. The exponentially increasing I_{OFF}, as the number of cycles increases for THALD FTJ, can be understood with D. Finally, it is concluded that both the I_{ON} and I_{OFF} could be improved using PEALD process for FTJ, which are correspondent to their ferroelectric properties, *i.e.*, P-V and I-V characteristics.

III. CONCLUSION

This work demonstrated two types of FTJ devices, where THALD and PEALD have been used in the fabrication process. The FTJ performance in terms of on/off ratio has been compared between the two FTJs with respect to the cycling endurance. It was found that the FTJ with PEALD exhibits boosted on/off ratio along with better cycling endurance. Most importantly, device performance of the FTJs was explained corresponding to not only P-V but also

2020 IEEE Silicon Nanoelectronics Workshop

the I-V characteristics. For the suppressed I_{OFF}, it could be explained with the stronger dielectric property of the more crystallized PEALD HZO layer having lower leakage and higher E_{BD}. The higher on/off ratio from the PEALD FTJ is attributed to the relatively its robustness against high E_{pulse}.

ACKNOWLEDGEMENT

This work was supported by ASCENT, one of the SRC/DARPA JUMP Centers.

REFERENCES

[1] S. Fujii, *et al.*, *2016 IEEE Symposium on VLSI Technology*, Jun. 2016, pp. 1–2.

[2] B. Max *et al.*, *IEEE Journal of the Electron Devices Society*, vol. 7, pp. 1175–1181, 2019.

[3] S. Starschich *et al.*, *Appl. Phys. Lett.*, vol. 110, no. 18, p. 182905, May 2017.

[4] M. Hyuk Park *et al.*, *Appl. Phys. Lett.*, vol. 102, no. 24, p. 242905, Jun. 2013.

[5] S. Shibayama *et al.*, *Journal of Applied Physics*, vol. 124, no. 18, p. 184101, Nov. 2018.

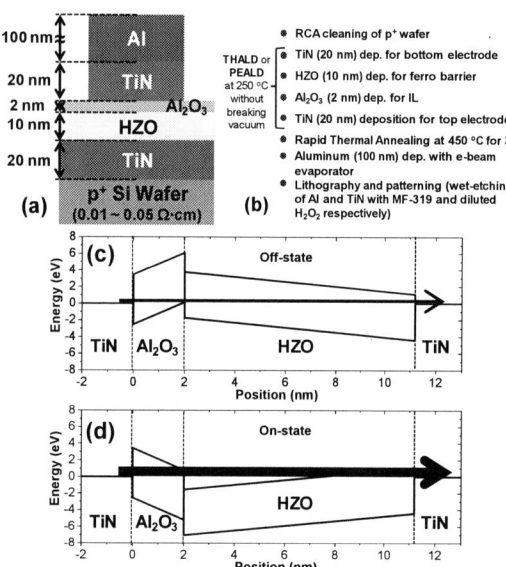

Fig. 1. (a) Device structure and (b) fabrication process flow of the HZO-based FTJ, (c) off-state and (d) on-state energy band diagram of the FTJ.

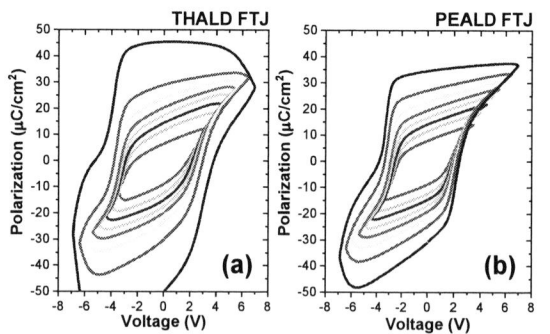

Fig. 2. P-V characteristics of (a) THALD FTJ and (b) PEALD FTJ (showing less leakage as sweeping voltage is large).

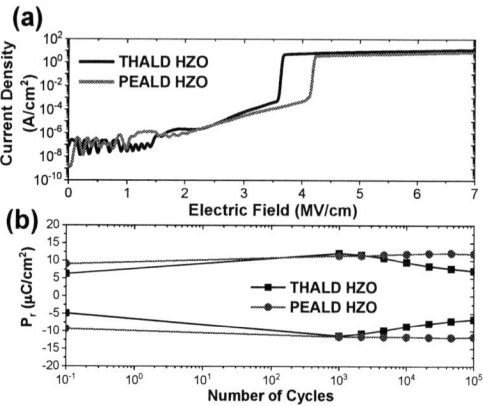

Fig. 3. (a) Current density-electric field characteristics and (b) P_r endurance of THALD and PEALD HZO capacitors.

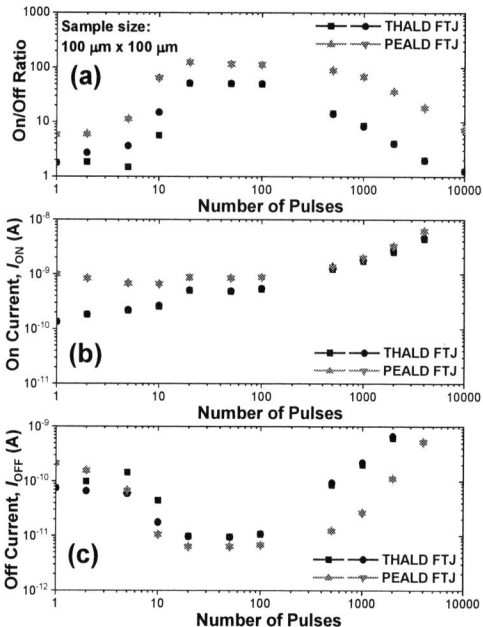

Fig. 4. (a) On/off ratio, (b) on-current and (c) off-current of the THALD and PEALD FTJs with two representative cells.

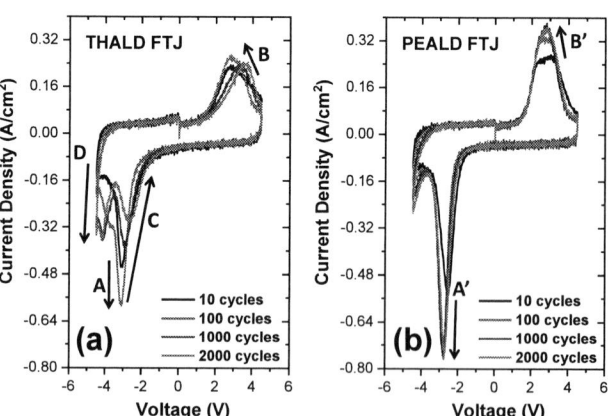

Fig. 5. I-V characteristics of (a) THALD FTJ and (b) PEALD FTJ with cycling endurance.

2020 IEEE Silicon Nanoelectronics Workshop

Process and Structure Considerations for the Post FinFET Era

Chun-Jung Su[1], Po-Jung Sung[1, 2], Kuo-Hsing Kao[3], Yao-Jen Lee[1], Wen-Fa Wu[1] and Wen-Kuan Yeh[1]

[1]Taiwan Semiconductor Research Institute, Taiwan; [2]Dept. Electrophysics, National Chiao Tung University, Taiwan; [3]Dept. Electrical Engineering, National Cheng Kung University, Taiwan

Email: cjsu@narlabs.org.tw

Abstract —Evolution of transistor structures, from planar, fin to gate-all-around (GAA) nanowire (NW)/nanosheet (NS), enables consecutive device scaling and performance boost. To further enhance the drive current per footprint, a vertically stacked configuration compatible with current CMOS technology may be a promising approach for extending Moore's Law. In this paper, we review the recent status of stacked FET architectures and beyond, as well as pointing out the challenges and perspectives.

Keywords: CMOS, FinFET, nanosheet, nanowire

I. INTRODUCTION

The commercial chipset using FinFET technology was firstly introduced at the 22 nm node in 2011, showing enhanced performance and reduced area over the planar FET architecture. Thanks to the better electrostatic control, this multigate structure enabled continued gate length (L_G) scaling. Yet, to further shrink device dimensions (including device feature size and cell height) and meet the IRDS requirement of performance beyond the 3 nm technology node (N3), new device architectures are needed.

While the tri-gated structure may fail to offer satisfactory electrostatic control for a FinFET beyond N3 as further scaling L_G, GAA transistors such as NW, nanoribbons or NS will need to be employed [1, 2]. They can be fabricated by current CMOS technology with minimal deviation from the standard FinFET processes and circumvent some of the patterning and design challenges as well as provide superior electrostatic and short channel control. In spite of the superior gate controllability over the channels, a single GAA NW or NS may not provide enough drive current due to the small cross-section and large series resistance. A vertically stacked structure can increase the effective device width per footprint without area penalty, and which has been experimentally demonstrated by different semiconductor channels, including Si, Ge, GeSn and InGaAs [3-6]. The key to the stacked NS formation and release is high-quality superlattice (such as Si/SiGe, Si/Ge and GeSn/Ge) epitaxy followed by dry and lightly selective etching. Accordingly, new process modules are introduced, including superlattice epitaxy, channel release and inner spacer formation. For the next-generation device miniaturization, a complementary FET (CFET) architecture is recently proposed to alleviate technology challenges for transistor evolution [7]. It features vertically stacked nFET and pFET (fin, NS or both) on a single device footprint, thus allowing for inverter area scaling. Despite the benefits of area [8], one of the challenges for CFETs is the formation of deep and shallow via for n/p FET interconnect, which may be the bottleneck of device performance due to parasitic resistance. This challenge can be relieved by introducing advanced middle-end of line (MEoL) contacts optimization [9, 10]. In this paper, the process and structure consideration for stacked NS and CMOS architectures, and their current status will be overviewed.

II. CONSIDERATION FOR STACKED NS

Fig. 1 shows the structures of FinFETs and stacked NSFETs. Conceptually, stacked NS can be considered as a bundle of vertical multi-fins laying down, and accordingly layout efficiency can be improved. The NS structure benefits the better SCE control and flexible width design. Fig. 2 depicts the performance matching of n/p FETs by adjusting the device width and number of NS [11].

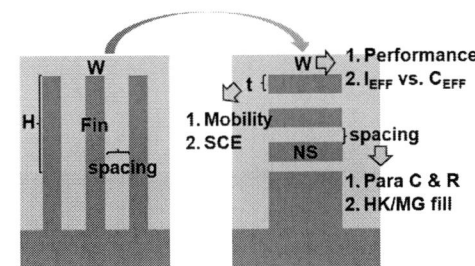

Fig. 1 A concept of converting a FinFET structure to an NSFET one. Parameters of an NS structure need to be redesigned regarding the device performance and integration challenges.

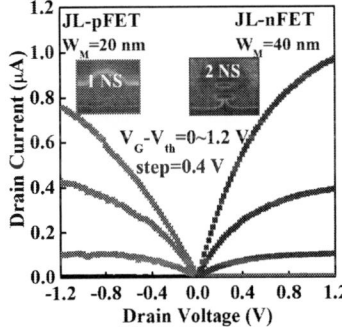

Fig. 2 Performance matching can be adjusted by width and number of NS.

There are three main structure parameters of NS that may impact the device performance and integration process as displayed in Fig. 1. NS thickness (t): thinner t guarantees stronger immunity to SCEs; however, mobility may be degraded [12]. Furthermore, the major surface orientation of a FinFET is (110) on sidewalls, where electron and hole mobilities are balanced [13]. However, the main conduction surface orientation of the NS

2020 IEEE Silicon Nanoelectronics Workshop

channel is (100), on which electron mobility is much higher than hole mobility. NS width (W): a smaller W limits the current due to the resistance, while a device with a larger W shows performance saturation owing to the trade-off between I_{EFF} and C_{EFF} [14]. NS spacing: small spacing poses process challenges of channel release and HK/MG filling, but large spacing causes higher C_{para} and R_{SD} limiting the currents. In addition, challenges of the performance loss on the bottom NS due to increased series resistance and formation of inner spacer to reduce parasitic gate-to-S/D capacitance are needed to be taken into account [15].

III. VERTICALLY STACKED CMOS

Fig. 3 shows the conceptual architecture of a CFET, in which a pFET is vertically placed onto an nFET or vice versa, and accordingly a CMOS footprint can be greatly reduced. A higher area efficiency and higher routing flexibility than FinFET technology have been evaluated by simulations [16]. On the other hand, from the process point of view, it is important to develop the advanced process for doping of S/D regions separately and forming the gates with dual work function (WF) for the stacked layers and the common gate configuration.

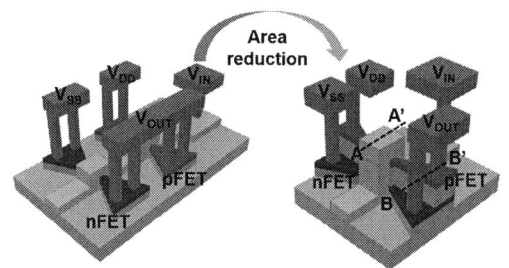

Fig. 3 A concept of a CFET scheme for area reduction by placing a pFET vertically onto an nFET.

A CMOS inverter and SRAM based on CFETs have been experimentally demonstrated by exploiting ultrathin poly-Si channels [17]. In the work, a junctionless (JL) scheme is employed for both n/p FETs to mitigate the S/D doping complexity, and manufacturing difficulties of vertically stacked S/D pads together with simplified lithographic steps. The cross-sectional TEM image along the gate direction (A-A' in Fig. 3) in Fig. 4(a) shows the vertically stacked n/p FETs. Fig. 4(b) displays the SEM image of contact regions for n/p FETs along B-B' direction in Fig. 3. The ultrathin channel (~8 nm) and thin isolation oxide layer (~30 nm) in-between minimize the series resistance of different contact depths. The voltage transfer curves (VTCs) of the inverter based on the optimized NS CMOS and NC CFETs are demonstrated in Fig. 5, depicting comparable properties while significant area reduction about 50 % is achieved by the CFET configuration [11, 17]. Widths and number of NS can be utilized for further performance matching, being similar to the strategy for the planar FET architectures, except no additional area requested. In addition, the requirement of multi-V_{th} engineering may be limited by the common gate structure. Typical dual WF metal gates of gate-last scheme may be implemented; however, more complicated process and techniques are demanded. Schottky

FETs without intentional doping may be another applicable approach to avoid S/D implant or *in-situ* doped epitaxy growth. The n/p FETs can be operated via appropriate S/D WF metal contacts to tune the polarity of conduction. A sub-gate structure can also be used to modulate channel potential for electrostatically tuning the V_{th} [18].

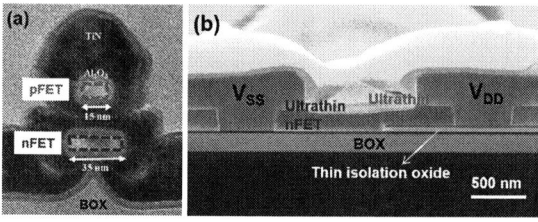

Fig. 4 (a) Cross-section TEM image of a CFET along A-A' in Fig. 3. (b) Cross-section SEM image of a CFET along B-B' in Fig. 3 to show the different contact levels of n/pFETs.

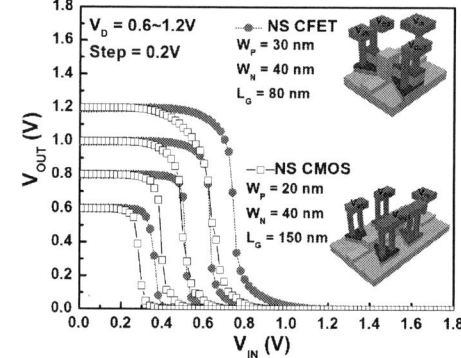

Fig. 5 VTCs of inverters based on the fabricated NS CMOS and NS CFETs.

ACKNOWLEDGMENTS

This work was supported in part by the Ministry of Science and Technology, Taiwan, under Grant MOST-107-2628-E-492-001-MY3, 109-2923-E-492-002-MY3, 109-2639-E-009-001, 109-2634-F-009-029 and 109-2636-E-006-004. Also in part by financially supported by the "Center for the Semiconductor Technology Research" from The Featured Areas Research Center Program within the framework of the Higher Education Sprout Project by the Ministry of Education (MOE) in Taiwan. The authors would also like to thank the support of Hitachi High-Technologies Corp.

REFERENCES

[1] H. S. P. Wong *et al.*, *Proc. of the IEEE* **87**, 537 (1999). [2]N. Singh *et al.*, *IEEE EDL* **27**, 383 (2006). [3] N. Loubet *et al.*, *VLSI*, 230 (2017). [4] C. L. Chu *et al.*, *IEEE EDL* **39**, 1133 (2018). [5] Y. H. Huang *et al.*, *IEDM*, 689 (2019). [6] J. J. Gu *et al.*, *IEDM*, 529 (2012). [7] J. Ryckaert *et al.*, *VLSI*, 141 (2018). [8] A. Mallik *et al.*, *VLSI*, 202 (2019). [9] C. Auth *et al*, *IEDM*, 673 (2017). [10] O. Varela Pedreira *et al.*, *IEEE IRPS*, 6B-2 (2017). [11] P. J. Sung *et al.*, *IEDM*, 504 (2018). [12] C. W. Yeung *et al.*, *IEDM*, 652 (2018). [13] M. Yang *et al.*, *IEDM*, 453 (2003). [14] J. Cai, *IEDM*, short course (2019). [15] H. Mertens *et al.*, *IEDM*, 828 (2017). [16] P. Schuddinck *et al.*, *VLSI*, 204 (2019). [17] S. W. Chang *et el.*, *IEDM*, 254 (2019). [18] H. C. Lin *et al.*, *IEDM*, 857 (2000).

Integration of ALD high-k dipole layers into CMOS SOI nanowire FETs for bi-directional threshold voltage engineering

Wonil Chung[1], Dongqi Zheng[1], Wei-E Wang[2], Mark Rodder[2] and Peide D. Ye[1]*

[1]School of Electrical and Computer Engineering, Purdue University, West Lafayette, IN 47907, USA
[2]Samsung Semiconductor, Inc., Advanced Logic Lab, Austin, TX 78746, USA
*Email: yep@purdue.edu

ABSTRACT

In this report, we successfully demonstrate V_T (threshold voltage) tuning in both n- and p- SOI nanowire FETs (NWFET) using ultrathin atomic layer deposition (ALD) dipole layers (Y_2O_3 and Al_2O_3) inserted directly under the ALD HfO_2 (3 nm). 0.7 nm of Y_2O_3 inserted between bottom SiO_x (< 1 nm) and top HfO_2 (3 nm) shifted the V_T by -138 mV and -58 mV for n- and p-NWFET, respectively while 0.7 nm Al_2O_3 shifted the V_T of n-NWFET by +219 mV and +134 mV for p-NWFET. V_T shift on planar SOI FETs were also observed with the similar trend only when such ALD dipole layers were inserted in direct contact with SiO_x. MOS capacitor's V_{FB} (flat band voltage) showed similar tunability with ALD dipole layer positioned near the SiO_x/Si channel interface. Intermixing multiple dipole layers can more precisely tune the V_T in desired direction and strengths by partially neutralizing each other.

INTRODUCTION

Abiding by the Moore's law, MOSFET dimension scaling has been aggressively executed in the past decades resulting in introduction of high-k oxides and metal gate (HKMG) process. However, controlling the threshold voltage (V_T) in scaled HKMG devices became more challenging due to unwanted V_T shifts which are thought to be accredited to dipole formation within the gate oxide stack [1]–[6]. The origin of dipole is not decisive yet; however, researchers suggest it could be related with oxygen vacancies [7], dopant electronegativity and ionic radius [8], ionic migration and reactivity at the interface [9] or oxygen areal density difference at the high-k/SiO_2 interface [2]. Multiple works employing high-k oxide on MOS capacitor structures have demonstrated dipole layer's V_T/V_{FB} tunability but such tuning in advanced 3D CMOS structures (fin or nanowire) for both n- and p-MOS were rarely reported. In this paper, we successfully integrate ultrathin ALD dipole layers to shift V_T bi-directionally in both nanowire and planar SOI FETs.

EXPERIMENT

Fabrication process flow of nanowire and planar SOI (Si thickness ~ 70 nm) FETs is shown in Fig. 1. Activation of ion implanted wafers was done in rapid thermal annealing (RTA) chamber at 1000 °C for 60 s. E-beam lithography was used throughout NWFET process. Buried oxide (BOX) under the defined fins was partially selectively etched with HF solution to release the nanowires from BOX. Dipole layers (Al_2O_3 or Y_2O_3) and the main HfO_2 dielectric (3 nm) were deposited in-situ using the same ALD. Trimethylaluminum (TMA), Tris(methylcyclopentadienyl)yttrium and Tetrakis (dimethylamino)hafnium (TDMAH) were used as precursors for Al_2O_3, Y_2O_3 and HfO_2, respectively at 250 °C. H_2O was used as oxidant. Capacitors studied in this report are fabricated on low-doped silicon bulk wafers using the same process as FETs except channel definition and S/D-related processes. Native SiO_x (< 1 nm) was used as an interfacial layer between the high-k and Si. Fig. 2 (a) and (b) show the false-colored SEM images of the fabricated nanowire structures immediately after step #7 of Fig. 1. Multiple parallel nanowires can be seen, 6 and 10 for the n- and p-NWFET, respectively. Fig. 2 (c) is the 3D illustration of a single nanowire with the underlying airgap. Fig. (d) is the microscope image of a planar SOI FET ready for electrical measurement.

RESULTS AND DISCUSSION

MOS capacitors were first fabricated to preliminarily test the effects of ALD dipole layer insertion on V_{FB} (extracted using C-V simulator [10]). Al_2O_3 dipole layers were inserted in 2 different positions, between SiO_x/HfO_2 and $HfO_2/$metal electrode as seen in Fig. 3 (a) and (b), respectively. Interestingly, the V_{FB} shifted negatively only when the dipole layer was inserted in direct contact with SiO_x (Fig. 3 (a)). Such trend can be more clearly observed in Fig. 4 where V_{FB} shifted more than +250 mV with $Al_2O_3 > 0.6$ nm. The ΔV_{FB} gradually

saturates when Al_2O_3 thickness (0 ~ 1.2 nm) reaches beyond 1 nm which coincides with previous reports [4], [11]. However, when the Al_2O_3 layer was placed on top of 3 nm HfO_2 away from SiO_x, V_{FB} was unaffected suggesting that the dominant interface that shifts V_T and V_{FB} is the SiO_x/high-k interface [11]. Similar trends were observed for Y_2O_3 dipole layer (0 ~ 1.5 nm) inserted under (Fig. 6 (a)) and above (Fig. 6 (b)) HfO_2. Fig. 7 depicts the extracted ΔV_{FB} with Y_2O_3 layer. Note that for Y_2O_3, the gradual V_{FB} saturation was not observed up to 1.5 nm. Unlike Al_2O_3, Y_2O_3 dipole layer shifts the V_{FB} negatively. This could be explained using the model proposed by K. Kita where the oxygen areal density difference between the SiO_x and dipole layer determines the strength and the direction of V_T shifts [2]. Planar SOI n-FETs were fabricated using the same dipole layers to validate the V_T shift trends. As presented in Fig. 5 and 8, 0.56 nm Al_2O_3 and 0.7 nm Y_2O_3 under 3 nm HfO_2 shifted the V_T by +118 mV and -224 mV, respectively.

To address the V_T engineering for the state-of-the-art 3D transistors, sub-1nm ALD dipole layers were integrated into complementary MOS (CMOS) nanowire structures. In order to place V_T values near 0 V symmetrically for both n- and p-NWFETs, it is important to be able to shift V_T in both positive and negative directions with desired strengths. Fig. 9 and Fig. 10 overlap the transfer curves acquired from multiple SOI n-NWFETs and p-NWFETs with either 0.7 nm of Al_2O_3 (blue circle) or Y_2O_3 (red triangle) ALD dipole layers under 3 nm HfO_2. It can be clearly seen from Fig. 11 (a) and (b) that Y_2O_3 shifted the V_T of both n-NWFETs by -139 mV and p-NWFETs by -58 mV while Al_2O_3 shifted the V_T of n-NWFET by +219 mV and p-NWFET by +134 mV. Shifts in V_T could also be resulted from various oxide interface related trap charges which are usually visible in the form of voltage hysteresis. To assess the effect of these charges on V_T shifts caused by the dipole layers, C-V hysteresis measurements on MOS capacitors fabricated simultaneously with NWFETs were acquired over 2.5 V of sweep range (Fig. 12). As seen in all 3 cases, voltage hysteresis caused by the wider voltage sweep range than transfer curves in Fig. 9 and 10 were observed to be negligible. This suggests oxide interface traps did not induce the significant V_T shifts observed after the insertion of dipole layers. Furthermore, although not shown in figures, mixing multiple ALD dipole layers weakened the V_T shift strength and even flipped the shift direction when the portion of the other dipole layer increased beyond certain point. Therefore, more precise V_T tuning is possible by optimizing the compositional ratio of intermixed ALD dipole layers and precursor pulsing strategies.

CONCLUSION

V_T shifts in n- and p- SOI NW/planar FETs and V_{FB} of MOS capacitors were studied with sub-nm ALD Y_2O_3 and Al_2O_3 dipole layers which shifted the V_T in negative and positive directions when inserted between the SiO_x and HfO_2. Such shifts were observed only when the dipole layers were deposited in direct contact with SiO_x suggesting that the dipole formed at SiO_x/high-k interface is responsible for the phenomenon. It was found that intermixing multiple dipole layers in different ratios can control the V_T shift strengths and directions. Dipole layer optimization strategies such as intermixing multiple ALD precursors per oxidation step or layer-by-layer intermixing of dipole oxides could be considered for more tailored V_T engineering.

This work is supported by Samsung Advanced Logic Lab.

References: [1] K. Iwamoto et al., VLSI, 2007. [2] K. Kita et al., IEDM, 2008. [3] Y. Yamamoto et al., IEDM, 2007. [4] Y. Kamimuta et al., IEDM, 2007. [5] P. Sivasubramani et al., VLSI, 2007. [6] S. Nittayakasetwat el at., J. Appl. Phys., vol. 125, no. 8, 2019. [7] S. Guha et al. APL, vol. 90, no. 9, 2007. [8] P. D. Kirsch et al., APL, vol. 92, no. 9, 2008. [9] J. Fei et al., JJAP, vol. 55, no. 4S, 2016. [10] G. Apostolopoulos et al., APL., vol. 84, no. 2, 2004. [11] K. Iwamoto et al., APL, vol. 92, no. 13, 2008.

2020 IEEE Silicon Nanoelectronics Workshop

3.2

1. SOI Wafer Wet Cleaning
2. Alignment mark/isolation dry etching
3. S/D region ion implantation
   N-type ($31P^+$): 15 keV, dose: $5 \times 15/cm^2$
   P-type ($49BF_2^+$): 15 keV, dose: $4 \times 15/cm^2$
4. Channel recess and dry etching (SF_6)
5. Activation: RTA, N_2, 1000 °C, 60 s, ATM
6. Fin patterning and dry etching (SF_6)
7. Nanowire release (HF solution)
8. ALD oxide deposition @250 °C
   Y_2O_3 and/or Al_2O_3 (< 1nm)+ HfO_2 (3 nm)
9. Forming Gas Annealing (FGA): 400 °C, 2 hr
10. S/D contact etching (BCl_3/Ar)
11. S/D contact metal dep./lift-off: Al, 150nm
12. Gate/S/D pad metal dep./lift-off: Al, 150nm
13. Post Metallization Annealing (PMA):
    RTA, FG, 250 °C, 5 min

Fig. 1. Key process flow of CMOS SOI nanowire FETs. Planar devices follow the same process except steps 4, 6 and 7.

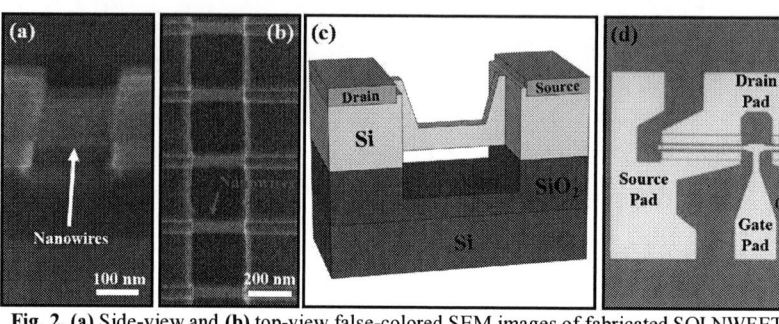

Fig. 2. (a) Side-view and **(b)** top-view false-colored SEM images of fabricated SOI NWFET with multiple parallel nanowires prior to step #8 in Fig. 1. **(c)** 3D illustration of the fabricated NWFET. **(d)** Microscope image of planar SOI device with channel width and length of 95 and 2 μm, respectively.

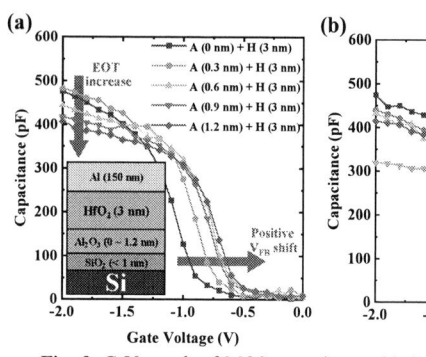
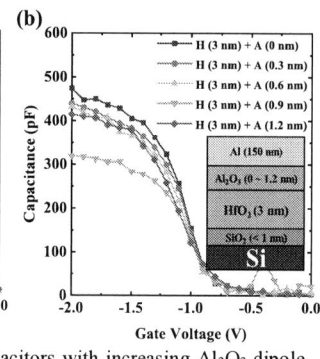
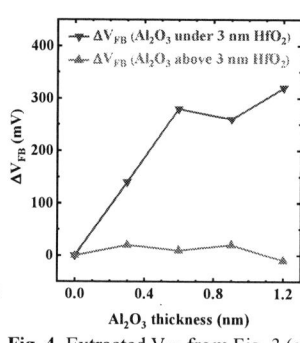
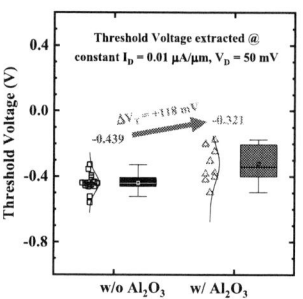

Fig. 3. C-V graph of MOS capacitors with increasing Al_2O_3 dipole layer thickness **(a)** under and **(b)** above the main ALD HfO_2 (3 nm). Al_2O_3 and HfO_2 are denoted as A and H, respectively. EOT increases and V_{FB} shifts positively with thicker Al_2O_3. However, the shift saturates gradually beyond 0.6 nm.

Fig. 4. Extracted V_{FB} from Fig. 3 (a) and (b). Note that if Al_2O_3 layer is deposited above the HfO_2, V_{FB} stays almost unchanged.

Fig. 5. Extracted V_T from planar SOI nFETs (**Fig. 2 (d)**), L/W = 40/95 μm) with and without 0.56 nm Al_2O_3 dipole layer under HfO_2 (3 nm).

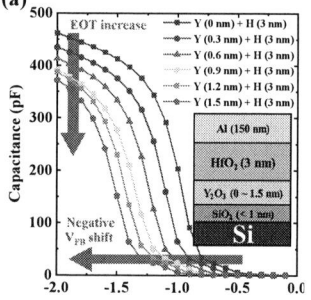
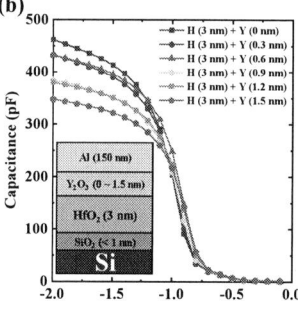
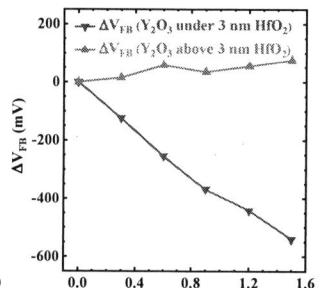
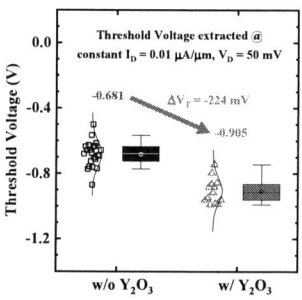

Fig. 6. C-V graph of MOS capacitors with increasing Y_2O_3 dipole layer thickness **(a)** under and **(b)** above the main ALD HfO_2 (3 nm). Y_2O_3 and HfO_2 are denoted as Y and H, respectively. Note the negative V_{FB} shift due to increasing Y_2O_3 dipole layer thickness.

Fig. 7. Extracted V_{FB} from Fig. 6 (a) and (b). Note that if Y_2O_3 layer is deposited above the HfO_2, V_{FB} stays almost unchanged.

Fig. 8. Extracted V_T from planar SOI nFETs (**Fig. 2 (d)**), L/W = 40/95 μm) with and without 0.7 nm Y_2O_3 dipole layer under HfO_2 (3 nm).

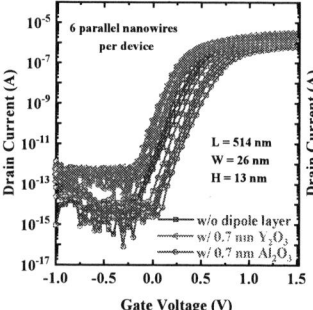
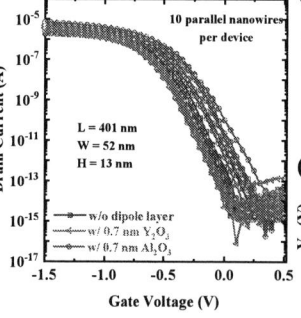
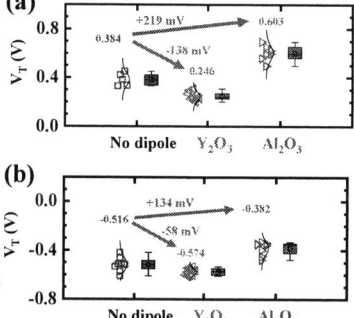
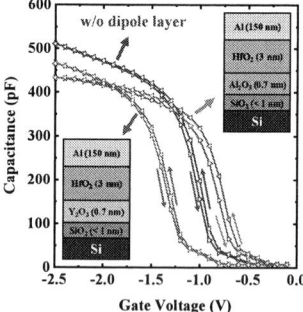

Fig. 9. Overlapped transfer curves of SOI n-NWFETs (**Fig. 2 (a)**, L/W/H = 514/26/13 nm) with 0.7 nm of dipole layers (Al_2O_3 or Y_2O_3) under 3 nm HfO_2.

Fig. 10. Overlapped transfer curves of SOI p-NWFETs (L/W/H = 401/52/13 nm) with 0.7 nm of dipole layers (Al_2O_3 or Y_2O_3) under 3 nm HfO_2.

Fig. 11. Extracted V_T from **(a)** n-NWFET (Fig. 9) and **(b)** p-NWFET (Fig. 10). 0.7 nm of Al_2O_3 and Y_2O_3 dipole layers shift the V_T in positive and negative directions, respectively for both n- and p-NWFETs.

Fig. 12. C-V hysteresis of MOS capacitors fabricated along with SOI n-NWFETs shown in Fig. 9. Hysteretic shifts in V_{FB} due to oxide charges are negligible after insertion of dipole layers even with 2.5 V sweep range.

2020 IEEE Silicon Nanoelectronics Workshop

978-1-7281-9736-4/20 $31.00 © 2020 IEEE

Vertical InAs/InGaAsSb/GaSb Nanowire Tunnel FETs on Si with Drain Field-Plate and EOT = 1 nm Achieving S_{min} = 32 mV/dec and g_m/I_D = 100 V^{-1}

Abinaya Krishnaraja, Johannes Svensson, Lars-Erik Wernersson
Lund University, Sweden. Email: abinaya.krishnaraja@eit.lth.se

Abstract

We present vertical InAs/InGaAsSb/GaSb nanowire tunnel FETs (TFETs) on Si demonstrating subthreshold swing (S) of 32 mV/dec with I_{ON} = 4 µA/µm for I_{OFF} = 1 nA/µm at V_{DS} = 0.3V. The demonstrated drive currents is the highest reported for a TFET with S below 40 mV/dec resulting in a transconductance efficiency as high as 100 V^{-1}. These results have been achieved by optimizing the source segment growth scheme and the device processing. The devices are compliant with low-power logic applications capable of operation at I_{OFF} = 100 pA/µm.

Introduction

Tunnel FETs, a promising candidate for future low-power electronics such as neuromorphic computing and Internet-of-Things (IoT), is estimated to provide advantages at the 3 nm node and beyond[1]. III-V TFETs have demonstrated sub-60 mV/dec slope operation along with better performance than their Si counterparts due to their ability to deliver higher drive currents [2]–[4]. However, achieving S much steeper than 60 mV/dec while maintaining technically relevant current levels has been a challenge. Among the reported devices with S < 40mV/dec, only operation at low drive currents (I_{60} = 10^{-7} – 10^{-4} µA/µm) has been achieved[5]–[7]. In this work, we present a TFET device with I_{60} = 26 nA/µm, the highest drive current reported for a sub-40mV/dec operating device. The modifications made to the composition and growth temperature of the source segment together with a scaled high-k, optimized bottom spacer, and introduction of a field-plate at the drain are key to the improvements in the overall performance.

Fabrication

The device schematic along with the process flow is illustrated in Fig. 1. The fabrication started with electron beam lithography definition of four differently sized Au dots (diameter 20, 24, 28, 32 nm) on a 260-nm-thick n$^+$-InAs layer deposited on a highly resistive Si (111) substrate. InAs/In$_{0.68}$Ga$_{0.32}$As$_{0.7}$Sb$_{0.3}$/GaSb vertical nanowires were then grown by VLS method as in [8] at a growth temperature of 440°C. After growth, the nanowires were digitally etched by ozone oxidation and citric acid to reduce the diameter to 17-27nm. A trilayer ALD 1nm/3nm/10nm Al$_2$O$_3$/HfO$_2$/Al$_2$O$_3$ high-k was deposited in which the gate oxide is Al$_2$O$_3$/HfO$_2$ with EOT ~1 and the third Al$_2$O$_3$ layer acts both as the bottom spacer to isolate the gate and drain contacts and also as a drain-side field-plate. After defining the field-plate length using a resist etch-back process, a 30-nm-thick W gate metal is deposited. The process is completed by depositing a resist spacer, forming via holes and depositing metal contacts.

Results

Transfer characteristics of a device with a nanowire diameter of 25 nm is shown in Fig. 2. The voltage sweeps in both bias directions exhibited sub 60-mV/dec behavior with a small hysteresis of 15 mV measured at I_{60} = 0.026 µA/µm. The device maintains S < 50 mV/dec over a current range of two orders of magnitude for V_{DS} = 0.05 – 0.3V reaching S_{min} of 32 mV/dec at V_{DS} = 0.15V and 0.3V. The drive current is I_{ON} = 4 µA/µm for I_{OFF} = 1 nA/µm at V_{DS} = 0.3V. The output characteristics in Fig. 3 confirms the band-to-band tunneling operation by the presence of an NDR with a Peak-to-Valley Current Ratio of 7.6. The statistical overview of the device performance in Fig. 4 shows that 32% of the devices operate below 45 mV/dec up to a V_{DS} of 0.3 V. Among the sub-60 mV/dec devices at V_{DS} = 0.3 V, a maximum I_{ON} of 8 µA/µm is achievable.

The present work is benchmarked in Fig. 5 against other III-V sub-60 performing TFETs in literature. Among the tunnel FET devices demonstrated so far, the nanowire devices show superior performance achieving the steepest slope at higher current levels. Fig. 6 shows this work benchmarked against our first generation TFETs [2]. The current devices show 10-15 mV/dec lower slope at V_{DS} = 0.1 – 0.3V and 2.5× higher I_{ON} for I_{OFF} = 1 nA/µm at V_{DS} = 0.1V and 2x higher transconductance efficiency. The improvements can be attributed to the altered source segment growth conditions and optimized device processing.

This work was supported by the Swedish Research Council and the European Union H2020 program SEQUENCE (Grant – 871764).

[1] IRDS 2017 Edition, IEEE Irds (2017) [2] E. Memisevic et al., IEDM, pp 19.1.1-19.1.4, 2016 [3] A. Alian et al., VLSI (2018) [4] G. Dewey et al., IEDM, pp 33.6.1-33.6.4 (2011) [5] E. Memisevic et al., Nanotechnology, vol. 29, p. 435201 (2018) [6] K. Tomioka et al., VLSI, pp 47-48 (2012) [7] K .Tomioka et al., Appl. Phys. Lett., 104, 073507 (2014) [8] E. Memisevic et al., Nano Lett., 17, p 4373-4380 (2017)

3.3

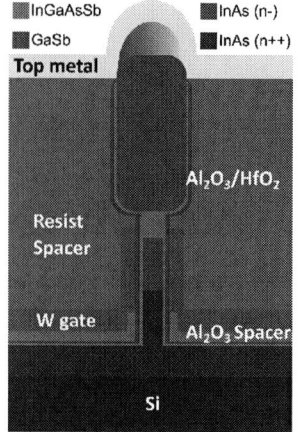

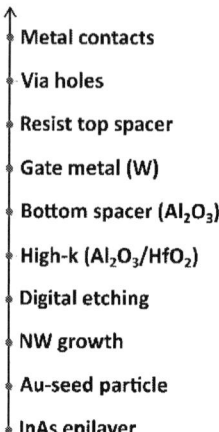

■ InGaAsSb ■ InAs (n-)
■ GaSb   ■ InAs (n++)

- Metal contacts
- Via holes
- Resist top spacer
- Gate metal (W)
- Bottom spacer (Al$_2$O$_3$)
- High-k (Al$_2$O$_3$/HfO$_2$)
- Digital etching
- NW growth
- Au-seed particle
- InAs epilayer

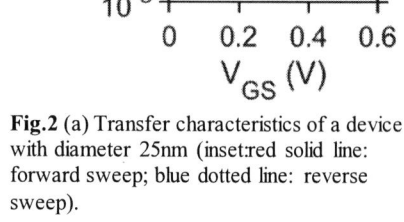

Fig.1 Illustration of the (a) device schematic and (b) process flow of the devices presented in this work.

Fig.2 (a) Transfer characteristics of a device with diameter 25nm (inset:red solid line: forward sweep; blue dotted line: reverse sweep).

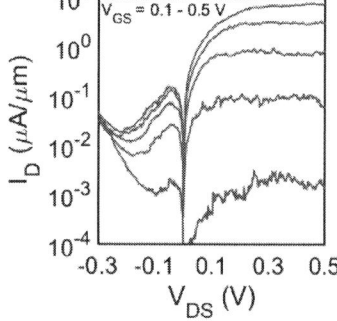

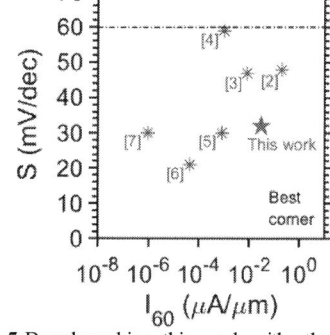

Fig.3 The output characteristics of the device in Fig 2. The forward bias clearly shows the negative differential resistance (NDR) behavior.

Fig. 4 I_{ON} of 32 sub-60 mV/dec performing devices plotted versus S. I_{ON} is determined for an I_{OFF} of 1 nA/μm at the corresponding drain bias.

Fig. 5 Benchmarking this work with other sub-60 mV/dec TFETs published by the authors and other reserach groups ($V_{DS} < 0.5$ V). The devices from this work shows the highest I_{60} for a sub-40 mV/dec device making it the steepest well-performing TFET.

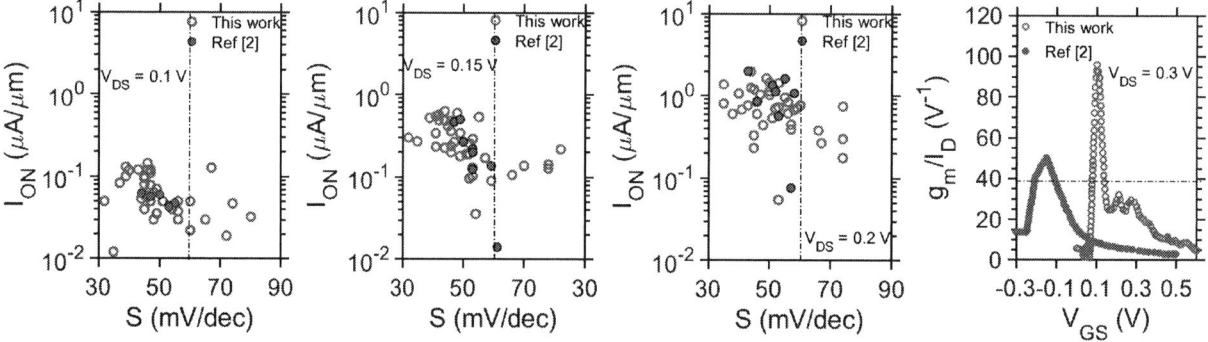

Fig. 6 Benchmarking of the devices from this work against devices from [2] which is the best III-V TFET demonstrated thus far. A significant improvement of nearly 10 mV/dec is observed in the slope while the current is over double that in [2] at $V_{DS} = 0.1$ V and similar to that in [2] at $V_{DS} = 0.15$ and 0.2 V. The transconductance efficiency is double that in [2] at $V_{DS} = 0.3$ V.

2020 IEEE Silicon Nanoelectronics Workshop

978-1-7281-9736-4/20 $31.00 © 2020 IEEE

Gap in pagination due to formatting issues.

Pages 19-20

4.1

Probabilistic computing based on spintronics technology

Shunsuke Fukami[1-5], William A. Borders[1], Ahmed Z. Pervaiz[6], Kerem Y. Camsari[6], Supriyo Datta, and Hideo Ohno[1-5]

[1]Laboratory for Nanoelectronics and Spintronics, Research Institute of Electrical Communication, Tohoku University, Japan
[2]Center for Innovative Integrated Electronic Systems, Tohoku University, Japan
[3]Center for Spintronics Research Network, Tohoku University, Japan
[4]Center for Science and Innovation in Spintronics, Tohoku University, Japan
[5]WPI-Advanced Institute for Materials Research, Tohoku University, Japan
[6]School of Electrical and Computer Engineering, Purdue University, USA
Email: s-fukami@riec.tohoku.ac.jp

Abstract — **There have been increasing demands on realizing computing hardware capable of addressing complex tasks that classical von-Neumann computers cannot readily execute. Here we show an unconventional computing scheme – probabilistic computing – based on a spintronics technology, which is promising to address various computationally hard problems. We present a proof-of-concept of the probabilistic computer with stochastic magnetic tunnel junctions, that can perform optimization problems at room temperature.**
Keywords: Probabilistic computing, Spintronics, Magnetic tunnel junction, Optimization problem

I. INTRODUCTION

Classical computers based on a concept of the Turing machine and von Neumann architecture are the heart of today's information society. Nonetheless, there are many computational problems, generally categorized into difficult classes of problems, which they cannot address efficiently. Optimization problem is a representative example of the difficult problems and development of dedicated hardware for optimization problems is one of the major challenges in recent electronics and computer engineering. For such computationally difficult problems, Richard Feynman gave a lecture in 1981 and suggested to actively utilize quantum mechanics [1], leading to enormous efforts to realize quantum computers. Nowadays, quantum annealing machines with thousands of quantum bits (qubits) are available to solve various optimization problems [2]. In addition, gated quantum computers with tens of qubits have also been intensively developed. Meanwhile, although not receiving as much attention, Feynman also suggested in the same lecture another concept of unconventional computing, a so-called probabilistic computing scheme, in which probabilistic behavior of physical systems is actively utilized to address difficult problems that are intrinsically probabilistic.

Here we show a proof-of-concept of the probabilistic computer based on a spintronics technology. We demonstrate integer factorization as an illustrative example of optimization problems [3].

II. RESULTS AND DISCUSSION

A. Stochastic Magnetic Tunnel Junction

Stochastic magnetic tunnel junction (MTJ) is an essential building block for the developed probabilistic computer. We prepared nanoscale stochastic MTJs by slightly modifying a current magnetoresistive random access memory (MRAM) technology [Fig. 1(a) left]. An increase in the free layer thickness from that for nonvolatile MRAM reduces the energy barrier between two energy minima. The reduced energy barrier provides a stochastic nature at ambient temperature, which can be controlled by the applied current as in the MRAM technology [Fig. 1(a) right].

B. Probabilistic Bit (p-bit)

The stochastic MTJ is incorporated into electronic circuits to form a probabilistic bit (p-bit), that behave as a binary stochastic neuron in neural networks. Right side of Fig. 1(b) shows a unit circuit of p-bit [4]. We confirm that the output of the prepared p-bit fluctuates in time between 0 and 1, and the probability to dwell at 0 and 1 can be controlled by input voltage. The time-averaged output voltage shows sigmoidal behavior with respect to the input [Fig. 1(b) left], indicating a capability to function as a stochastic neuron.

C. Integer Factorization using Probabilistic Computer

A number of developed p-bits are connected with a microcontroller and digital-to-analog converter to construct a rudimentary probabilistic computer [Fig. 1(c) left]. The microcontroller is programmed to give interactions among p-bits with synaptic weights to solve given problems, based on an algorithm inspired by the adiabatic quantum computing [2] and stochastic neural network. Using the constructed system, we demonstrate integer factorizations of 35 into 5 and 7 by 4 p-bits and 945 into 63 and 15 by 8 p-bits [Fig. 1(c) right]. From a circuit simulation, we also confirm that the demonstrated scheme is advantageous over the simulated annealing using digital CMOS circuits in terms of footprint and energy consumption.

III. CONCLUSION

In conclusion, we show a proof-of-concept of a probabilistic computer that Feynman envisioned. Since its building block, the p-bit, is prepared on the basis of current Mb- to Gb-scale MRAM technology and can be updated without any forced sequence unlike classical computers, it is possible to realize parallel and asynchronous operation of a large number of p-bits, leading to a new paradigm of efficient computation of difficult problems.

2020 IEEE Silicon Nanoelectronics Workshop

978-1-7281-9736-4/20 $31.00 © 2020 IEEE

ACKNOWLEDGMENTS

A portion of this work has been supported by ImPACT Program of CSTI, JST-CREST JPMJCR19K3, JSPS KAKENHI 19J12206, Cooperative Research Projects of RIEC, and ASCENT, one of six centers in JUMP, an SRC program sponsored by DARPA.

REFERENCES

[1] R. P. Feynman, "Simulating physics with computers," Int. J. Theor. Phys., vol. 21, pp. 467-488 (1982).

[2] T. Albash and D. A. Lidar, "Adiabatic quantum computation," Rev. Mod. Phys., vol. 90, 015002 (2018).

[3] W. A. Borders, A. Z. Pervaiz, S. Fukami, K. Y. Camsari, H. Ohno, and S. Datta, "Integer factorization using stochastic magnetic tunnel junctions," Nature, vol. 573, pp. 390-393 (2019).

[4] K. Y. Camsari, S. Salahuddin, and S. Datta, "Implementing p-bits With Embedded MTJ," IEEE Electron Device Letters, vol. 38, pp. 1767-1770 (2017).

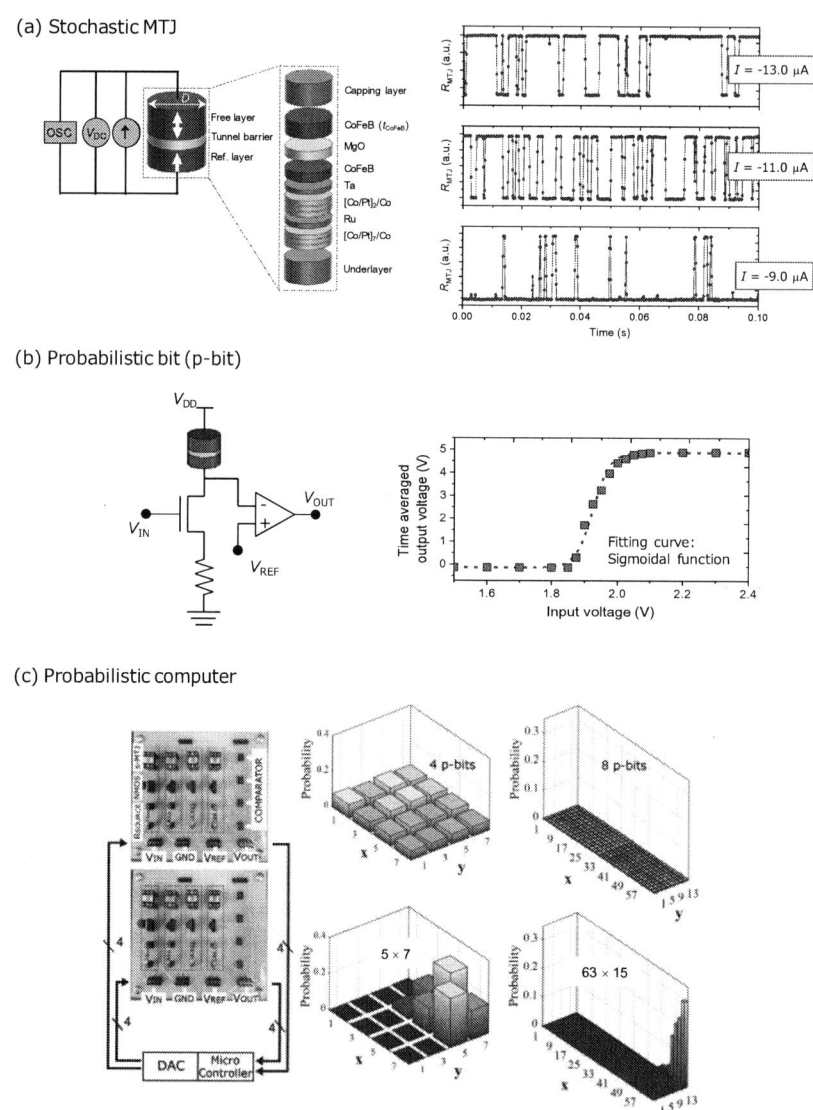

Fig. 1. Developed (a) stochastic magnetic tunnel junction, (b) probabilistic bit (p-bit), and (c) probabilistic computer with eight p-bits that address integer factorization.

A Self-align Gate-last Resistive Gate Switching FinFET Nonvolatile Memory[4.2]
Feasible for Embedded Applications

W. Y. Yang[1], E. R. Hsieh[2], C. H. Cheng[1] and Steve S. Chung[1]

[1]Department of Electronics Engineering, National Chiao Tung University, Taiwan, [2] Dept. of Electrical Engineering, National Central University, Taiwan

Abstract- In this work, we propose a FinFET resistance gate switching nonvolatile memory (RG-FinFET) which comprises a simple RRAM structure on top of a HKMG FinFET gate. The readout is taken from the FinFET V_T or I_D and its operation is based on the resistance switching instead of the conventional charge storage. The SET/RESET operation of the memory is made by the edge tunneling between top gate and source. The RG-FinFET shows ultra-low switching current, FORMing-free and ultra-fast SET/RESET speed. Comparing to conventional drain-type 1T1R, proposed gate-type 1T features low power consumption, smaller size in layout and larger window. It also exhibits excellent reliabilities, e.g., a very large window with highly stable retention, no sneak path, immunity to disturbances etc. Moreover, RG-FinFET is fully compatible with the logic CMOS technology and well-suited for NOR type memories, showing great potential for the future embedded applications.

1. Introduction

As Moore's law continues, conventional charge-based 1T memory like floating gate or SONOS has faced an inevitable scaling problem and reliability issues such as over-erase and slow erasing time by hot hole injection and data retention[1-2]. Besides, the latency between floating gate memory and DRAM is getting much larger. Among them, resistance changing, such as RRAM evolved as one of the promising solutions to meet the requirement. However, there are issues of sneak path, forming, and uniformity issues, in RRAM[3-4]. which need to be solved.

In 1T floating gate memory, the changes of the stored charges represent the information of "1" or "0". Here, we proposed a different concept by connecting an MIM to the top of a MOSFET, Fig. 1, named as resistive-gate FinFET NVM. If the resistance(R) of MIM is high resistance (reset), the connection of gate electrode can been seen as "open"; otherwise, if it is a low resistance state (set), this connection will be short. The performance of RG-FinFET will be demonstrated as follows. Endurance, retention, power consumption and disturbance test will then be examined.

2. Device Preparation and Experimental Model

TiN/HfO$_2$/Al$_2$O$_3$/TiN RRAM samples were prepared. By using an n-channel FinFET core transistor with its gate connected to RRAM as the unit cell (RG-FinFET). The process flows of RG-FinFET and conventional 1T1R structure are given in Fig. 2, where the RG-FinFET can effectively reduce redundant processes and improve layout efficiency (Fig. 3).

3. Results and Discussions
A. The Characteristics of RG-FinFET Cells

The basic characteristic of RRAM operation is shown in Fig. 4, by operating compliance current set to 500nA, we can operate RRAM with ultra-low current switching with FORMing-free characteristic. In DC sweep operation, SET can be done at 3V and RESET at -2V. As aforementioned, the architecture in Fig. 1, the change of gate resistance is used to differentiate "1" and "0". The readout from the drain current can be used to identify the two states. In other words, the cell of RG-FinFET can be switched between "1" and "0" states, or "OFF" and "ON" states respectively. Fig. 5 compares the I_D-V_G of RG-FinFET at different resistance states. At high resistance state, most of V_{gate} is developed across the RRAM so a low drain current was measured; at low resistance state, part of V_{gate} is across the gate dielectric, which gives rise to a large drain current. The two states, SET and RESET, provide two distinct levels of drain current.

B. Design of RRAM and FinFET Dimension Matching

In the RG-FinFET memory, FinFET with longer length has larger threshold voltage shift compared to that with smaller device. This is because gate leakage current is proportional to one with smaller area.

Fig. 6 shows a correlation between S.S. and channel length. We found that average S.S. degrades while gate length of FinFET increases. It is because the increase of gate length causes more gate leakage current, i.e., the influence from resistance will be more significant. S.S. curves at different resistance levels are also shown in the figure, and it was observed that S.S. degrades in the larger device. From I_d-V_g curves, the larger device has more voltage shift but ON current degrades. Devices having various widths are plotted in Figs. 7 and 8, in which we can see more current leaks in OFF state, and also degradation of ON current is found for the same reason of more gate leakage in larger width device. In gate-type 1T structure, the larger area between the gate and channel leads to more gate leakage current and thus loosing controllability. Read drain current distribution is shown in Fig. 9, in which smaller width shows more narrower ON state distribution and a tendency with smaller OFF state current.

C. Reliability Evaluation of RG-FinFET

Using pulse transient measurement, Fig. 10 shows that SET and RESET can be completed within 70 and 200ns respectively. Fig. 11 shows the endurance test, a significant 2,000x window can be achieved which is much larger than that of pure RRAM, and endurance over 10,000 set/reset cycles is also obtained. In Fig. 12, the retention can maintain over 100,000 seconds baked at 125 °C, excellent flat window in 10-year lifetime prediction. To introduce RG-FinFET into a memory array, we must consider carefully the mutual influence between adjacent cells. Therefore, two worse cases (SET and RESET) of RG-FinFET array have been considered in Fig. 13, and the results show very good immunity of the disturbances.

D. The Comparison with Conventional Drain-Type 1T1R Structure

Here, we will compare the FinFET and RRAM device pair in two types of structure, the novel 1T gate-type and conventional drain-type 1T1R. As shown in Fig. 14, the HRS and LRS are controlled by the drain and gate biases of the transistor. For example, at V_D= 0.7V, the drain current of 1T1R is about 0.01mA at HRS; 0.1 mA at LRS, from which the window is around 10. This is about the on/off ratio of single MIM RRAM. In Fig. 15, its operating current is larger than the RG-FinFET ones, in Fig. 4 where the proposed gate-type is self-limited by the FinFET. The comparison between RG-FinFET and drain-type 1T1R structure is shown in Table 1. It shows great advantages of the gate-type, especially very low energy efficiency, ultra-low current operation, and very large window.

In summary, we propose a FinFET resistance gate switching nonvolatile memory which comprises a simple RRAM structure and a HKMG FinFET. This memory features a FORMing-free, ultra-fast program speed (70ns), and good reliabilities (endurance, retention, disturbance) as shown in Table 2. In comparison to the RRAM in gate-type 1T1R structure, the smaller the FinFET, the better performance can be achieved, because the degradation of S.S is small and the ON/OFF ratio is large. Different from conventional drain-type 1T1R structure, gate-type 1T exhibits fairly *large window* featuring *low power* consumption. Its smaller cell area with the 1T structure shows great potential for embedded applications.

Acknowledgments This work was support in part by the Ministry of Science and Technology, Taiwan, under *Research of Excellence program* contract No. 109-2639-E009-001 and JDP-109-Y1-050, NDL, Taiwan.

References: [1] D. Kahng and S. Sze, *BSTJ*, p. 1288,1967. [2] M. White, *Ckts & Dev. Magazine*, p. 22, 2000. [3] H. D. Lee et al., *VLSI*, p. 151, 2012. [4] A. Kalantarian et al., *IRPS*, p. 6C.4.1., 2012. [5] Y. X. Liu et al., *IEEE SOI*, p. 113, 2013. [6] S. Tsuda et al., *IEDM Tech*, p. 11.1.1, 2016. [7] H. W. Pan et al., *IEDM*, p. 10.5.1, 2015. [8] E. R. Hsieh et al., *IEEE Trans. Electron Devices*, p. 4910, 2017.

4.2

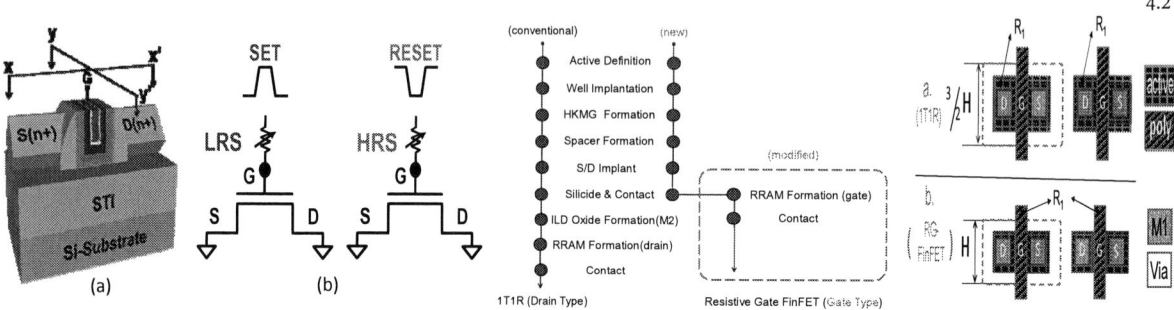

Fig. 1 The schematic of (a) gate-type 1T structure (RG-FinFET) NVM and (b) its operation for two logic states.

Fig. 2 Process flows of drain type 1T1R (left) and RG-FinFET (right). By integrating an RRAM on top of gate, the process flow can be effectively reduced.

Fig. 3 Enhance layout efficiency by gate type 1T structure(RG-FinFET).

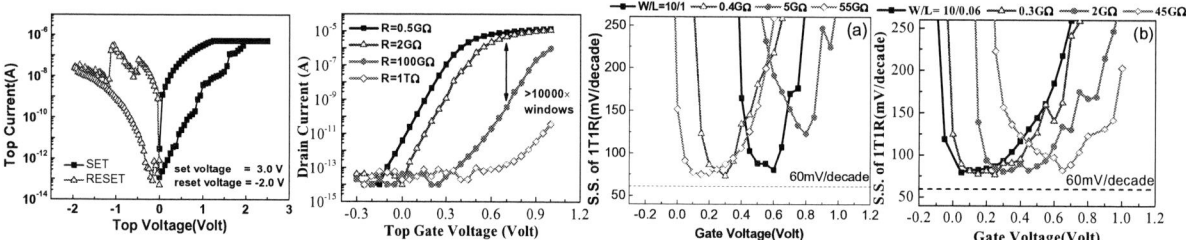

Fig. 4 The bipolar switching characteristics of RRAM show ultra-low current operation (<1 μA) under $|V_{gs}|$<3.0V with FORMing-free property.

Fig. 5 The I_D-V_G curves of RG-FinFET in LRS and HRS exhibit a large ON/OFF ratio (>10^{4}x). Owing to modulation of RRAM, two distinctive states can be readout.

Fig. 6 The S.S. characteristics of RG-FinFET with different gate lengths at different resistance values. S.S. degrades while gate length of FinFET increases.

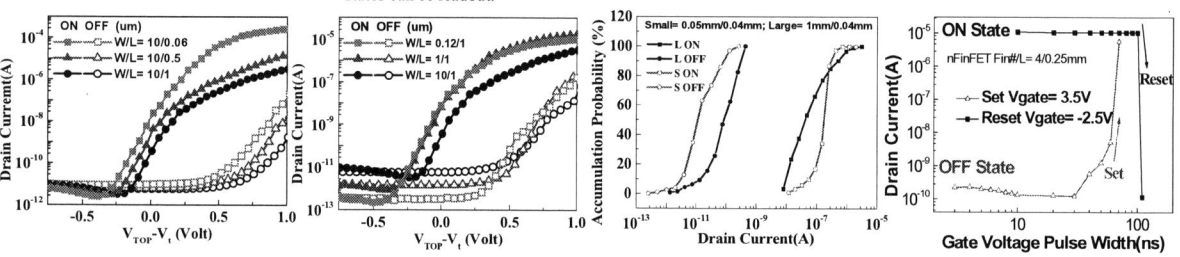

Fig. 7 The I_D-V_G curves of RG-FinFET with different gate lengths. ON/OFF ratio degrades while gate length of FinFET increases.

Fig. 8 The I_D-V_G curves of RG-FinFET with different widths. More leakage current in OFF state and degradation of ON current for the same reason of larger device in width.

Fig. 9 The readout drain current distributions of RG-FinFET with different widths, a large ON/OFF ratio with small widths is achieved.

Fig. 10 SET and RESET pulse transients of RG-FinFET show ultra-fast speed with <70ns SETTing time and <200ns RESETTing time.

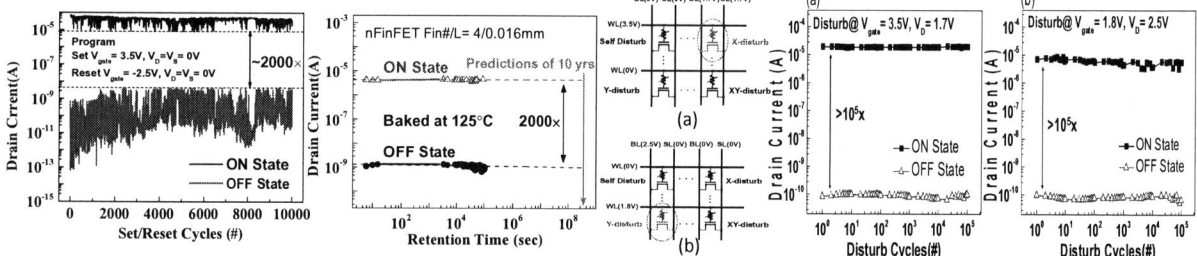

Fig. 11 The endurance of RG-FinFET shows more than 10^4 cycles with fairly large 2,000x window.

Fig. 12 Retention test of RG-FinFET shows 2,000x ON/OFF ratio over 100,000 seconds at 125 ˚C.

Fig. 13 Two-case studies of the disturbances in the RG-FinFET for (a) SET and (b) RESET show excellent immunities to the disturbances.

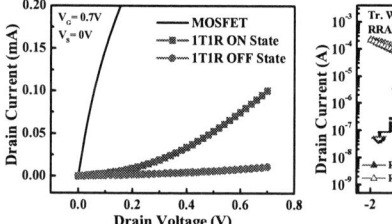

Fig. 14 I_D-V_D curve of drain-type 1T1R structure. ON and OFF states are controlled by the drain and gate biases of FinFET transistor.

Fig. 15 Switching characteristics of drain-type 1T1R. Almost the same curves implied that the same window for 1T1R and MIM. Also, operating current is much larger than the RG-FinFET ones, **Fig. 4**.

	Gate-Type (New)	Drain-Type
Unit Cell	1T	1T1R
Read Method	ΔV$_t$	ΔI$_D$
Switching	Electric field	Current driving
Min. ON Current	20 nA	100 μA
Max. OFF Current	0.1 nA	10 μA
ON/OFF Ratio	>2000	>10 (same as MIM RRAM)
Energy Efficiency(fJ)	30	6500
Memory Array	NOR, NAND	NOR

Table 1 The comparison between drain type and RG-FinFET. New gate-type 1T cell exhibits much larger window, ultra-low operation current and better energy efficiency.

	SOI FinFET SONOS [5]	Bulk FinFET MONOS [6]	HKMG FinFET [7]	HKMG FinFET [8]	This Work
Mechanism	Charge Trapping	Charge Trapping	Resistance Unipolar	Resistance Bipolar	Resistance Bipolar
Tech. Node	65 nm	16/14 nm	16 nm	14 nm	14 nm
Operation Voltage(V)	10	6	2-3	2-3	2.5-3.5
Speed(sec)	50 m	10 m	SET: 50 n RESET: 1 m	SET: 50 n RESET: 100 n	SET: 70 n RESET: 200 n
Endurance (#)	N/A	10^5	10^6	10^5	10^4
Retention(s)	10 years @25°C	10 years @150°C	10 years @150°C	10 years @125°C	10 years @125°C
Feature Size	N/A	11.3@14nm	43.3@16nm	N/A	13.75@14nm

Table 2 Benchmarks of key features in this work with other reported NVMs. RG-FinFET shows ultra fast program speed(70ns), excellent reliabilities.

978-1-7281-9736-4/20 $31.00 © 2020 IEEE

2020 IEEE Silicon Nanoelectronics Workshop

Design and Analysis of Core-Gate Shell-Channel 1T DRAM

Md. Hasan Raza Ansari[1], Jae Yoon Lee[1], Seongjae Cho[1,*], and Byung-Gook Park[2]

[1]Department of Electronic Engineering, College of IT Convergence Engineering, Gachon University, Seongnam-si, Gyeonggi-do 13120, Republic of Korea, [2] Department of Electrical and Computer Engineering with Inter-university Semiconductor Research Center (ISRC), Seoul National University, Seoul 08826, Republic of Korea
*E-mail: felixcho@gachon.ac.kr

ABSTRACT

The work showcases the utility of core-gate shell-channel (CGSC) architecture for one-transistor dynamic random-access memory (1T DRAM). The advantage of gate-all-around (GAA) is that the structure has less variability issue compared with other multi-gate devices. CGSC in GAA helps to achieve a fully-depleted channel and form deeper potential well for effective charge storage. The proposed 1T DRAM cell achieves retention time (T_{ret}) of ~3.5 s at 85 °C for a gate length of 100 nm and ~5 ms at 125 °C with gate length of 10 nm, even at elevated temperatures. The device demonstrates low power (25.18 nW for write "1") and energy (0.02 fJ for read "0") consumptions for DRAM operations.

I. INTRODUCTION

Increase in the demand on high-speed, larger data capacity, and more reliable memory devices for applications in emerging electronics, cloud computing, and internet-of-things (IoT) calls for ultimate device scaling [1,2]. Downscaling of cell transistor and capacitor in a DRAM unit cell has been hindered by short-channel effects (SCEs) and reduction in the storage capacitance, which degrades the DRAM performances [1]. Floating-body (FB) effect in the 1T DRAM cell based on silicon-on-insulator (SOI) can lead to a memory effect and has been a platform for 1T DRAM [3]. However, there is still room to resolve the SCEs and random dopant fluctuation (RDF) issues for the SOI-based 1T DRAM [4,5]. In this work, a novel device featuring core gate and shell channel is proposed and optimized to improve the memory performance metrics.

II. RESULTS AND DISCUSSION

CGSC device structure (Figs. 1(a) and (b)) for 1T DRAM application is validated by TCAD device simulation [6]. For activating the realistic models, 2-D simulation works have been conducted in a rotated structure (many of the models are not functional in the 3-D simulation) [6]. In order to validate the physical models, the simulation results have been also calibrated with the experimental ones from GAA device as shown in Fig. 2 [7]. The incorporated models include temperature and concentration-dependent mobility model, Lombardi's mobility model, Shockley-Read-Hall (SRH) recombination and generation model, impact ionization model, bandgap barrowing model, Auger recombination model, and band-to-band tunneling (BTBT) model [6]. Carrier generation and recombination in a 1T DRAM device are responsible for device metrics largely, and the optimal performances can be obtained through precisely adjusted voltage and timing scheme as shown in Table I [3,8]. A kink effect is observed in the output characteristics of the designed 1T DRAM due to the FB effect as shown in Fig. 3. Fig. 4 shows the transient analysis of the device along with voltage waveform over a period in sequence, W1−5×R1−W0−5×R0 (W: write, R: read). BTBT mechanism is used to perform write "1" (W1), which warrants lower power consumption and higher reliability compared with the write operation by impact ionization [9]. It is evident from Fig. 5(a) that applying a positive bias on the drain and a negative bias on the core gate reduces the effective tunneling width and generates holes for state "1" (higher read current) [9]. Evacuation of holes from the potential well (write "0") is done by increasing the potentials at the core and the outer gates, which allows the holes to recombine with electrons for state "0" (lower read current) (Fig. 5(b)). Sensing margin (SM) was defined as the difference between state currents and retention time (T_{ret}) was estimated when SM reaches down to 50% of its initial value [3],[8],[9]. T_{ret} is affected by device structure, bias,

dimensions, and temperature, and is guidelines to be > 64 ms at 85 °C, irrespective of L_g, by IRDS [10]. Fig. 6 shows the variation of state current (I_{D1} and I_{D0}) with hold time. A larger negative core gate hold voltage ($V_{GC,H}$) deteriorates I_{D0} due to increase of thermal generation and BTBT, which generates the holes and I_{D0} starts approaching to I_{D1}. However, a bias toward positive $V_{GC,H}$ perturbs I_{D1} due to increase of potential, which leads to recombination of the stored holds and I_{D1} goes to I_{D0}. Therefore, an optimal bias condition is required to control the generation and recombination of holes to increase T_{ret}. The maximum T_{ret} is achieved to be ~3.5 s with SM = 0.33 μA (L_g = 100 nm) at $V_{GC,H}$ = -0.3 V and 85 °C. Increase in underlap length (L_{un}) reduces the electric field (Fig. 7(a)), which reveals the trade-off relation between write time (WT) and T_{ret} (Fig. 7(b)). Reduction in electric field by employing the underlap structure helps to increase the T_{ret} owing to increased effective storage area and reduces BTBT rate during hold "0" operation. However, increase in L_{un} requires a longer time (higher voltage) to reach state "1" for a voltage. Thus, speed of the memory is degraded with L_{un} for a given write voltage. Thanks to the core-gate structure, gate controllability and deeper potential well for better data retention are acquired [5]. It is evident from Fig. 8(a) that reduction in L_g reduces the effective storage area, which results in source barrier lowering due to the SCEs. Reduction of L_g and elevation of temperature increases both state currents and SM (Fig. 8(b)). However, the shaded area in Fig. 8(b) indicates the regions where the SCEs are prominent. Here, a drastic change in I_{D0} is observed, and thus, SM is subject to decrease. Reduction of storage area and increase of temperature affects the storage capacitance and boosts the thermal generation and recombination, respectively, which degrades the retention characteristics of the memory cell [9]. Fig. 9 shows the variation of state currents over hold time with and without trap-assisted tunneling (TAT) model for L_g = 100 nm at 85 °C. It is explicit from Fig. 9 that the simulation with TAT model shows a fast degradation of state currents compared with the case turning on the model, due to increases in SRH recombination and generation. T_{ret} > 64 ms is achieved even at 85 °C for L_g = 100 nm. Table I also summarizes the power and energy consumptions during the DRAM operations. The device consumes an extremely low power of 25.18 nW and energy of 1.51 fJ during W1 by utilizing BTBT mechanism. During read operation, the device consumes even lower power and energy with a small drain bias. Fig. 10 demonstrates the comparison between DRAM performance metrics of inversion mode (IM) and junctionless (JL) devices in the same structure. JL 1T DRAM shows the advantage in terms of fabrication and speed over the IM device [9]. However, the JL device is not capable of achieving a long T_{ret} (only ~ 7 ms even at L_g = 100 nm) due to shallower potential well and shorter carrier lifetime. In comparison to a double-gate JL device in an existing literature [9], the core-gate shell-channel JL 1T DRAM cell accomplishes 5 times longer T_{ret}. It is revealed that the advantageous features brought by the novel structuring are more highlighted in the JL structure.

III. CONCLUSION

In this work, a novel 1T DRAM with core-gate shell-channel structure with better performance metrics and higher temperature tolerance has been proposed and analyzed. The device shows an elongated T_{ret} = ~ 3.5 s and the data is effectively retained with T_{ret} > 64 ms at 105 °C (L_g = 20 nm) and T_{ret} = ~ 5 ms at 125 °C (L_g = 10 nm). Its high applicability to both stand-alone and embedded DRAMs has been verified.

ACKNOWLEDGEMENT This work was supported by the Ministry of Trade, Industry and Energy through KSRC support program (Grant No. 10080513).

4.3

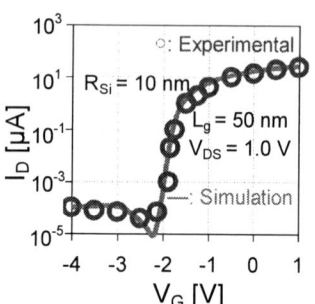

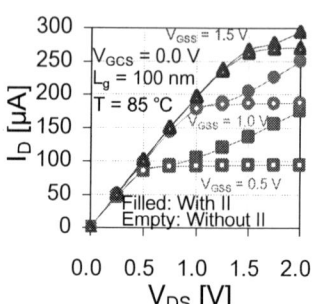

Fig. 1. Schematic of the core-gate shell-channel 1T DRAM. (a) Cross-sectional and (b) aerial views. Channel doping is $N_A = 10^{15}$ cm^{-3}, and source and drain doping concentrations are both $N_D = 10^{19}$ cm^{-3}. The equivalent oxide thickness (EOT) is 1 nm, Si channel thickness (R_{Si}) is 10 nm, and underlap length (L_{un}) is 10 nm, while gate length (L_g) is varied from 100 nm down to 10 nm. p^+ poly-Si gate was assumed so that the gate workfunction can make a deeper potential well for a range of elevated temperature from 85 °C to 125 °C.

Fig. 2. Comparison of transfer characteristic (I_D-V_{GS}) curves for calibration of physical simulation models with the experimental values [7].

Fig. 3. I_D-V_{DS} of the core-gate shell-channel 1T DRAM with $L_g = 100$ nm (floating-body effect is witnessed) at different gate voltages.

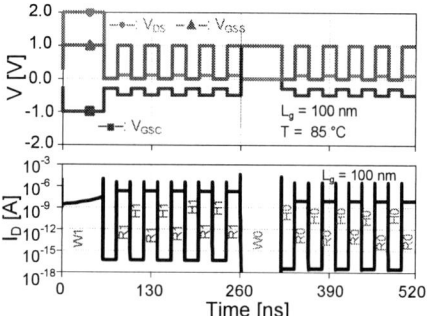

Fig. 4. Voltage waveform and transient analysis of the core-gate shell-channel 1T DRAM device.

Table I. Optimized voltage and timing schemes for operation of the core-gate shell-channel 1T DRAM cell.

Operation	V_{GS} [V]	V_{GSC} [V]	V_{DS} [V]	T [ns]	P [nw]	E [fJ]
W1	1.0	-1.0	2.0	60	25.18	1.51
Hold	0.0	-0.3	0.0	--	--	--
R1	1.0	-0.1	0.1	20	18.74	0.37
R0					0.87	0.02
W0	1.0	1.0	0.0	10	--	

* W1/W0: Write "1"/Write "0"
* H1/H0: Hold "1"/Hold "0"
* R1/R0: Read "1"/Read "0"

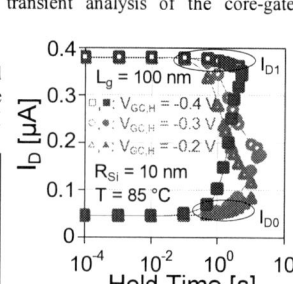
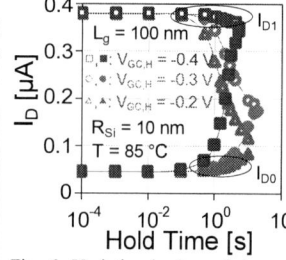

Fig. 6. Variation in I_{D1} and I_{D0} over hold time with different core hold voltage ($V_{GC,H}$) for $L_g = 100$ nm at 85 °C.

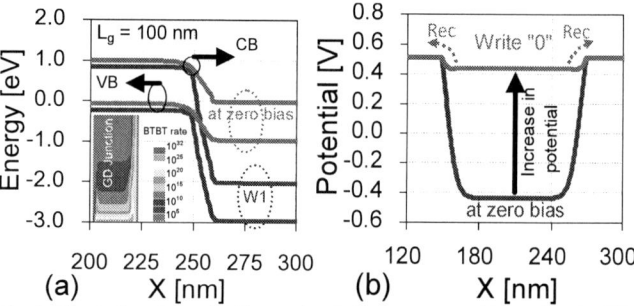

Fig. 5. Physical analyses. (a) Energy band diagram at zero bias and write "1" condition. (b) potential profile at zero and write "0" for $L_g = 100$ nm at 85 °C (CB: conduction band, VB: valence band).

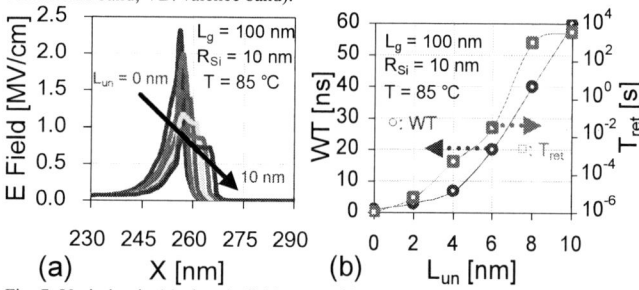

Fig. 7. Variation in (a) electric field at zero bias condition with different L_{un} and (b) WT and T_{ret} with L_{un} in the 1T DRAM cell with $L_g = 100$ nm at 85 °C

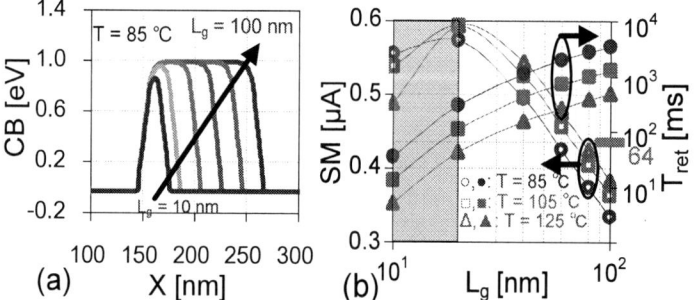

Fig. 8. Variation in (a) CB minimum energy contour at zero bias condition with different L_g's and (b) SM and T_{ret} with L_g at different temperatures.

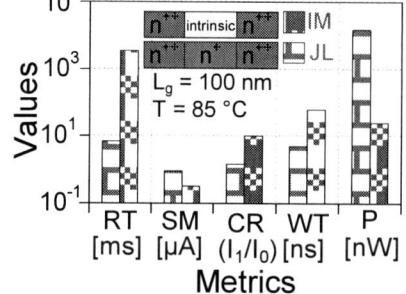

Fig. 10. Comparison between inversion-mode (IM) and junctionless (JL) MOSFET-based 1T DRAM cells in the same structure with the same critical dimensions at 85 °C.

REFERENCES

[1] S. K. Kim *et al.*, *MRS Bull.*, vol. 43, no. 5, 2018. [2] R. H. Dennard, *Nat. Electron.*, vol. 1, no. 6, 2018. [3] S. Okhonin *et al.*, in *IEEE SOI Conference*, 2001. [4] J. P. Colinge, *Microelectron. Eng.*, vol. 84, no. 9–10, 2007. [5] H. M. Fahad *et al.*," *Nano Lett.*, vol. 11, no. 10, 2011. [6] Atlas User's Manual. Silvaco Int., 2016. [7] S.-J. Choi *et al.*, *IEEE EDL.*, vol. 32, no. 2, 2011. [8] G. Giusi *et al.*, *IEEE TED*, vol. 57, no. 8, 2010. [9] M. H. R. Ansari *et al.*, *IEEE TED*, vol. 65, no. 3, 1205, 2018. [10] IRDS 2018 online available at http://www.irds.net.

2020 IEEE Silicon Nanoelectronics Workshop

Flash Memory based Computing-In-Memory to Solve Time-dependent Partial Differential Equations

Yang Feng, Xuepeng Zhan, Jiezhi Chen[*]

School of Information Science and Engineering, Shandong University, Qingdao, P. R. China

*Email: chen.jiezhi@sdu.edu.cn

Abstract—To construct a time-dependent partial differential equation (PDE) solver to improve the computation efficiency with ultra-high accuracy, a flash memory based computing-in-memory (CIM) hardware system is proposed in this work. On the basis of spice simulations on 65nm NOR flash memory technology together with a precision-extension technique, 64-bit precision accuracy can be achieved successfully. This is important to design PDE solvers with high efficiency and precision that required in scientific research and engineering.

1. Introduction

Computing-in-memory (CIM) has become a rising concern as its potential to break the "memory gap" of conventional Von Neumann architecture. Many emerging non-volatile memories (NVM) represented by resistive random access memory (RRAM) have the ability to implement CIM [1-2]. Time-dependent partial differential equations (PDEs) are widely used in many scientific researches, such as the wave equation, the heat-conduction equation and the Schrödinger equation. Recently, Wei D. Lu *et al.* proposed an algorithm of memristor-based CIM system as PDE solvers [3]. IBM group proposed an architecture by designing a mix-precision CIM system as a linear equation solver, using PCRAM arrays to perform approximate calculations while the high-precision processing is processed in central processing unit (CPU) [4]. However, RRAM needs to work over a few uA to keep sufficient on/off ratio, which caused large accumulation currents when the RRAM array is large. Compared to RRAM, flash memory is a memory technology satisfying the requirements of non-volatile, ultra-high density and low cost. Besides, Flash memory enables to realize vector-matrix multiply-and-accumulate (MAC) operation with high accuracy and good tolerance to device error [5-7].

In this work, we proposed a flash memory based CIM hardware system to solve time-evolving problems that requiring high-precision and accurate solutions. Based on finite difference method and Jacobi iteration algorithm, we show the solution of the wave equation. Matrix partition methods utilize the structural characteristics of flash memory array effectively. Also, in order to get better accuracy, we use precision extension technology to extend the accuracy from low precision (4-bit) to high precision (64-bit).

2. The CIM hardware system Scheme and Results

The core processing unit is constructed with flash memory arrays of 65nm NOR flash memory technique in which gate length (L_g) and width (W_g) of the cell are 130/80nm, as shown in Fig. 1 (b). Characterization of the saturation regime are shown in Fig. 1 (a). We use the threshold voltages and gate voltage pulse time as two multipliers, and the results can be acquired by measuring the integral of currents. It should be noted that, in our flash array, 5 cells in the same position of the 900×1 array are commonly used as the input of an integrator since they are used to acquire the same unknown.

The time-dependent PDE we solved is a wave equation as:

$$\frac{\partial^2 u}{\partial t^2} = 0.5 * \left(\frac{\partial^2 u}{\partial x^2} + \frac{\partial^2 u}{\partial y^2}\right)$$

Following FDM, we discretized it as:

$$\frac{u_{i,j}^{k+1} + -2u_{i,j}^k + u_{i,j}^{k-1}}{(\Delta t)^2} = 0.5 * \frac{u_{i+1,j}^k + u_{i,j+1}^k - 4u_{i,j}^k + u_{i-1,j}^k + u_{i,j-1}^k}{h^2}$$

where u represents the wave height; i is the x-axis grid index, j is the y-axis grid index; h is the distance between two neighboring grid points along the x-axis or y-axis; k is the time index; Δt is the numerical time step. Equation can be re-written in a five-point stencil format as:

$$u_{i,j}^{k+1} = 2u_{i,j}^k - u_{i,j}^{k-1} + 0.5(\frac{\Delta t}{h})^2 * (u_{i+1,j}^k + u_{i,j+1}^k - 4u_{i,j}^k + u_{i-1,j}^k + u_{i,j-1}^k)$$

We solved the wave equation in a 30×30 grid, with spatial steps h=0.15 and time step $\Delta t = 0.15$. We divided the solution of the unknown vector into 30×30 2D distributed grids, indicating the coefficient matrix is a 900×900 matrix. The coefficient matrix is mapped into the flash memory array as threshold voltages, then the unknown vector is transformed into pulse time as input. The large coefficient matrix magnifies the hardware challenges, and the matrix is a sparse matrix with only a few non-zero elements. To make the most of the array, we extract five non-zero columns in the matrix (each column is a 900×1 vector). To represent 900×1 vector, five 900×1 array is constructed, and the output is the sum of the charge in the same position of five array. Fig. 2 shows this method of 3×3 grid as an example. By inputting the $u^{(k)}$ obtained from the last iteration into the next iteration, the $u^{(k+1)}$ is obtained. The process is then repeated iteratively by feeding $u^{(k+1)}$ to the array as the next input vector and the new state of the system is computed iteratively. Fig. 4 shows the initial condition and the result of floating-point solver, respectively. In order to improve the solution accuracy of the hardware system, the precision-extension technique is adopted, as shown in Fig. 5 and Fig. 6.

3. Conclusions

This work proposed a novel flash memory based nonvolatile computing-in-memory (CIM) hardware system to solve time-evolving problems that requiring high-precision and accurate solutions. By using the matrix features after FDM, we optimize the matrix structure to utilize flash memory arrays more efficiently. Moreover, by using the precision-extension technique, 64-bit precision accuracy is also realized, which is important for its future applications in scientific research and engineering.

Acknowledgment: This work is supported by China Key Research and Development Program (2016YFA0201802), the Fundamental Research Funds of Shandong University and the National Natural Science Foundation of China (91964105, 61874068).

References: [1] M. Prezioso, et al., Nature, 2015, 521(7550): 61-64.; [2] L-X. Xia, et al., DAC 2016; [3] W-D. Lu, et al., Nature Electronics,2018,1(7), 411-420; [4] M. Le Gallo, et al., Nature Electronics, 2018, 1(4): 246-253; [5] S-T. Lee, et al., IEDM 2019: 38.4.1-38. 4.4; [6] K. Ishimaru, IEDM, 2019. [7] X. Guo, et al., IEDM, 2017: 6.5. 1-6.5. 4.

2020 IEEE Silicon Nanoelectronics Workshop

4.4

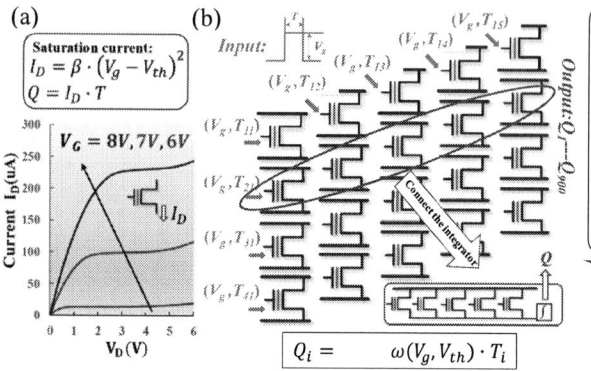

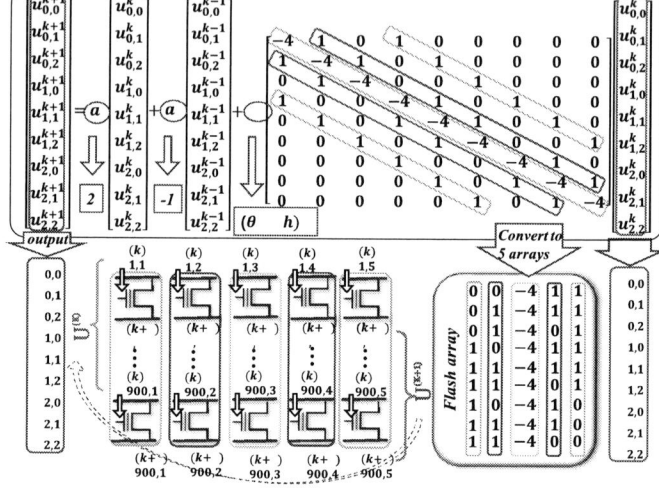

Fig. 1. (a) Cell saturation current formula and cell currents as sweeping VD (Vth=5V). (b) The values of the elements are mapped as pulse time and threshold voltages. Vector–matrix multiplication operations of the slice are performed by supplying the input vector as voltage pulses to the rows and reading out the charge outputs at each.

Fig. 2. In an example where the equation's domain is discretized into a 3×3 grid. Iterative relations of time-dependent PDE is transformed into matrix form for the purpose of mapping matrix into flash memory array. The output charge represents a new estimate next time.

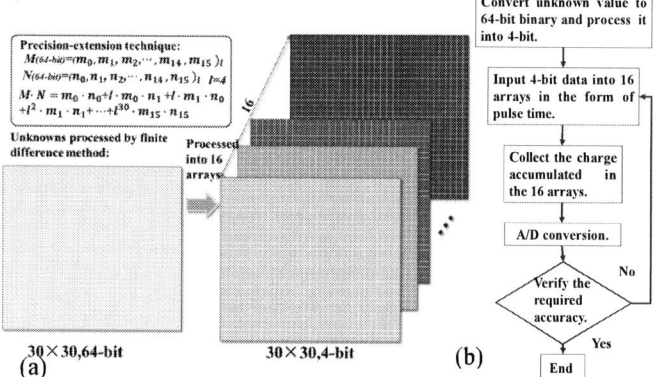

Fig. 3. (a) Illustration of precision extension technology. Operations of 64-bit numbers can be processed in a physical system based on 4-bit devices. Original 64-bit numbers can be represented by 4-bit numbers (b) Flow chart of the 900×5 matrix iteration process with precision extension technology applied.

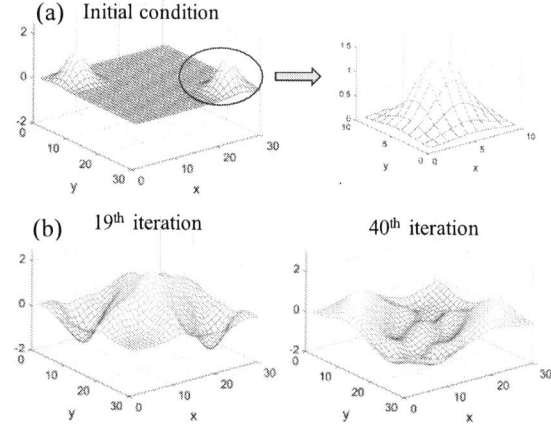

Fig. 4. (a)The initial condition of time-dependent PDE. (b) The measured outputs at iteration numbers 19 and 40, where the z-axis represents the wave height.

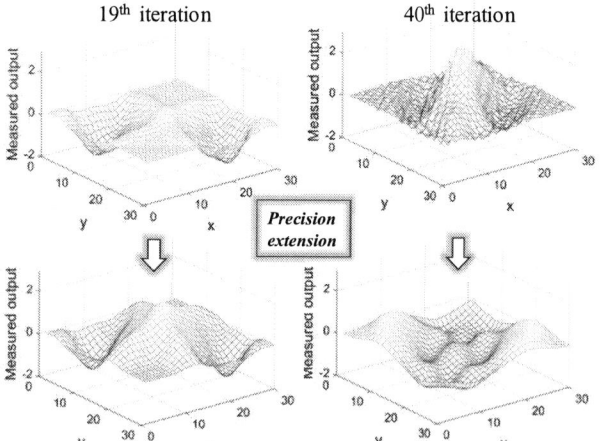

Fig. 5. Improvement of the accuracy of the measured outputs at iteration numbers 19 and 40 after using the precision-expansion technique.

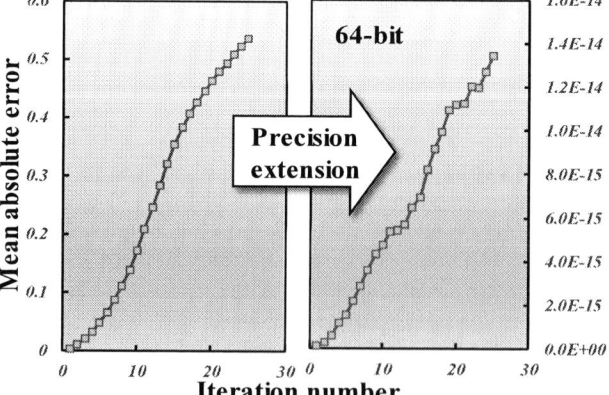

Fig. 6. Evolution of the mean absolute error for precision-extension technique, measured against the exact numerical solution.

2020 IEEE Silicon Nanoelectronics Workshop

978-1-7281-9736-4/20 $31.00 © 2020 IEEE

A Novel High-Density and Low-Power Ternary Content Addressable Memory Design Based on 3D NAND Flash

H. Z. Yang[1], P. Huang[1†], R. Z. Han[1], Y. C. Xiang[1], Y. Feng[2], B. Gao[3],
J. Z. Chen[2], L. F. Liu[1], X. Y. Liu[1], J. F. Kang[1]

[1]Institute of Microelectronics, Peking University, Beijing 100871, China
[2]School of Information Science and Engineering, Shandong University, Qingdao 250100, China
[3]Institute of Microelectronics, Tsinghua University, Beijing 100084, China
Email: [†]phwang@pku.edu.cn

Abstract — A novel ternary content addressable memory (TCAM) design based on 3D NAND Flash is proposed, in which two adjacent flash cells consist of one TCAM cell. HSPICE simulations show that the proposed TCAM has the ultra-low energy consumption of 0.298 fJ/bit/search for a 64-bit word and the cell density with 96-layer 3D NAND is reduced ~582 times compared to the conventional CMOS-based TCAMs.

I. INTRODUCTION

Ternary content addressable memory (TCAM) is widely used in various applications that need fast data searching, such as pattern recognition [1], network router [2], and approximate computing [3]. However, conventional CMOS-based TCAM consumes large circuit area and its standby power consumption increases when scaling down. To overcome these challenges, TCAM designs based on emerging non-volatile memories (NVMs) such as RRAM [4], PCM [5], STT-MRAM [6] are proposed to eliminate the standby power and enlarge storage capacity. Yet, the fabrication technology of most emerging NVMs are still not mature enough for large-scale integration. Among all the NVMs, 3D NAND Flash may be a potential candidate thanks to its industrial technology and invincible high-density [7]. In this work, we propose a novel 3D NAND based TCAM design, in which two flash cells are used to consist of one TCAM cell. After verifying NAND transistor model with experimental data, the data search sensing margin, speed, power consumption and density are evaluated based on the HSPICE simulations. The simulated results imply the 3D NAND based TCAM is a promising technology for high density data search with large capacity. The detailed simulations for the impacts of 3D NAND's characteristics on the performance of the proposed TCAM design are performed to guide future design optimizations.

II. PROPOSED TCAM DESIGN

Fig. 1 shows the structure of the proposed TCAM design based on 3D NAND Flash array. The search data is converted into voltage pulses and applied to the bit lines (BLs). The data to be searched is stored in the array beforehand by program and erase operations. The search results are read out in parallel in the form of current flowing in the source lines (SLs). The number of BLs should be twice of the word width. The TCAM cell design is shown in Fig. 2 (a). The store and search scheme are illustrated in Fig. 2 (b) and (c), respectively.

III. FEASIBILITY DEMONSTRATION

To demonstrate the feasibility of the proposed TCAM, we establish the compact model of NAND transistor for circuit simulation. Fig. 3 shows the I_d -V_g curves fitted with BSIM model on the basis of the measured data in Ref. [8]. To consider parasitic effects, we use the unit cell as shown in the inset of Fig. 3 and the values are calculated according to Ref. [7]. From the simulation results of Fig. 4, we can see that the current of fully matched case is much lower than the mismatched currents, which verifies the feasibility. Sensing margin, defined as the ratio of 1-bit mismatched current with the fully matched case, is ~200x. As for the currents tend to rise first and then descend to a stable level, we think it's from the gate's coupling capacitance. Considering the sensing margin may be affected by V_{th} variation, we add the distribution and it's still adequate for the sensing margin as shown in Fig. 5.

IV. PERFORMANCE EVALUATION

To analyse impacts of various factors on the performance of the proposed TCAM, we carried out further simulations. Fig. 6 shows the sensing margin at different V_{th} variations and it's usually possible to tighten the variation under $200mV$ [9]. Fig. 7 indicates the negative relationship of sensing margin with the word width, which is because the sum of I_{off} gets comparable with I_{on} as the increasing word width. The influence of parasitic effects on the searching speed is shown in Fig. 8, as a call for the improvement of technology. Fig. 9 and Fig. 10 show the effects of V_{pass} on the searching speed, energy consumption and sensing margin. Fig. 11 and Fig. 12 indicate the impacts of 3D NAND layers on these target values. A trade-off among the three targets should be noticed when setting conditions. The energy consumption is 0.298 fJ/bit/search for a 64-bit word at the lowest-power condition. It should be noticed that we can further reduce the energy consumption by lowering V_{BL}. Fig. 13 shows the 3D NAND based TCAM's density is about ~582x higher than the SRAM-based when the number of layers is 96.

V. CONCLUSION

A 3D NAND-based TCAM design with high density and low power consumption has been presented, in which two flash cells are used to consist of one TCAM cell The simulation results indicate that the proposed design can reduce energy consumption, with 0.298 fJ/bit/search for a 64-bit word. In 96-layer 3D NAND, the cell density is ~582 times higher than that of the traditional CMOS-based TCAMs. The design will be beneficial for the search applications which require low energy consumption and large volume data.

ACKNOWLEDGMENTS

This work was supported in part by National Natural Science Foundation of China under Grant 61841404, and National Key Research and Development (2016YFA0201804, 2019YFB2205100).

REFERENCES

[1] D. R. B. Ly et al., IEDM 2019. [2] J. Huang et al., IEEE Trans on Networking, 26, p.976, 2018. [3] A. Kumar et al., NVMTS 2016. [4] R. Z. Han et al., Jpn. J. Appl. Phys, 57, 04FE02, 2018. [5] J. Li el al., JSSC, 49, p.896,2014. [6] B. Yan et al., NVMTS 2016. [7] P. Wang et al., IEEE Trans on VLSI, 27, p.988, 2019. [8] H. Lue et al., IEDM 2017. [9] L. Cui et al., ICCD 2019.

2020 IEEE Silicon Nanoelectronics Workshop

4.5

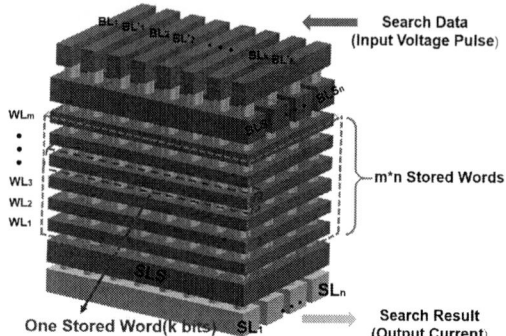

Fig. 1 Structure of TCAM design based on 3D NAND Flash array with (m*n) k-bit words.

Fig. 2 (a) TCAM cell design based on 3D NAND Flash. (b)Definition of TCAM logic with NAND cells. (c) Scheme of searching operation with '0', '1' and 'X' in TCAM logic.

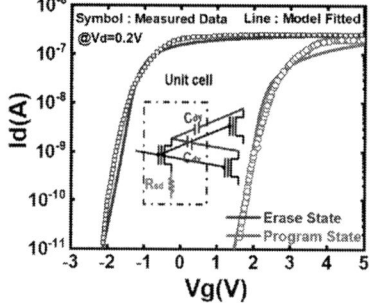

Fig. 3 I_d-V_g curve fitted with BSIM model on the basis of the measured data in Ref. [8]. The inset is the unit cell used in the simulation of 3D NAND array. The value of Cdx, Cdy and Rsd is 0.03fF, 0.02fF and 4Ω. R_{BL} and R_{SL} is set as 0.5Ω and 10Ω.

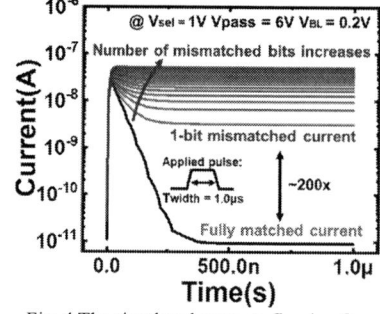

Fig. 4 The simulated currents flowing from source lines with sixteen 16-bit words. The simulation results show that the ratio of mismatched currents to the fully matched current is large enough.

Fig. 5 The distribution curves of the fully matched case and 1-bit mismatched case considering V_{th} variation. Gray lines are raw data and solid lines are the mean current. The mismatched currents are so close together that they look like only one curve.

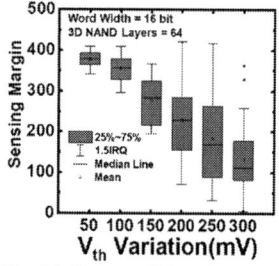

Fig. 6 Influence of V_{th} variation in 3D NAND array on sensing margin of the TCAM system.

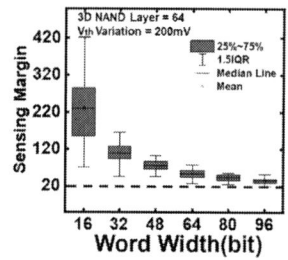

Fig. 7 Influence of word width on sensing margin of the TCAM system, setting σ_{Vth} = 200mV.

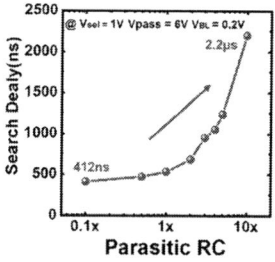

Fig.8 The impacts of parasitic resistance and capacitance in 3D NAND array on the search delay of the TCAM system.

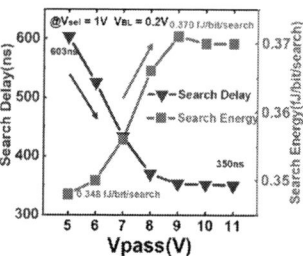

Fig. 9 Search delay and energy consumption vs the WL voltage applied to the unselected layers. Increased V_{pass} reduce unselected layers' cell resistance.

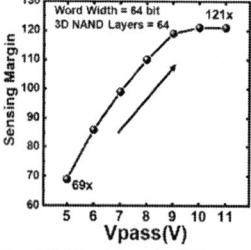

Fig. 10 The impact of V_{pass} on the sensing margin. Lower unselected layers' cell resistance causes higher V_{ds} of selected layers and therefore, higher sensing margin.

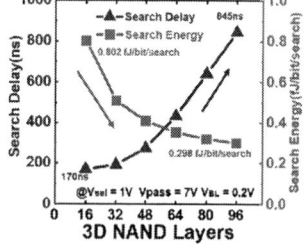

Fig. 11 Search delay and energy consumption vs 3D NAND layers. The adding layers bring more RC on the signal path and as a result, lower V_{ds} on selected layers.

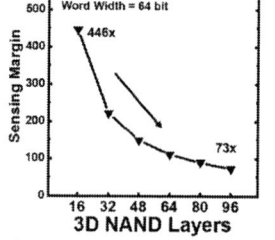

Fig. 12 The relationship of the proposed TCAM's sensing margin with 3D NAND layers .

Fig. 13 Cell area of the 3D-NAND-based TCAM and cell density ratio of the SRAM-based TCAM with the number of 3D NAND layers.

2020 IEEE Silicon Nanoelectronics Workshop

978-1-7281-9736-4/20 $31.00 © 2020 IEEE

5.1

Silicon quantum dot devices for spin-based quantum computing

Tetsuo Kodera

[1]Department of Electrical and Electronic Engineering, Tokyo Institute of Technology, Japan
Email: kodera.t.ac@m.titech.ac.jp

Abstract — Demonstration of long spin coherence time in isotopically-enriched silicon (Si) qubits have accelerated research of spin-based quantum computing. The next major issue is large scale integration of qubits. Physically-defined quantum dot (QD) on Si-on-insulator (SOI) is one of the promising candidates for multiple scaled qubits. This work demonstrates the fabrication of physically-defined QDs with control gates and charge sensor, and the observation of Pauli spin blockades in triple QDs and p-channel QDs.

Keywords: Si qubits, quantum dots, spin

I. INTRODUCTION

Scalable architectures such as solid state qubits are necessary to perform useful computations. Silicon quantum dots (Si QDs) are investigated intensively because of the compatibility of the current technologies in very large scale integration. Si atoms contain only 4% of isotope with nuclear spins and hence electron-spin coherence time is long in Si QDs. Moreover, using isotopically-engineered Si [1], qubits with long electron spin coherence time and high fidelity of gate control have been demonstrated [2,3]. By the demonstration of 2-qubit logic gates [4], Si qubit researches have been accelerated [5]. The next major remaining issue is large scale integration of qubits.

The advantage of Si based systems is that they can utilize CMOS technologies for large-scale integration. At Tokyo Tech, we have studied physically-defined QDs fabricated on Si-on-insulator (SOI) substrates [6-12], which are promising for high density integration. Here, we describe the fabrication and transport properties of physically-defined Si QDs, with an emphasis on Pauli spin blockade (PSB) [13], which is essential for spin-based qubits.

II. FABRICATION

The physically-defined Si QD structures are fabricated as follows: A 40-nm-thick Si-on-insulator (SOI) substrate is etched to form patterned trenches defining QDs, side gates (SGs), and a charge sensor (CS) using electron beam lithography and reactive ion etching techniques. After cleaning, the Si surface was thermally passivated. A 75-nm-thick SiO_2 layer and a poly-Si top gate are deposited by low-pressure chemical vapor deposition. The top gate covers an area of the order of 100 mm^2 covering the QDs. Phosphorus ions are implanted to make the SOI metallic at cryogenic temperature except for the area covered by the top gate, followed by thermal activation. A 300-nm-thick aluminum layer is thermally evaporated for Ohmic contacts after contact holes have been made in the SiO_2 layer using

hydrogen fluoride. Finally, the sample is annealed in a forming gas atmosphere.

III. PAULI SPIN BLOCKADE IN TRIPLE QDS

Figure 1(a) shows scanning electron micrograph (SEM) of a triangularly-positioned triple QD (TTQD) [10,11]. Three QDs (QD1, QD2, and QD3: the numbered white dotted circles) form a triangle. This is the minimum QD configuration for a two-dimensional qubit array, which may enable large-scale qubit integration. A set of SG structures is used to tune the potential around each TQD structure. A CS utilizing a QD is integrated to detect changes in the number of electrons in the TTQD.

In the measurement, the magnetic field dependence of the current through QD1 and QD3 in the TTQD was studied at a temperature of 300 mK. Charge triple point is a current peak observed in a region where energy levels of the QDs are aligned in bias window. The charge triple point without applying magnetic field is shown in Fig. 1(b), and the charge triple point for in-plane magnetic field $B_{//} = 1$ T is shown in Fig. 1(c) [11]. The areas enclosed by the green dashed lines and the yellow dashed lines in both figures have the same size. It was found that the magnitude of the current at the charge triple point decreased when the in-plane magnetic field was applied, especially in the rectangular region surrounded by the yellow dashed lines. This current suppression is attributed to the phenomenon called PSB. PSB is a phenomenon in which the tunneling between QDs is suppressed by Pauli exclusion principle. This phenomenon can be used for initialization and readout of spin state.

IV. PAULI SPIN BLOCKADE IN P-CHANNEL QDS

We have also studied spin-related hole transport in physically-defined p-channel Si double QD (DQD) [7,9,12]. Moderate spin-orbit interaction in the valence band can be used for controlling hole spins. This can lead to avoid complicated systems such as micro-magnet for electron spin resonance using a slanting magnetic field, which is advantageous for future integration. Schematic of device structure and SEM image are shown in Fig. 2 (a) and (b), respectively. Poly-crystalline Si (poly-Si) top gate with gate oxide (SiO_2) covering QD area is used to induce two-dimensional hole gas (2DHG) and adjusts charge accumulation. SGs and QDs denoted by yellow structures are capacitively coupled with each other. The single QD (SQD) can work as CS, which is used to detect hole transport and charge state in the DQD system.

Charge triple points are indicated by yellow lines obtained for $B_{//}=0$ in Fig. 2(c) and for 1 T in Fig. 2(d),

2020 IEEE Silicon Nanoelectronics Workshop

978-1-7281-9736-4/20 $31.00 © 2020 IEEE

respectively [12]. The bottom of the charge triple points indicated by yellow dashed lines is the PSB region where hole transport is suppressed due to Pauli's exclusion principle. The width of Δ_{ST} showing PSB region became smaller in (d) than in (c) because of increasing Zeeman energy with increasing $B_{//}$. From the in-plane magnetic field dependence of Δ_{ST}, g-factor of DQD is obtained as ~2.0.

V. CONCLUSION

In this work, we prepared physically defined QDs on SOI using electron beam lithography and reactive ion etching. We observed PSB by controlling SGs both in TTQD and p-channel DQDs. These results are promising basis for spin-based qubits with large scale integration.

ACKNOWLEDGMENTS

This work was financially supported by JST CREST (JPMJCR1675), MEXT Quantum Leap Flagship Program (MEXT Q-LEAP) Grant Number JPMXS0118069228, and JSPS-KAKENHI Grants-in-Aid (No. 20H00237).

REFERENCES

[1] K. M. Itoh and H. Watanabe, "Isotope engineering of silicon and diamond for quantum computing and sensing applications". *Mater. Res. Soc. Commun.* **4**, 143–157 (2014).

[2] M. Veldhorst, et al., "An addressable quantum dot qubit with fault-tolerant control-fidelity", *Nat. Nanotechnol.*, **9**, 981 (2014).

[3] J. Yoneda, et al., "A quantum-dot spin qubit with coherence limited by charge noise and fidelity higher than 99.9%" *Nat. Nanotechnol.*, **13**, 102–106 (2018).

[4] M. Veldhorst, et al., "A two-qubit logic gate in silicon" *Nature* **526**, 410–414 (2015)

[5] J. Yoneda, et al., "Quantum non-demolition readout of an electron spin in silicon" *Nat. Commun.*, **11**, 1144-1-7 (2020).

[6] G. Yamahata, et al., "Magnetic field dependence of Pauli spin blockade: A window into the sources of spin relaxation in silicon quantum dots" *Phys. Rev. B* **86**, 115322 (2012).

[7] K. Yamada, et al., "Fabrication and characterization of p-channel Si double quantum dots" *Appl. Phys. Lett.* **105**, 113110 (2014).

[8] K. Horibe, et al., "Lithographically-defined few-electron silicon quantum dots based on a silicon-on-insulator substrate" *Appl. Phys. Lett.* **106**, 083111 (2015).

[9] Y. Yamaoka, et al., "Charge sensing and spin-related transport property of p-channel silicon quantum dots" *Jpn. J. Appl. Phys.* **56**, 04CK07-1-4 (2017).

[10] R. Mizokuchi, et al., "Physically defined triple quantum dot systems in silicon on insulator" *Appl. Phys. Lett.* **114**, 073104-1-4 (2019).

[11] M. Tadokoro, et al., "Pauli spin blockade in a silicon triangular triple quantum dot" *Jpn. J. Appl. Phys.* **59** SGGI01-1-3 (2020).

[12] H. Wei, et al., "Estimation of hole spin g-factors in p-channel silicon single and double quantum dots towards spin manipulation" *Jpn. J. Appl. Phys.* **59** SGGI10-1-5 (2020).

[13] K. Ono, et al., "Current rectification by Pauli exclusion in a weakly coupled double dot system", *Science* **297**, 1313-1317 (2002).

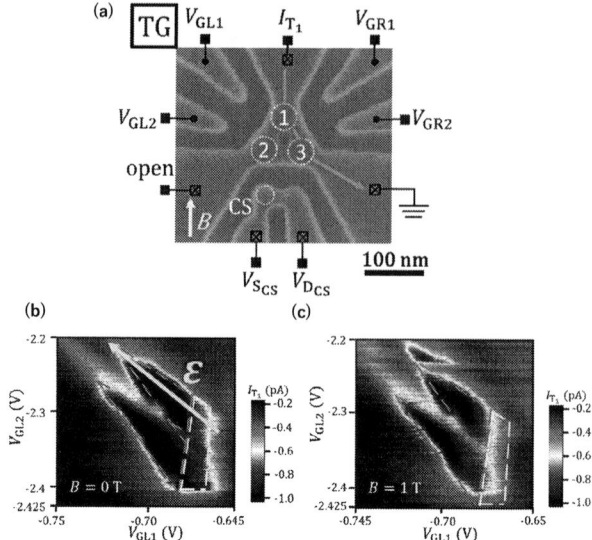

FIG. 1 (a) Scanning electron micrograph (SEM) of the triangularly-positioned triple QD (TTQD). Dark black area corresponds to buried oxide (BOX), and light black area corresponds to SOI. The QDs in the TTQD are indicated by white dotted circles. Each QD has its own reservoir. Yellow arrow indicates the direction of applied magnetic field. (b) Charge triple points without magnetic field. (c) Charge triple points when the magnetic field $B = 1$ T. In green dashed triangles, current flows through TTQD while the current is suppressed in yellow dashed polygon. By comparing (b) and (c), it turns out that the magnetic field enhances the suppression. The light blue arrow in (b) shows a detuning ε of the charge triple point. [11]

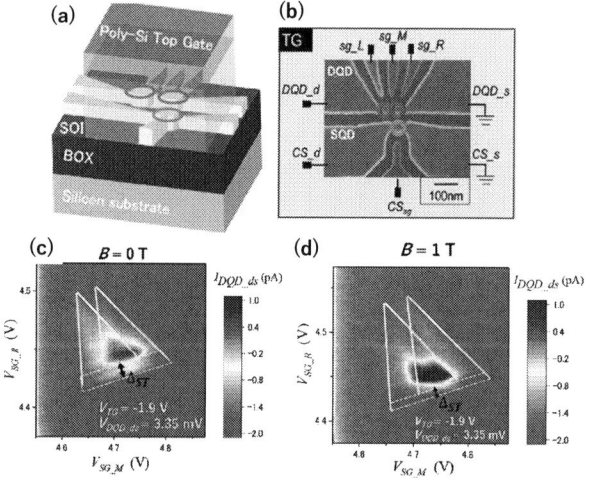

FIG. 2 (a) Schematic of p-channel QD structure. Single QD (SQD) and double QD (DQD) are indicated by blue and red circles, respectively. (b) SEM image focusing on QD area (denoted by red and blue circles) on device chip. (c, d) Charge triple points of DQD indicated by yellow lines obtained for $B_{//}$=0 (c) and 1 T (d). The bottom of the charge triple points indicated by yellow dashed lines is the PSB region where hole transport is suppressed due to Pauli's exclusion principle. PSB was observed in both (c) and (d), although the width of Δ_{ST} became smaller in (d) than in (c). [12]

Integrated Circuits Composed of Nanowire and Single-Electron Transistors Operating at Room Temperature

Tomoko Mizutani[1], Kiyoshi Takeuchi[1], Takuya Saraya[1], Masaharu Kobayashi[1,2], Toshiro Hiramoto[1]

[1]Institute of Industrial Science, The University of Tokyo, Tokyo, Japan

[2]Systems Design Lab, School of Engineering, The University of Tokyo, Tokyo, Japan

Email: mizutani@nano.iis.u-tokyo.ac.jp

Abstract — **Integrated circuits composed of nanowire transistors and a silicon single-electron transistor on a chip have been fabricated. The circuit operations of current/voltage conversion and voltage amplification have been demonstrated at room temperature and at low operation voltage by nanowire MOS/SET circuits for the first time.**

Keywords: single-electron transistor, nanowire transistor, integrated circuit, low-voltage

I. INTRODUCTION

A silicon-based single-electron transistor (SET) [1], whose characteristics are controlled with a single electron, is promising as a future ultra-low-power device. Although the operating temperature of a SET depends on the size of the dot shape called "Coulomb Island", the realization of the nano-scale process has enabled stable operation at room temperature [2]. Generally, it is difficult to put SETs to practical use by replacing MOS transistors because the drive current of a SET is ultra-small. CMOS/SET circuits have been reported in which CMOS circuits are integrated on a chip to compensate for disadvantages of a SET [3-5].

On the other hand, a nanowire transistor, which is effective for suppressing short channel effects due to excellent electric field controllability of the gate electrode, is attracting attention as next-generation devices. Since a SET is usually in the form of nanowire channel structure, the nanowire MOS/SET configuration would be very promising. However, it has been still very difficult to integrate a room-temperature operating SET with nanowire FETs because the fabrication process is not mature enough to form a dot as small as 3nm and uniform nanowire structures with small variability [6].

In this study, we fabricated MOS/SET circuits, in which MOS circuits composed of nanowire transistors and a SET operating at room temperature are integrated on a chip as novel low-voltage circuits. The circuit operations of current/voltage conversion and voltage amplification have been demonstrated at room temperature by the nanowire MOS/SET circuits for the first time.

II. DEVICE STRUCTURE

Intrinsic channel all-around (GAA) silicon nanowire nFETs included in the circuits were fabricated on an SOI substrate [5,7]. Fig.1 shows a schematic of the device structure of a SET with an ultra-small dot less than 3 nm in the nanowire channel. Fig.2 shows measured I_d-V_g curves of the SET at room temperature. The gate voltage (V_{gs}) at the peak position of Coulomb blockade oscillation is 0.46 V, which is independent of the drain voltage (V_{ds}). As shown in Fig.3, Peak-to-valley current ratio (PVCR) decreases as V_{ds}

increases but the peak current increases and tends to saturate around V_{ds} = 150 mV.

III. CIRCUIT DEMONSTRATIONS

Fig.4 shows the nanowire MOS/SET circuits designed in this study, consisting of a current/voltage converter and an inverter. The total number of transistors is 55, including a SET. Fig.5 shows I_d-V_d curves of a load nanowire nFET (NW1) in the converter. NW1 has high resistance and fast saturation, so it is suitable for a load MOS. As shown in Fig.6, the converter successfully converted the ultra-small current of the SET to voltage. VDD1 is set to 200 mV or less, based on Fig.3. The output voltage ranges from 0.123V to 0.183V at VDD1 = 200 mV.

The invertor consists of driver nanowire nFETs (NW2) and a load nanowire nFET (NW3). NW2 is composed of 52 nanowire transistors connected in parallel to achieve a steep output slope. Figs.7-8 show I_d-V_d curves of NW2 and NW3 in the inverter. The dotted line in Fig.7 divides the operating region of NW2 into a saturated region (I) and a linear region (II). The load curve at VDD2 = 2 V, taken from the dotted line plotted the drain current at V_{gs} = V_{ds} in Fig.8, is also plotted in Fig.7. The input of the invertor (the output of the converter) is 0.123V - 0.183V at VDD1 = 200 mV and NW2 operates almost in the region (I), so the output voltage of the inverter would have a steep slope. Fig.9 shows the transfer characteristics of the inverter at VDD2 = 2 V. The output voltage drops by V_{OFF} from 2 V, as shown by the load curve in Fig.8, but a very steep sloop is obtained. Fig.10 shows the slope of the transfer characteristics. The slope of -8 V/V or steeper can be secured within the input voltage range of VDD1 = 200 mV.

Finally, fig.11 shows the output voltage of the nanowire MOS/SET circuit versus the gate voltage of the SET. The large voltage oscillation, which is almost the same as the Coulomb blockade oscillations of SET (Fig.2) and the voltage range is approximately from 0.2V to 1V, was successfully obtained.

CONCLUSION

Low-voltage integrated circuits composed of a nanowire transistors and a silicon single-electron transistor on a chip have been fabricated. The circuit operations of current/voltage conversion and voltage amplification have been demonstrated at room temperature by nanowire MOS/SET circuits, indicating that the nanowire MOS/SET integration is promising for novel functional integrated circuits.

ACKNOWLEDGMENT

This work was partly supported by a Grant-in-Aid for Scientific Research of MEXT, Japan.

2020 IEEE Silicon Nanoelectronics Workshop

REFERENCES

[1] K. K. Likharev, Proc. IEEE 87, p. 606, 1999.
[2] M. Saitoh el al., IEEE TNANO, vol. 2, p. 241, 2003.
[3] K. Uchida el al., IEEE TED, vol. 50, p. 1623, 2003.
[4] M. Vinet el al., IEDM, p. 645, 2013.
[5] R. Suzuki el al., JJAP, vol. 52, 04CJ05, 2013.
[6] T. Mizutani et al., SNW, p. 21, 2015.
[7] R. Suzuki et al., JJAP, vol. 52, 104001, 2013.

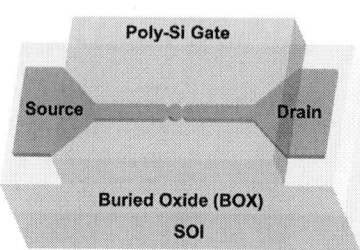

Fig.1. A schematic of a SET fabricated on a SOI substrate. The SET has an ultra-small dot less than 3 nm in the channel.

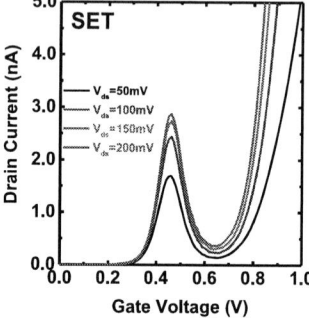

Fig.2. Measured I_d-V_g curves of the SET at room temperature.

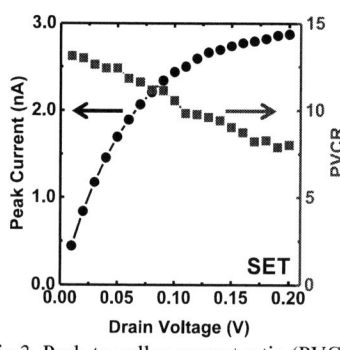

Fig.3. Peak-to-valley current ratio (PVCR) and peak current as a function of the drain current (V_{ds}).

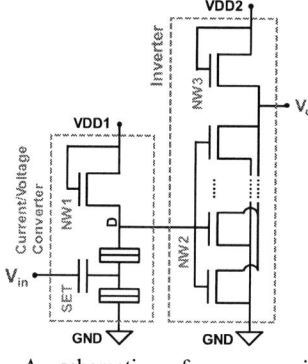

Fig.4. A schematic of a nanowire MOS/SET circuit. It consists of a current/voltage converter and an inverter.

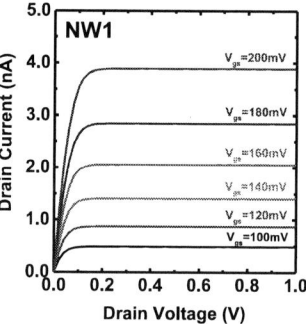

Fig.5. Measured I_d-V_d curves of a load nanowire nFET (NW1) in the converter.

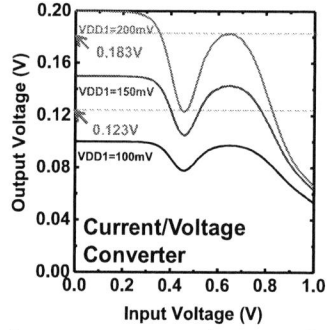

Fig.6. Output voltage of the current/voltage converter (D node) as a function of gate voltage of the SET (V_{in} node).

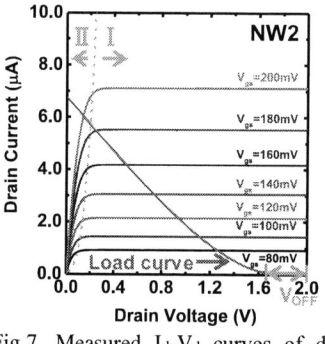

Fig.7. Measured I_d-V_d curves of driver nanowire nFETs connected in parallel (NW2) in the inverter.

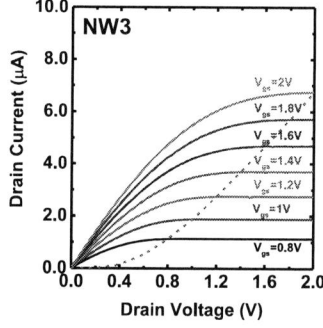

Fig.8. Measured I_d-V_d curves of a load nanowire nFET (NW3) in the inverter.

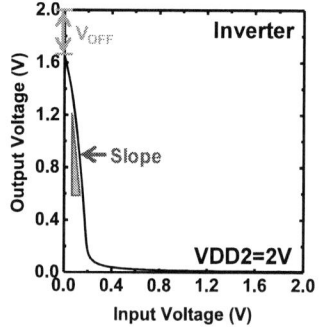

Fig.9. The transfer characteristics of the inverter at VDD2 = 2 V.

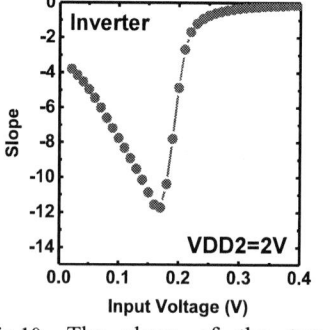

Fig.10. The slope of the transfer characteristics in Fig.9.

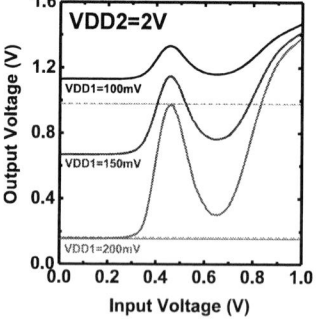

Fig.11. Measured output voltage (V_{out} node) of the nanowire MOS/SET circuit as a function of the gate voltage of the SET (V_{in} node).

2020 IEEE Silicon Nanoelectronics Workshop

A Study of Single-Electron Tunneling Functionalities in Highly-Doped Silicon-on-Insulator Junctionless Transistors

T. Teja Jupalli,[1,2] G. Prabhudesai,[2] M. Hasan,[2] A. Debnath,[2] P. Jeevan Kumar,[2] M. Tabe,[2] and D. Moraru[1,2*]

[1]Graduate School of Science and Technology, Shizuoka University, 3-5-1 Johoku Hamamatsu 432-8011, Japan
[2]Research Institute of Electronics, Shizuoka University, 3-5-1 Johoku Hamamatsu 432-8011, Japan
*Email: moraru.daniel@shizuoka.ac.jp

Abstract As an extension of the continuous miniaturization trend for Si transistors, single-electron effects related to dopants in the nanoscale transistor channels have been recently considered for low-power and fundamental applications. Research on dopant-based single-electron tunneling (SET) functionality extends from single-donor quantum dots (QDs) to multiple-donor QDs. Different from using complex techniques for doping, we show that a simpler, uniform doping technique at high-concentration of nanoscale silicon-on-insulator (SOI) transistors can offer the statistical conditions for the formation of isolated multiple-donor clusters that allow SET functionality, even at room temperature.

I. INTRODUCTION

As transistors are continuously downscaled, the role of discrete dopants becomes critical. Single-donor-atom transistors have been studied over the past decade for single-electron tunneling (SET) functionality using many approaches [1], including atomistic control by scanning tunneling microscope (STM) technology [2]. However, it was previously demonstrated that, even if the nanoscale silicon-on-insulator (SOI) channel is uniformly doped at relatively low concentrations (~10^{18} cm^{-3}), single P-donors can still work as distinct quantum dots (QDs) for SET operation [3]. Nevertheless, this functionality is limited to low temperatures ($T \lesssim 100$ K) due to low tunnel barriers [4]. In order to solve this problem, research has been done to enhance the tunnel barriers by implementing a selective-doping technique in a nanoscale channel, with P-donors doped at moderate concentration (~10^{19} cm^{-3}); thus, several P-donor atoms may form a cluster if they are close to each other within the Bohr radius, while high tunnel barriers are likely preserved in the undoped regions. This device design allowed the observation of Coulomb-blockade behavior even at elevated temperatures, with SET functionality visible at $T \approx 150$ K [5].

Despite this, it remains difficult to control the selective doping in nanoscale at higher concentration, due to side diffusion and complex electron-beam technology. It is, therefore, preferable to obtain such multiple-donor clusters in uniformly-doped Si nano-channels, i.e., in so-called *junctionless* transistors [6]. In such heavily-doped Si junctionless transistors, SET functionalities have not been sufficiently analyzed because it is expected that they exhibit basically a metallic behavior. Nevertheless, it can be also expected that, in junctionless transistors with highly-reduced dimensionality, SET functionalities re-emerge. The purpose of this study is, thus, to statistically analyze, by simulation and experiment, the possibility of SET functionality in heavily-doped SOI junctionless transistors with high doping in the channels, $N_D > 1 \times 10^{20}$ cm^{-3}.

II. DEVICE STRUCTURE: HIGHLY-DOPED SOI-FETS

For experiment, silicon-on-insulator (SOI) FETs were fabricated using electron-beam lithography and dry etching process, to form nanoscale channels from a thin highly-doped SOI layer ($N_D \approx 1.8 \times 10^{20}$ cm^{-3}). For this N_D, distance between neighboring P-donors is ~2 nm. Typical dimensions of the channel under a broad gate are: $t_{SOI} \approx 10$ nm, $W \lesssim 200$ nm (by design) and $L \approx 30$ nm (in this study). Device structure is shown in **Fig. 1**.

III. STATISTICS OF QUANTUM DOT FORMATION IN HIGHLY-DOPED NANO-CHANNELS

In order to understand the possibility for SET functionality in such devices, we first make a statistical analysis of heavily-doped nanostructures, by mapping donor clusters that can work as QDs, i.e., that contain coupled P-donors while remaining isolated from source and drain leads. For that, we consider random distributions of donors in a Si matrix, with substitutional positions. Then, for different nanostructure dimensions, we evaluate the number of donor-clusters (**Fig. 2**) and the average number of donors in the clusters containing the largest number of donors in each case (**Fig. 3**). As a typical example of length, we show data for $L=30$ nm. From **Fig. 2**, it is found that the number of clusters increases almost linearly with width, suggesting the formation of parallel paths for wider structures; in order to isolate a unique-QD path, low-dimensionality (with width $< \sim 20$ nm) is necessary. It is also seen from **Fig. 3** that the number of donors in the largest clusters varies significantly with dimensions, but for structures of width $< \sim 20$ nm, most likely clusters have ~7 ± 2 donors.

IV. ELECTRICAL CHARACTERISTICS AND SET FUNCTIONALITY

Electrical measurements were carried out in a probing system with variable temperature, in the range of 8~300 K. I_D-V_G characteristics were measured typically for V_D=5~100 mV, while stability diagrams were measured by changing V_D as a parameter. We measured SOI-FETs with different designed channel widths, and found that current flows only for those with $W \geq 40$ nm, suggesting a consumption of the finer channels by the final dry oxidation for gate oxide; this allows a rough estimation of real channel width by simply subtracting this value from W_{des}. **Figure 4(a)** shows I_D-V_G characteristics in a wide range of voltages for a wide-channel device (W_{des}=200 nm, L=50 nm) at low temperature (T=8.3 K), while **Fig. 4(b)** shows the stability diagram. This behavior is representative for most devices with wider channels, that exhibit

2020 IEEE Silicon Nanoelectronics Workshop

effectively metallic behavior, i.e., no control of the current level over a wide range of V_G. For SOI-FETs with intermediate dimensions (W_{des}=100 nm, L=30 nm), as shown in **Fig. 5**, weak SET signatures can be found in several devices, i.e., small current peaks in I_D-V_G characteristics [**Fig. 5(a)**] and rough Coulomb diamonds [**Fig. 5(b)**]. However, the Coulomb diamonds are blurred and a large charging effect can be observed for V_G≈2 V, which suggests the presence of multiple interacting ODs in such channels. It should be noted though that this device preserves current oscillations even at room temperature (T=300 K).

Finally, **Fig. 6(a)** shows the I_D-V_G characteristics for one of the narrowest-channel devices (W_{des}=50 nm, L=10 nm), which clearly exhibits SET functionality observed by the broad, large-period current peaks with superimposed fine-pitch peaks (especially visible at T=50 K). This can be explained by the interplay between a small, confined OD and a larger, more extended OD, as illustrated in **Fig. 6(b)**. From the calculation of the period, ΔV_G, and assuming that the OD has a circular parallel-plate structure, the radius is estimated to be roughly 3.0±0.4 nm, which is consistent with a OD formed by a several P-donors strongly coupled to each other. Most interestingly,

such OD exhibits SET functionality even towards room temperature, as seen from T-dependence data.

V. CONCLUSIONS

In this study, junctionless transistors with P-donor concentration N_D≈1.8×10²⁰ cm⁻³ were characterized as a function of channel dimensions. While wide-channel SOI-FETs exhibit a metallic behavior, SOI-FETs with lower dimensionality exhibit SET behavior, with current peaks surviving even up to T=300 K for the smallest devices. This SET functionality is ascribed to the effect of a low number of isolated clusters of several P-donors, statistically formed in nanoscale channels even at such high doping concentrations. This result provides a pathway for practical donor-based SET functionalities.

ACKNOWLEDGMENTS We thank K. Sakazaki for device fabrication. This work was supported by MEXT Kakenhi (I19K045290) and Coop. Res. of RIE (Shizuoka Univ.).

REFERENCES
[1] D. Moraru *et al.*, Nanoscale Res. Lett. **6**, 479 (2011).
[2] M. Fuechsle *et al.*, Nature Nanotechnol. **7**, 242 (2012).
[3] M. Tabe *et al.*, Phys. Rev. Lett. **105**, 016803 (2010).
[4] E. Hamid *et al.*, Phys. Rev. B **87**, 085420 (2013).
[5] A. Samanta *et al.*, Appl. Phys. Lett. **110**, 093107 (2017).
[6] J. P. Colinge *et al.*, Nat. Nanotechnol. **5**, 225 (2010).

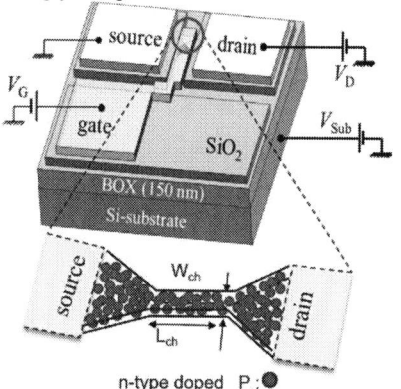

Fig. 1. SOI-FET device structure and biasing circuit. Inset: nano-channel region with P-donors distributed between source and drain, in a region with sizes of L_{ch} and W_{ch}.

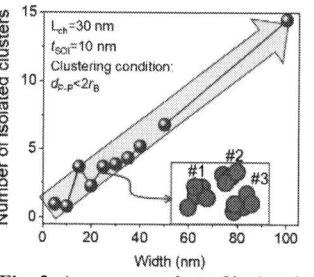

Fig. 2. Average number of isolated clusters of P-donors in a nanostructure (10 nm × 30 nm × width), showing a linear dependence (marked by an arrow). Inset: Schematic illustration of counting of donor-clusters in a nanostructure.

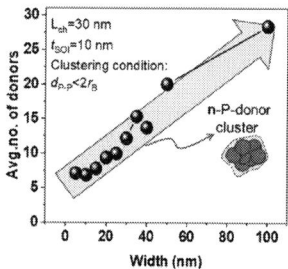

Fig. 3. Average number of P-donors/largest cluster in a nanostructure (10 nm × 30 nm × width), showing a clear dependence on width (indicated by a broad arrow). Inset: Schematic illustration of **n** P-donors in a (largest) cluster.

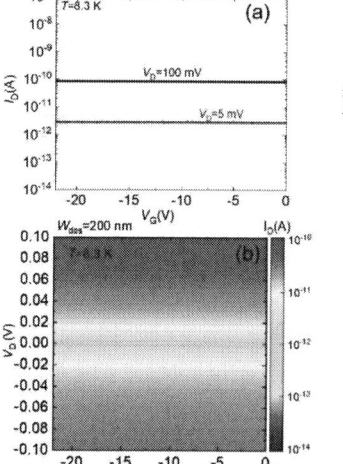

Fig. 4. Electrical characteristics for a wide SOI-FET (W_{des}=200 nm) at LT (T=8.3 K) as: (a) I_D-V_G characteristics; (b) stability diagram in wide range of V_G.

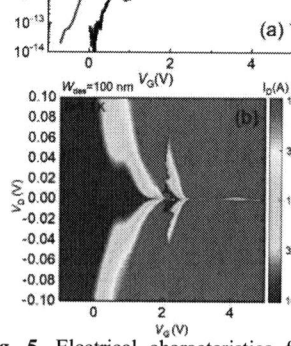

Fig. 5. Electrical characteristics for a moderate-size SOI-FET (W_{des}=100 nm) as: (a) I_D-V_G characteristics at T=8.3 K and 300 K; (b) stability diagram at 8.3 K.

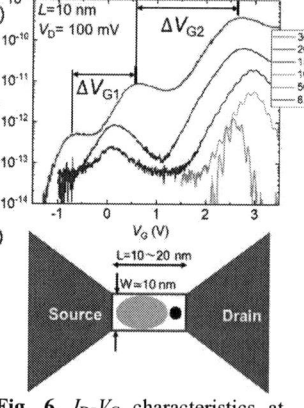

Fig. 6. I_D-V_G characteristics at T≥8.3 K as a function of T, for a small-size SOI-FET (W_{des}=50 nm). Current peaks are observed (even at 300 K), with different pitches at LT. (b) Model to tentatively explain SET current oscillations by interacting QDs.

2020 IEEE Silicon Nanoelectronics Workshop

5.4

Double-gate single-electron devices formed by single-layered Fe nanodot array

Takayuki Gyakushi, Yuki Asai, Beommo Byun, Ikuma Amano, Atsushi Tsurumaki-Fukuchi,
Masashi Arita and Yasuo Takahashi

Graduate School of Information Science and Technology, Hokkaido University
Email: gyakushi.takayuki.d8@elms.hokudai.ac.jp

Abstract — **Single-electron devices (SEDs) composed of nanometer-scale dots have been attracted due to their low-power consumption and high functionality. In this paper, we fabricated double-gate SEDs formed by Fe nanodot array which showed periodic Coulomb blockade oscillation characteristics derived from a single dot. As a result, we confirmed that the charge state of the single dot could be controlled by two gates. In addition, we also found an interesting phenomenon that the two gates, which attached parallel to the nanodot array, unevenly affected the nanodots.**

Keywords: Single-electron device, Self-assembled nanodot array, Coulomb blockade, Disordered system, Conductance oscillation, Mesoscopic physics

I. INTRODUCTION

Single-electron devices (SEDs) have been extensively studied as a candidate for future integrated devices because of their low-power consumption and high functionality [1]-[3]. We have already established a method to fabricate SEDs by the use of self-assembled Fe nanodot array [4],[5]. Many dots are aligned with a small gap to the neighbouring dots where the average dot size and tunnel gap are controlled by the average thickness of the Fe film. In general, many nanodots aligned in series and in parallel between the source and drain electrodes provide complicated current oscillations. However, if the average Fe thickness is controlled to the certain value in which the Fe nanodots are just connected with each other, the devices sometimes showed almost periodic current oscillations as a function of gate voltage (Fig. 1) which is thought to come from a single-electron island between the electrodes [5].

In this paper, we fabricated double-gate SEDs by attaching a metal top-gate to the Fe nanodot array and evaluated Coulomb blockade oscillation characteristics. Consequently, we could control the charge state of a single dot by using the two gates.

II. EXPERIMENTAL METHOD

The fabrication processes used for the Fe nanodot array SEDs are almost the same as those reported in Ref. [5]. A schematic cross section and a plane-view scanning electron microscope image of the fabricated SEDs are shown in Fig. 2. A top-gate is formed over the array and a Si substrate is used as a back-gate. Average Fe thickness t_{Fe} was controlled from 1.8 to 2.8 nm. The length L of the Fe nanodot array was 50 to 400 nm, while the width W was 600 nm. All measurements were performed in a closed cycle cryogenic probe station at $T = 8$ K.

III. RESULTS AND DISCUSSION

Almost all the devices with $t_{Fe} > 1.8$ nm showed clear Coulomb blockade oscillations as a function of top-gate voltage V_T and back-gate voltage V_B. Contour plots of the drain current I_D of the samples which showed periodic Coulomb blockade oscillation characteristics as a function of V_T versus V_B are shown in Fig. 3(a) and (b). These samples are denoted as sample A ($t_{Fe} = 2.4$ nm, $L = 50$ nm, $V_D = 20$ mV) and sample B ($t_{Fe} = 2.4$ nm, $L = 400$ nm, $V_D = 5$ mV) (V_D, drain voltage). In both samples A and B, the charge state of a single dot could be controlled by applying gate voltages to the top-gate and the back-gate electrodes. From the slope of the observed current peak (yellow dotted-line), the gate-capacitance ratio of the back-gate capacitance C_B and the top-gate capacitance C_T, C_B/C_T were evaluated to be ~ 1.2 for sample A and ~ 2.0 for sample B, respectively. Similarly, in the four samples which showed periodic Coulomb oscillation characteristics, the gate-capacitance ratio spreads over from 1.2 to 2.7. This phenomenon seems strange because the Fe nanodot array was sandwiched between the gate electrodes via thick gate insulators, which is thought to provide a uniform electric field. Since the distance between dots is also small, inhomogeneous size and dispersion of the dots cause inhomogeneity of electric field owing to the 3D structure of Fe nanodot array. A schematic image of uneven distribution of electric field from nanodots to both gate electrodes is shown in Fig. 4. As an example, the electric fields toward the top-gate electrode from a small dot placed between large dots should be shielded by the large size dots, which causes the decrease of the top-gate capacitances. This means that, in the case of the multiple dot structure, the gate capacitance of the dot does not only depend on the insulator thickness but also on the 3D structure. These phenomena were confirmed by the use of electromagnetic field simulation. This causes randomly distributed gate controllability depending on each dot, which is applicable to flexible logic gates [6].

III. SUMMARY

Double-gate SEDs formed by the Fe nanodot array were successfully fabricated. It was confirmed that the charge state of the single dot could be controlled by two gates. We also found an interesting effect that two gates unevenly affected the nanodots. This finding may be useful to achieve higher functionality such as reservoir computing.

ACKNOKLEDGMENT

This work is supported by JSPS KAKENHI (15H01706, 16H0433906,16K18073, and 19K04484), the cooperative Research Project of the Research Center for Biomedical Engineering with Research Institute of Electronics, Shizuoka University, Nanotechnology Platform (Hokkaido Univ. and

2020 IEEE Silicon Nanoelectronics Workshop

978-1-7281-9736-4/20 $31.00 © 2020 IEEE

Kyushu Univ.) and Microscopic Analysis for Nano materials science & Bio science Open Unit (MANBOU) programs organized by the Ministry of Education, Culture, Sports and Technology (MEXT) Japan.

REFERENCES

[1] K. K. Likharev, *et al.*, Proc. IEEE, **87,** pp. 606-632 (1999).

[2] Y. Takahashi, *et al.*, J. Phys. Condens. Matter., **14,** pp. 995-1033 (2002).

[3] S. Cotofana, *et al.*, IEEE Trans. Comput., **54,** pp. 243-256 (2005).

[4] H. Hosoya, *et al.*, Matter. Sci. Eng. B, **147,** pp. 100-104 (2008).

[5] T. Gyakushi, *et al.*, Thin Solid Films, **704,** 138012 (2020).

[6] T. Kaizawa, *et al.*, IEEE Trans. Nanotechnol., **8,** pp. 535-541 (2009).

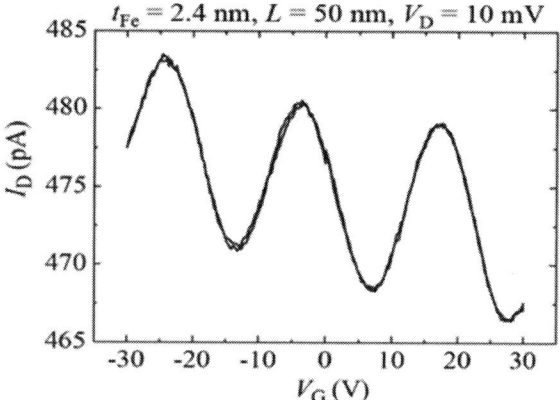

Fig. 1 The almost periodic Coulomb blockade oscillation characteristic measured in forward and reverse gate voltage scan. The device parameters are t_{Fe} = 2.4 nm, L = 50 nm, V_D = 10 mV, respectively. [5]

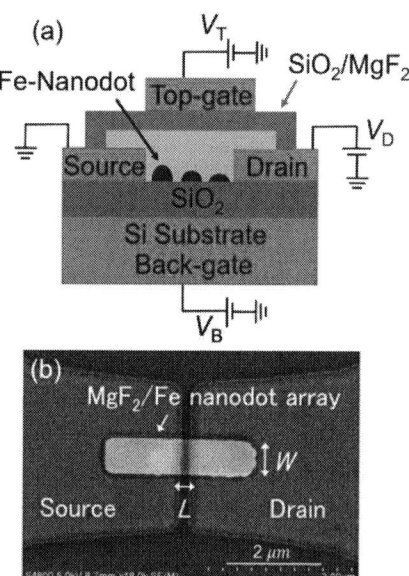

Fig. 2 (a) Schematic cross-sectional view of the fabricated device. (b) SEM image of the device directly after forming the nanodot array. The top-gate electrode is formed over the entire area.

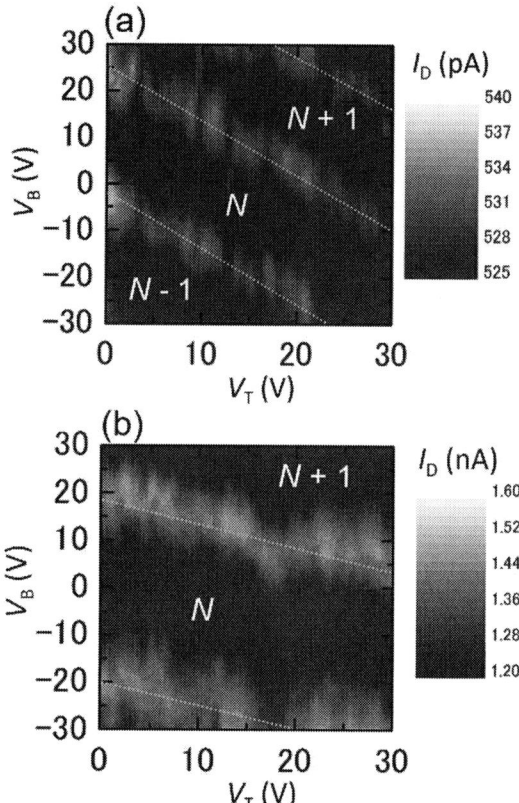

Fig. 3 Contour plot of drain current I_D as a function of the top-gate (V_T) versus the back-gate (V_B) voltage. (a) sample A and (b) sample B. N is the number of electrons in a single dot.

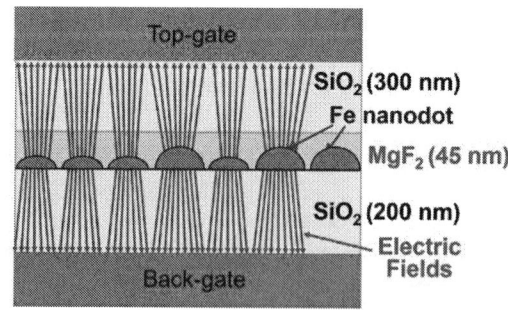

Fig. 4 A schematic image of uneven distribution of electric field from nanodots to both gate electrodes.

A Novel High-Density and Low-Power Ternary Content Addressable Memory Design Based on 3D NAND Flash

H. Z. Yang[1], P. Huang[1†], R. Z. Han[1], Y. C. Xiang[1], Y. Feng[2], B. Gao[3],
J. Z. Chen[2], L. F. Liu[1], X. Y. Liu[1], J. F. Kang[1]

[1]Institute of Microelectronics, Peking University, Beijing 100871, China
[2]School of Information Science and Engineering, Shandong University, Qingdao 250100, China
[3]Institute of Microelectronics, Tsinghua University, Beijing 100084, China
Email: [†] phwang@pku.edu.cn

Abstract — A novel ternary content addressable memory (TCAM) design based on 3D NAND Flash is proposed, in which two adjacent flash cells consist of one TCAM cell. HSPICE simulations show that the proposed TCAM has the ultra-low energy consumption of 0.298 fJ/bit/search for a 64-bit word and the cell density with 96-layer 3D NAND is reduced ~582 times compared to the conventional CMOS-based TCAMs.

I. INTRODUCTION

Ternary content addressable memory (TCAM) is widely used in various applications that need fast data searching, such as pattern recognition [1], network router [2], and approximate computing [3]. However, conventional CMOS-based TCAM consumes large circuit area and its standby power consumption increases when scaling down. To overcome these challenges, TCAM designs based on emerging non-volatile memories (NVMs) such as RRAM [4], PCM [5], STT-MRAM [6] are proposed to eliminate the standby power and enlarge storage capacity. Yet, the fabrication technology of most emerging NVMs are still not mature enough for large-scale integration. Among all the NVMs, 3D NAND Flash may be a potential candidate thanks to its industrial technology and invincible high-density [7]. In this work, we propose a novel 3D NAND based TCAM design, in which two flash cells are used to consist of one TCAM cell. After verifying NAND transistor model with experimental data, the data search sensing margin, speed, power consumption and density are evaluated based on the HSPICE simulations. The simulated results imply the 3D NAND based TCAM is a promising technology for high density data search with large capacity. The detailed simulations for the impacts of 3D NAND's characteristics on the performance of the proposed TCAM design are performed to guide future design optimizations.

II. PROPOSED TCAM DESIGN

Fig. 1 shows the structure of the proposed TCAM design based on 3D NAND Flash array. The search data is converted into voltage pulses and applied to the bit lines (BLs). The data to be searched is stored in the array beforehand by program and erase operations. The search results are read out in parallel in the form of current flowing in the source lines (SLs). The number of BLs should be twice of the word width. The TCAM cell design is shown in Fig. 2 (a). The store and search scheme are illustrated in Fig. 2 (b) and (c), respectively.

III. FEASIBILITY DEMONSTRATION

To demonstrate the feasibility of the proposed TCAM, we establish the compact model of NAND transistor for circuit simulation. Fig. 3 shows the I_d -V_g curves fitted with BSIM model on the basis of the measured data in Ref. [8]. To consider parasitic effects, we use the unit cell as shown in the inset of Fig. 3 and the values are calculated according to Ref. [7]. From the simulation results of Fig. 4, we can see that the current of fully matched case is much lower than the mismatched currents, which verifies the feasibility. Sensing margin, defined as the ratio of 1-bit mismatched current with the fully matched case, is ~200x. As for the currents tend to rise first and then descend to a stable level, we think it's from the gate's coupling capacitance. Considering the sensing margin may be affected by V_{th} variation, we add the distribution and it's still adequate for the sensing margin as shown in Fig. 5.

IV. PERFORMANCE EVALUATION

To analyse impacts of various factors on the performance of the proposed TCAM, we carried out further simulations. Fig. 6 shows the sensing margin at different V_{th} variations and it's usually possible to tighten the variation under $200mV$ [9]. Fig. 7 indicates the negative relationship of sensing margin with the word width, which is because the sum of I_{off} gets comparable with I_{on} as the increasing word width. The influence of parasitic effects on the searching speed is shown in Fig. 8, as a call for the improvement of technology. Fig. 9 and Fig. 10 show the effects of V_{pass} on the searching speed, energy consumption and sensing margin. Fig. 11 and Fig. 12 indicate the impacts of 3D NAND layers on these target values. A trade-off among the three targets should be noticed when setting conditions. The energy consumption is 0.298 fJ/bit/search for a 64-bit word at the lowest-power condition. It should be noticed that we can further reduce the energy consumption by lowering V_{BL}. Fig. 13 shows the 3D NAND based TCAM's density is about ~582x higher than the SRAM-based when the number of layers is 96.

V. CONCLUSION

A 3D NAND-based TCAM design with high density and low power consumption has been presented, in which two flash cells are used to consist of one TCAM cell The simulation results indicate that the proposed design can reduce energy consumption, with 0.298 fJ/bit/search for a 64-bit word. In 96-layer 3D NAND, the cell density is ~582 times higher than that of the traditional CMOS-based TCAMs. The design will be beneficial for the search applications which require low energy consumption and large volume data.

ACKNOWLEDGMENTS

This work was supported in part by National Natural Science Foundation of China under Grant 61841404, and National Key Research and Development (2016YFA0201804, 2019YFB2205100).

REFERENCES

[1] D. R. B. Ly et al., IEDM 2019. [2] J. Huang et al., IEEE Trans on Networking, 26, p.976, 2018. [3] A. Kumar et al., NVMTS 2016. [4] R. Z. Han et al., Jpn. J. Appl. Phys, 57, 04FE02, 2018. [5] J. Li el al., JSSC, 49, p.896,2014. [6] B. Yan et al., NVMTS 2016. [7] P. Wang et al., IEEE Trans on VLSI, 27, p.988, 2019. [8] H. Lue et al., IEDM 2017. [9] L. Cui et al., ICCD 2019.

5.5

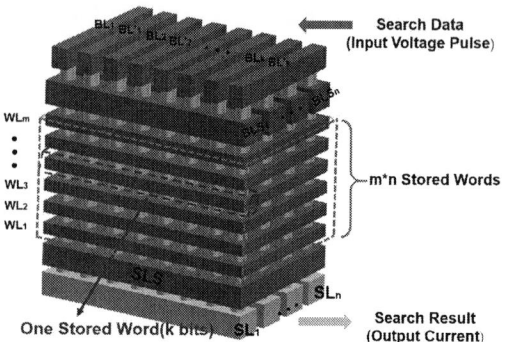

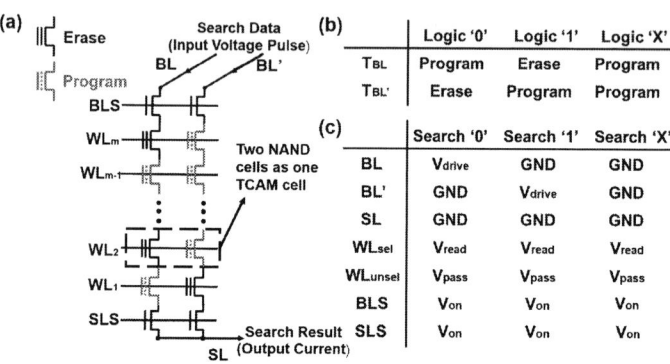

	Logic '0'	Logic '1'	Logic 'X'
T_{BL}	Program	Erase	Program
$T_{BL'}$	Erase	Program	Program

	Search '0'	Search '1'	Search 'X'
BL	V_{drive}	GND	GND
BL'	GND	V_{drive}	GND
SL	GND	GND	GND
WL_{sel}	V_{read}	V_{read}	V_{read}
WL_{unsel}	V_{pass}	V_{pass}	V_{pass}
BLS	V_{on}	V_{on}	V_{on}
SLS	V_{on}	V_{on}	V_{on}

Fig. 1 Structure of TCAM design based on 3D NAND Flash array with (m*n) k-bit words.

Fig. 2 (a) TCAM cell design based on 3D NAND Flash. (b)Definition of TCAM logic with NAND cells. (c) Scheme of searching operation with '0', '1' and 'X' in TCAM logic.

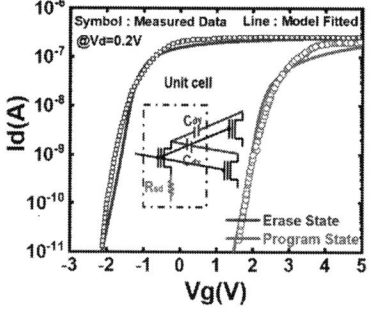

Fig. 3 I_d-V_g curve fitted with BSIM model on the basis of the measured data in Ref. [8]. The inset is the unit cell used in the simulation of 3D NAND array. The value of Cdx, Cdy and Rsd is 0.03fF, 0.02fF and 4Ω. R_{BL} and R_{SL} is set as 0.5Ω and 10Ω.

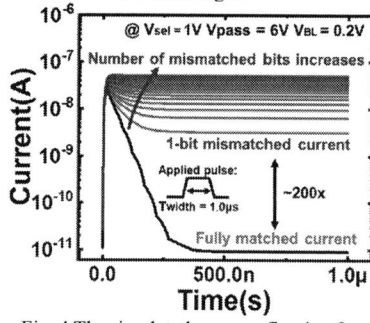

Fig. 4 The simulated currents flowing from source lines with sixteen 16-bit words. The simulation results show that the ratio of mismatched currents to the fully matched current is large enough.

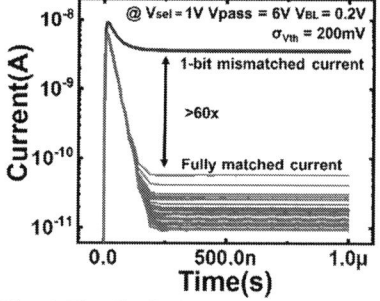

Fig. 5 The distribution curves of the fully matched case and 1-bit mismatched case considering V_{th} variation. Gray lines are raw data and solid lines are the mean current. The mismatched currents are so close together that they look like only one curve.

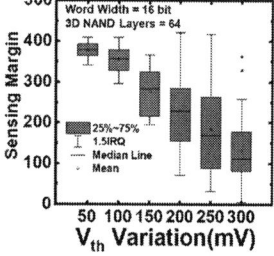

Fig. 6 Influence of V_{th} variation in 3D NAND array on sensing margin of the TCAM system.

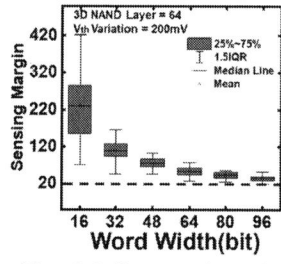

Fig. 7 Influence of word width on sensing margin of the TCAM system, setting σ_{Vth} = 200mV.

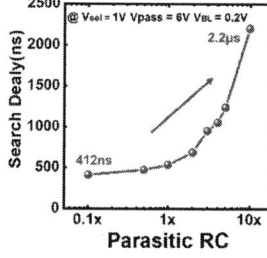

Fig.8 The impacts of parasitic resistance and capacitance in 3D NAND array on the search delay of the TCAM system.

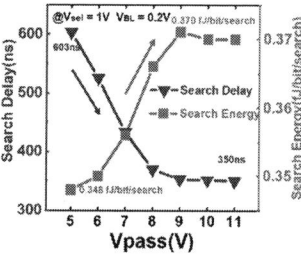

Fig. 9 Search delay and energy consumption vs the WL voltage applied to the unselected layers. Increased V_{pass} reduce unselected layers' cell resistance.

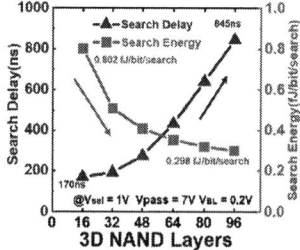

Fig. 10 The impact of V_{pass} on the sensing margin. Lower unselected layers' cell resistance causes higher V_{ds} of selected layers and therefore, higher sensing margin.

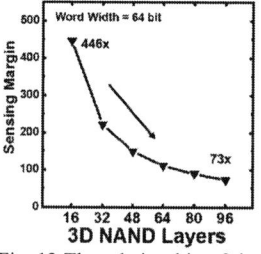

Fig. 11 Search delay and energy consumption vs 3D NAND layers. The adding layers bring more RC on the signal path and as a result, lower V_{ds} on selected layers.

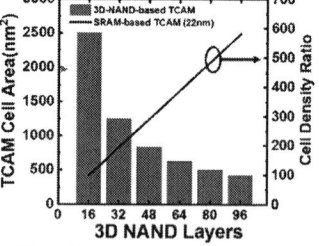

Fig. 12 The relationship of the proposed TCAM's sensing margin with 3D NAND layers .

Fig. 13 Cell area of the 3D-NAND-based TCAM and cell density ratio of the SRAM-based TCAM with the number of 3D NAND layers.

2020 IEEE Silicon Nanoelectronics Workshop

978-1-7281-9736-4/20 $31.00 © 2020 IEEE

6.1

Charge-assisted Recovery and Degradation in Charge-trapping 3D NAND Flash Memory, Experimental Evidences and Theoretical Perspectives

Xiaolei Ma[1], Rui Cao[1], Fei Wang[1], Xuepeng Zhan[1,2], Jiezhi Chen*

[1]School of Information Science and Engineering, Shandong University, Qingdao, P. R. China
[2]State Key Laboratory of High-end Server & Storage Technology, Jinan, P. R. China
*Email: chen.jiezhi@sdu.edu.cn

1. Abstract

In this work, high temperature (HT) baking impacts on the charge-trap (CT) type 3D NAND flash memory are studied systematically. It is found that HT baking impacts are directly correlated to cell states. On the one side, memory cells can be partly recovered when baking at the erase state while serious degradations occur when baking at the program state. Specially, after 2k Program/Erase (P/E) cycling in TLC 3D NAND chip, ~25% error bits can be cured by HT baking after cells' programming. To study the underlying mechanisms, theoretial investigations are done by the first principle calculations, indicating that charge impacts are weak on the defects at the interface but effective to modulate the defect trap levels in the tunneling oxide. It is concluded that electrons de-trapping from charged Vo defects make Vo defects inactive and "cure" the tunneling oxide. However, electrons accumulation will not only inhibit electrons detrapping but also assist electrons charging to the neutral Vo defects, which "generate" more active Vo defects and cause the degradation.

2. Introduction

Device recovery is an eternal topic for device and system engineering because it is the most fundamental solution to prolong the lifetime of electrical devices, such as logic devices and memory cells. Previously, some works showed that degradations caused by the electrical stressing can be cured after high temperatures (HT) annealing [1-4]. Similarly, same effects of HT annealing can be observed in BE-SONOS charge-trapping (CT) flash memories after 100K P/E cycling. It is concluded that the intrinsic mechanism for the cured memory cells originates from the tunneling dielectric recovery by thermal baking [5]. For the same reason, in [6], special circuits are designed and integrated as a local heater to achieve ultra-high endurance flash memory. However, some other reports showed contradict results that device or dielectric recovery cannot be observed at all [7-9]. In a word, it is still not clear why different results can be observed even the device structures are almost same. Therefore, it is reasonable to believe that some other impact factors exist and contribute to the dielectric recovery, like the impacts of hydrogen diffusion [10] and charge accumulation, which are strongly related to the device process and operation schemes.

In this work, by characterizing TLC 3D NAND flash memory chips, it is found that erasing cells before baking helps memory cell recovery while programming cells before baking will cause serious degradation. Atomistic studies are done by the first principle calculations, showing that charge-assisted memory recovery and degradation originates from charge trapping and detrapping in previously generated Vo defects in the tunneling oxide.

3. Experimental Setup and Calculation Approaches

TLC 3D NAND flash memory chips are characterized by using a FPGA-based raw NAND tester. Error bits together with cells' threshold voltage distributions can be measured and analyzed. As shown in Fig. 1, before placing the NAND chip into the oven, various P/E cycling were done in several blocks at room temperatures (RT) and then divided those blocks to three groups, G-level programmed blocks, random programmed blocks, and erased blocks. Three hours baking was performed at 200℃ in a special oven for memory chip characterizations. Then, after baking, we continued the P/E cycling and studied error bits changing of each block. The first-principle calculations are done to study trap energy levels with different charging states, including Si_3N_4/SiO_2 interface and Si/SiO_2 interface. The geometry optimization and DFT self-consistent electronic structures calculations of Si_3N_4/SiO_2 interface structure are performed using generalized gradient approximation (GGA) function to the exchange and correlation interactions, and the SG15 pseudopotential. The atomic configuration of each defect structure with different charging sates are relaxed till the forces converge at 0.05 eV/Å.

4. Results and Discussions

As shown in Fig.1, using the TLC NAND chip, we can program cells from the erase state to seven different states, wherein G-level is the highest with the most number of electrons in the silicon nitride CT layer. Baking effects of various blocks are summarized in Fig. 2. For G-level programmed blocks, obvious degradations can be observed for 0.5k~2K P/E cycled blocks; for random data programmed blocks, baking effects are difficult to be distinguished; while for erased blocks, blocks recovery can be clearly observed. As shown in Fig. 3, baking at the erase state will be more effective to assist memory recovery when the blocks experience more P/E cycling. For 2k P/E cycled blocks, ~25% error bits can be reduced by the baking. Detail information of error bits changing in different program levels are extracted in Fig. 4, it is interesting to find that, error bits' degradation happens in all A~G states, while reduced error bits mainly happens in A~D middle-low states.

It should be noted that higher program levels indicate more electrons are stored in the CTL of memory cell while a large number of hole injection happens during the erasing process. To study to underlying mechanisms of experiment observations, atomistic calculations are done to investigate the charge effects on generated defects in the tunneling layer (TNL). Firstly, we injected electrons and holes into the defective Si_3N_4/SiO_2 interface to understand the influence of charges on trap level shown in Fig. 6. Compared with the case of electron injection, the energy difference(ΔE) from trap level to $Ec(Si_3N_4)$ is smaller than the ΔE after hole injection, which means that electrons accumulation can "generate" more active Vo defects in the CT layer bandgap and cause cells degradation. We can also draw this conclusion at the Si/SiO_2 interface, as shown in Fig. 7. The extra electrons around the defect can drive the trap level close to $Ec(Si)$, while the extra holes around the defect can drive the trap level away from $Ec(Si)$. It is noticed that charge impacts on the defects are weak at the interface. Next, one electron is removed to simulate the de-trapping process in HT baking. It is found that electrons de-trapping from charged Vo defects make Vo defects inactive and "cure" the tunneling oxide. The results at Si_3N_4/SiO_2 interface and Si/SiO_2 interface are summarized in Figs. 8 and 9, respectively. Furthermore, hydrogen impacts should not be ignored [10] because the applied electric field can cause the hydrogen ions diffusion and easily combined with oxygen vacancy. Once the H combine with the Si dangling bond left by the oxygen vacancy, the defect energy level appears in the band gap of Si_3N_4(Fig. 10(a)) and in the band gap of Si (Fig. 10(b)), which could be the major reason for reliability issues. However, holes accumulate can make trap levels move out from the band gap of Si or Si_3N_4(Fig. 11), which can make Vo defects inactive and not to contribute to the charge loss or weak charge injection in P/E disturb.

5. Conclusions

Charge impacts on the recovery and degradation of CT 3D NAND flash memories are systematically studied in this work. By characterizing TLC 3D NAND chips with 2k P/E cycling, it is found that ~25% error bits can be recovered by baking the blocks at erase state while large error bits' degradation happens by baking the blocks at G-level state. By using the first principle calculations, it is considered that electrons de-trapping from the charged Vo defects is the underlying reason that cures the tunneling oxide. However, electrons accumulation will not only inhibit electrons detrapping but also assist electrons charging to the neutral Vo defects, which "generate" more active Vo defects and cause the degradation.

Acknowledgment: This work is supported by China Key Research and Development Program (2016YFA0201802), the National Natural Science Foundation of China (91964105, 61874068), the Joint fund for Intelligent Computing of Shandong Natural Science Foundation (ZR2019LZH009), and the Fundamental Research Funds of Shandong University.

2020 IEEE Silicon Nanoelectronics Workshop

6.1

Reference: [1] N. Mielke et al., IRPS 2006, p.29; [2] J. C. King et al., IEEE electron device letters 15, 475 (1994); [3] P. Riess et al., Applied physics letters 72, 3041 (1998); [4] R. Moazzami et al., IEDM 1992, p.973; [5] C.C. Hsieh et al., IEDM 2010, p.114; [6] H-T Lue et al., IEDM 2012, p.1, session 9.1.1; [7] J. De Blauwe et al., IEDM 1997, p.93; [8] R. Scott et al., IRPS 1995, p.131; [9] K. Sakakibara et al., IRPS 1996, p.100; [10] Y. Mitani et al., IRPS 2018, p.3A.4.

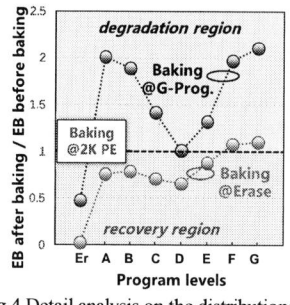

Fig.1 The characterization method in this work. V_{th} distributions are measured by a NAND tester.

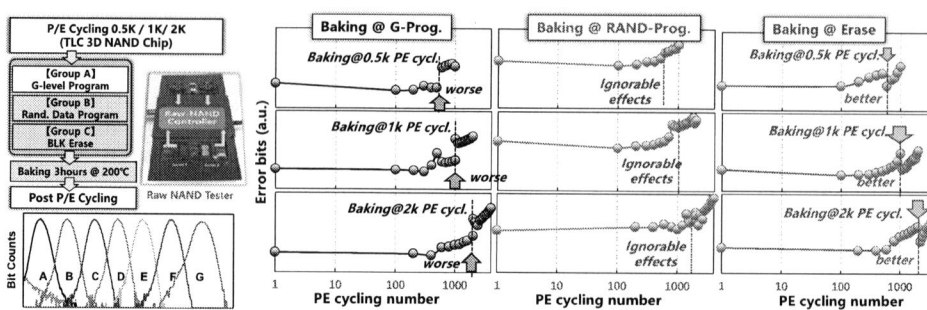

Fig.2 Measured error bits with HT annealing at 0.5k/1k/2k P/E cycling points. Before baking the chip, block states are separated into three groups, (left) G-level blocks, (middle) random data programmed blocks and (right) erased blocks.

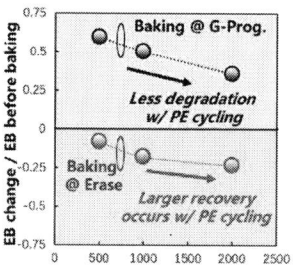

Fig.3 Error bits' modulations at different baking conditions, erase baking reduce error bits while G-level baking degrades error bits.

Fig.4 Detail analysis on the distributions of error bits' recovery and degradation at different program levels.

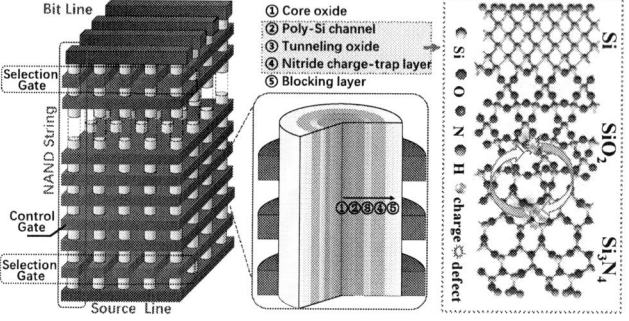

Fig.5. A typical CT 3D NAND structure and atomic interface, electron trapping from the Si_3N_4 to the oxygen vacancy can cause charge loss.

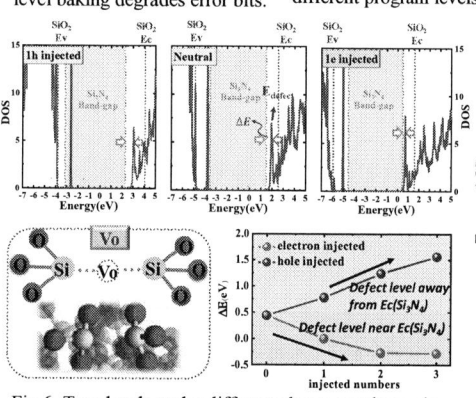

Fig.6. Trap levels under different charge numbers when the defect locates at Si_3N_4-SiO_2 interface. ΔE represents the energy difference between defect level and $E_c(Si_3N_4)$. The defect type is oxygen vacancy.

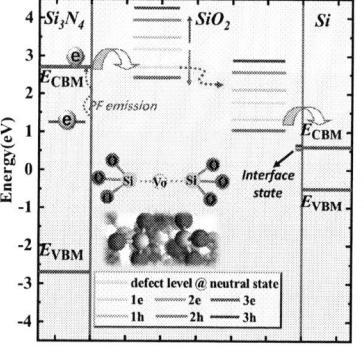

Fig.7. Summarized charge impacts on the trap energy modulations of Vo defects at Si_3N_4-SiO_2-Si interface.

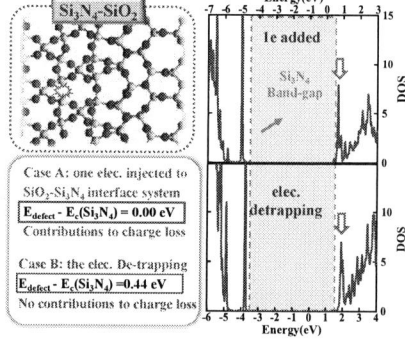

Fig.8. Modulations of trap level after electrons detrapping when the defect locates at Si_3N_4-SiO_2 interface.

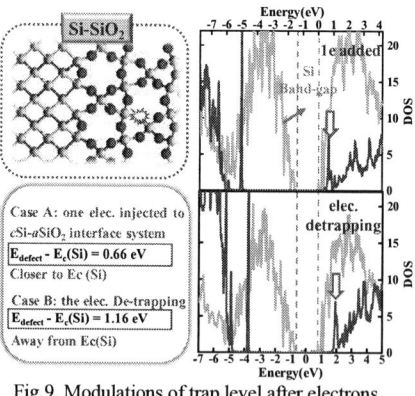

Fig.9. Modulations of trap level after electrons detrapping when the defect locates at the Si-SiO_2 interface.

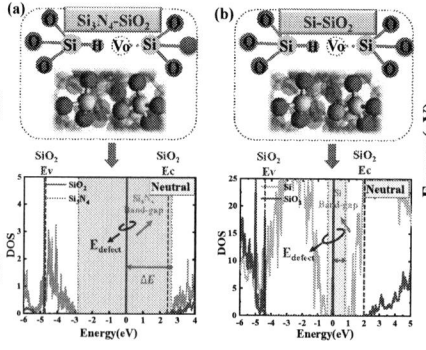

Fig.10. A schematic diagram of the combination of hydrogen ions with oxygen vacancy at (a) Si_3N_4-SiO_2 interface and (b) Si-SiO_2 interface, and the determination of the trap level by DOS.

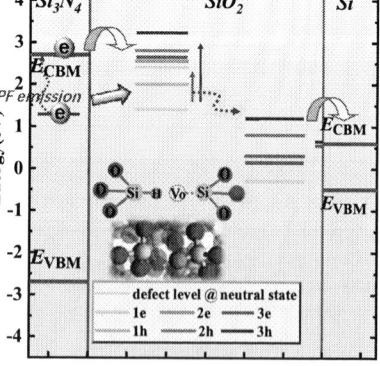

Fig.11. At Si_3N_4-SiO_2-Si interface, gathered hydrogen ions around the defects have large impacts on the trap levels of defects.

2020 IEEE Silicon Nanoelectronics Workshop

6.2

Approximate 3D-TLC NAND Flash Write with Initial Error Injection for Application-level Reliability Improvement of Machine Learning-based Computing

Shun Suzuki[1], Hiroki Aihara[1], Keita Mizushina[1], Shin Yamaguchi[1] and Ken Takeuchi[2]

[1]Chuo University, [2]The University of Tokyo, E-mail: takeuchi@co-design.t.u-tokyo.ac.jp

Abstract

In conventional SSD controller, error correcting code (ECC) is required to correct errors in 3D-TLC NAND flash memories. However, in machine learning (ML)-based computing, some errors can be tolerated and ECC is not necessary. By utilizing the error tolerance, this paper proposes Approximate Initial Error Injection (AIEI) which initially injects errors to memory cells to improve the application-level reliability after the data-retention (D.R.). AIEI initially injects 16.7% of errors to 3D-TLCNAND flash cells. Then, after the D.R., measured bit-error rate (BER) decreases by 34%. Measured acceptable D.R. time is extended by 4.9-times.

Introduction

ML automatically recognizes texts, images and voices based on enormous datasets collected by various sensors and then stored in NAND flash storages. However, the reliability of data is degraded by memories with long D.R. time and high write and erase (W/E) cycles. Generally, in conventional SSD controller, strong ECC such as LDPC code is required to correct errors after the D.R. but strong ECC degrades SSD performance. In ML-based computing, some errors do not affect ML accuracy [1]. Therefore, it is necessary for ML-based computing to decrease BER below the acceptable values after the D.R. because the reliability of data in non-volatile memories is degraded as D.R. time increases. In this paper, by utilizing the error tolerance of ML-based computing, proposed AIEI initially injects V_{TH}-up errors to decrease V_{TH}-down errors after the D.R. (Fig. 1). This is because V_{TH}-down errors are over 4000-times larger than V_{TH}-up errors and worsen recognition accuracy at the long D.R. time. For example, as shown in Fig. 2, about 20% and 37% BER of training data are acceptable for image recognition of Cifar-10 [2] and MNIST [3], respectively. Thus, in the case of Cifar-10, AIEI initially injects V_{TH}-up errors within 20% BER and then the reliability of data after the D.R. is improved because V_{TH}-down errors are largely reduced. In addition, because AIEI has no complex calculation, proposed data modulation does not require data overhead or SSD performance penalty, compared with conventional reliability enhancement techniques (e.g. Asymmetric Coding (AC) [4], Word-line Batch V_{TH} Modulation (WBVM) [5], etc.). Furthermore, AIEI can be widely applied to various error tolerant applications, not only training data of ML.

Error Characteristics of 3D-TLC NAND Flash Memories

Fig. 3(a) shows the structure of 3D triple-level cell (TLC) NAND flash memories which store 3 bits per memory cell [6]. In 3D-TLC NAND flash, reliability problems of vertical charge de-trap [7] and lateral charge migration [8] affect recognition accuracy of ML (Fig. 3(b)). By vertical charge de-trap, during the D.R., V_{TH} of higher V_{TH}-states largely decrease and BER increases (Fig. 4). V_{TH}-down error from "P7" to "P6"-state ("P7"→"P6" error) is 69-times higher than "P1"→"P0" error by large charge loss. Contrarily, in lateral charge migration, electrons are migrated to neighboring word-lines (WLs) because of lateral electric field. Thus, in AIEI, electrons lost by vertical charge de-trap are compensated by lateral charge migration between adjacent higher V_{TH}-states. As shown in Fig. 5, when adjacent WLs are "P7"-states, measured BER decreases by 91% than when those are "P0"-states after the D.R.

Approximate Initial Error Injection

In this paper, during the D.R., proposed AIEI decreases V_{TH}-down errors which degrade the recognition accuracy of ML. Fig. 6 describes AIEI and measured V_{TH}-distribution. AIEI initially injects V_{TH}-up errors to some V_{TH}-states with any percentage. For instance, when AIEI injects 100% "P6"→"P7" error, initial BER is about 4%, but after the D.R., AIEI decreases errors thanks to Advantage 1-3 (Fig. 7). First, lateral charge migration of adjacent higher V_{TH}-states compensates the charge loss of vertical charge de-trap (V_{TH}-down error) (Fig. 8). As shown in Fig. 5, when adjacent WLs are "P7"-states, measured BER after the D.R. is smaller than when those are "P6"-states. Therefore, measured BER of "P2", "P3" and "P4"-state

decreases by 9.0% by injecting 100% "P6"→"P7" error. Second, V_{TH} window between "P6" and "P5" increases and thus V_{TH}-down error largely decreases as shown in Fig. 9. If AIEI initially injects 100% "P6"→"P7" error, "P6"→"P5" error decreases by 99% after the D.R. because "P6"-state is eliminated by AIEI. Advantage 2 becomes more effective by eliminating higher V_{TH}-states such as "P6" because original reliability of higher V_{TH}-states is worse (Fig. 4). Finally, the initial error injection decreases BER after the D.R. because initial error cells eventually return to the original V_{TH} region by vertical charge de-trap (Fig. 10(a)). These cells are defined as "recovery cells," that is, after injecting errors initially, V_{TH} recovers to the correct values as shown in Fig. 10(b). If AIEI injects 100% "P6"→"P7" error, recovery cells increase by 3.8-times than 25% "P6"→"P7" error injection. Much more aggressive error injection increases initial BER but after the long D.R. BER largely decreases. In addition, similar to Advantage 2, AIEI is more effective for higher V_{TH}-state whose BER is naturally higher.

Experimental Results of Proposed AIEI

Thanks to Advantage 1-3, AIEI enhances the reliability of 3D-TLC NAND flash after the D.R. AIEI is experimentally demonstrated by changing error injection ratio and target injected V_{TH}-state as shown in Fig. 6. Fig. 11 shows the measured BER when error injection ratio of AIEI is changed. When AIEI injects 100% "P6"→"P7" error, measured BER after the D.R. decreases by 23% with a drawback of 4.2% initial BER. Furthermore, as shown in Fig. 12, BER is more significantly reduced when AIEI injects errors to higher V_{TH}-states. This is because higher V_{TH} is more likely to cause errors after the D.R. and Advantage 1-3 become more effective. Figs. 11 and 12 mean that AIEI is more effective with larger error injection ratio and for higher target injected V_{TH}-states. Next, more aggressively, AIEI is evaluated by injecting errors to multiple V_{TH}-states in the order of higher V_{TH}. Fig. 13(a) shows measured BER with 100% error injection to multiple V_{TH}-states. In short D.R., measured BER of error injection of "P6"→"P7", "P5"→"P6", "P4"→"P5" and "P3"→"P4" largely decreases because more cells recover rapidly as described in Fig. 13(b). As a result, measured BER after the D.R. decreases by 34% whereas initial BER increases by 16.7%. That is, AIEI recovers more error cells by aggressive error injection. Fig. 14 compares conventional AC and WBVM. Conventional techniques modulate input data and enhance the reliability of memory cells by bit-flip techniques where flag cells are data overhead. Conventional AC, WBVM and proposed AIEI decrease measured BER by 14%, 19% and 34%, respectively. Considering ML-based computing can tolerate 20% BER, acceptable D.R. time is enhanced by 1.5, 1.7 and 4.9-times.

Conclusion

Table 1 summarizes this work. Proposed AIEI initially injects errors to improve application-level reliability of 3D-TL NAND flash. Data for ML-based computing tolerate some errors and thus AIEI is effective. If AIEI injects errors of "P6"→"P7", "P5"→"P6", "P4"→"P5" and "P3"→"P4", after the D.R. measured BER decreases by 34%. Measured acceptable D.R. time is extended by 4.9-times. Although initial BER increases by 16.7%, those errors are acceptable for ML. Furthermore, AIEI has no data overhead and encode/decode time penalty because of no complex calculation.

Acknowledgment

The authors thank Y. Mori for his support. This paper is based on results obtained from a project commissioned by the New Energy and Industrial Technology Development Organization (NEDO).

References

[1] T. Hirtzlin et al., arXiv:1904.03652v1 [cs.ET], Apr. 2019. [2] CIFAR-10 dataset, Available: https://www.cs.toronto.edu/~kriz/cifar.html [3] MNIST dataset, Available: http://yann.lecun.com/exdb/mnist/ [4] S. Tanakamaru et al., ISSCC, Feb. 2011. [5] Y. Deguchi et al., A-SSCC, Nov. 2017. [6] H. Tanaka et al., VLSI Tech., Jun. 2007. [7] R. Yamada et al., IRPS, Apr. 2000. [8] Y. Fukuzumi et al., IEDM, Dec. 2007.

2020 IEEE Silicon Nanoelectronics Workshop

978-1-7281-9736-4/20 $31.00 © 2020 IEEE

6.2

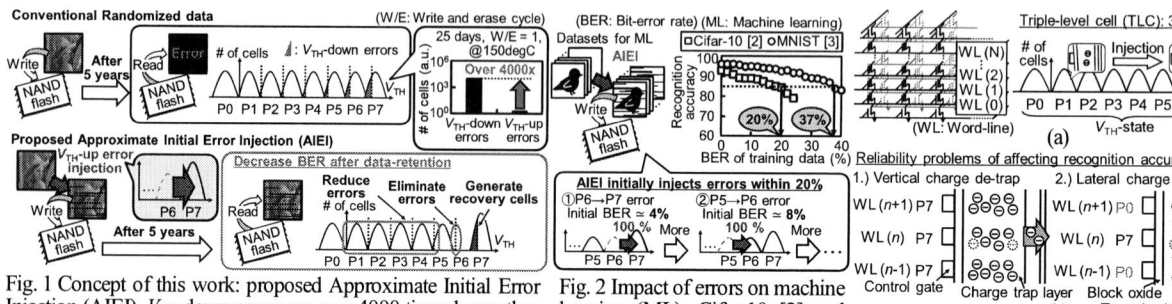

Fig. 1 Concept of this work: proposed Approximate Initial Error Injection (AIEI). V_{TH}-down errors are over 4000-times larger than V_{TH}-up errors. Thus, V_{TH}-up errors are initially injected to higher V_{TH}-states to compensate V_{TH}-down errors and improve the application-level reliability of machine learning after the data-retention (D.R.).

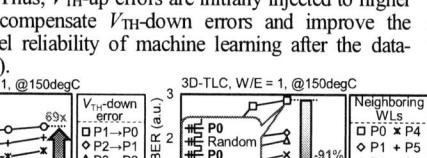

Fig. 2 Impact of errors on machine learning (ML). Cifar-10 [2] and MNIST [3] tolerate high bit-error rate (BER) (20% and 37%) of training data, and therefore initial errors by AIEI are acceptable.

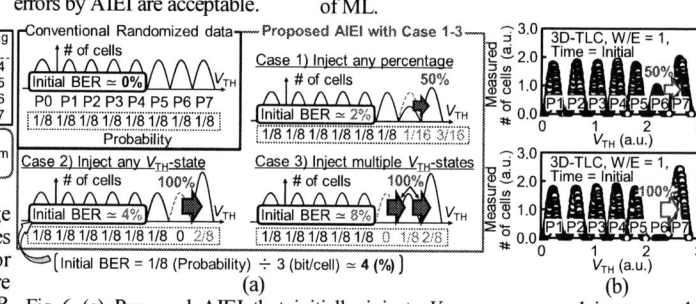

Fig. 3 (a) Structure of 3D triple-level cell (TLC) NAND flash memories. (b) Reliability problems that impact on recognition accuracy of ML.

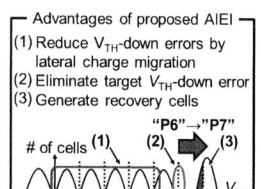

Fig. 4 Measured vertical charge de-trap which is the cause of recognition accuracy degradation. After the D.R., "P7"→"P6" error increases by 69-times than "P1"→"P0" error.

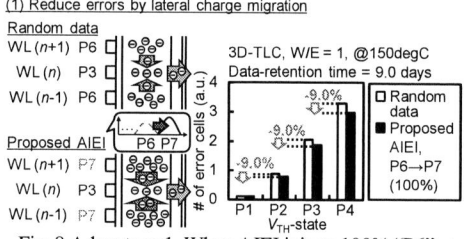

Fig. 5 Measured lateral charge migration which compensates V_{TH}-down errors and is useful for AIEI. When adjacent WLs are "P7"-states, measured BER decreases by 91% after the D.R. than when those are "P0"-states.

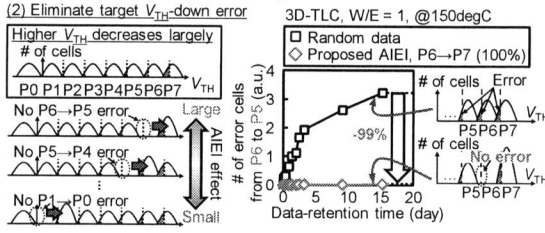

Fig. 6 (a) Proposed AIEI that initially injects V_{TH}-up errors and improves the application-level reliability of ML after the D.R., and (b) measured V_{TH}-distributions with 50% and 100% "P6"→"P7" error injection by AIEI.

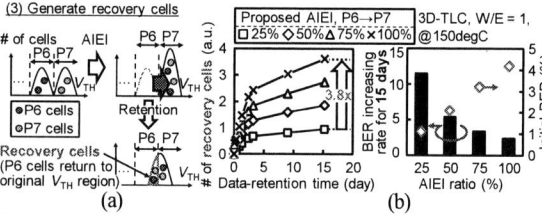

Fig. 7 Advantage 1-3 of AIEI which suppresses V_{TH}-down errors to improve recognition accuracy of ML after the D.R.

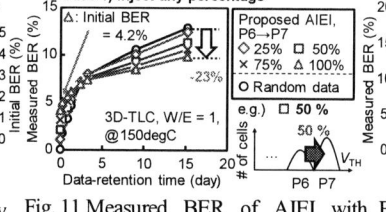

Fig. 8 Advantage 1. When AIEI injects 100% "P6"→"P7" error, error cells of "P2", "P3" and "P4"-states decrease by 9.0% thanks to lateral charge migration (V_{TH} increase) that compensates for vertical charge de-trap (V_{TH} decrease) during the D.R.

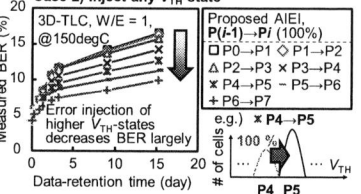

Fig. 9 Advantage 2. V_{TH}-down error of "Pi"-state is eliminated by removing "Pi"-state by AIEI, and it is more effective for higher V_{TH}-states. When AIEI injects 100% "P6"→"P7" error, "P6"→"P5" error cells decrease by 99% after the D.R.

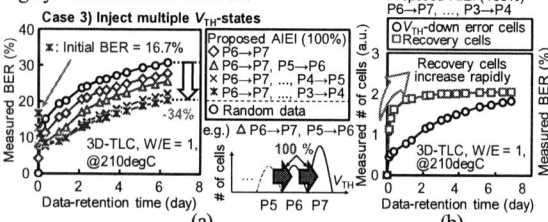

Fig. 10 Advantage 3. (a) AIEI recovers initial error cells by returning error cells to original V_{TH} region, which decrease BER after the D.R. (b) If AIEI injects 100% "P6"→"P7" error, recovery cells increase by 3.8-times compared with 25% error injection. Much higher error injection increases initial BER, while BER largely decreases after the D.R.

Fig. 11 Measured BER of AIEI with various error injection ratios. When AIEI injects 100% "P6"→"P7" error, measured BER decreases by 23% after the D.R. whereas initial error increases by 4.2%.

Fig. 12 Measured BER of AIEI with 100% "P(i-1)"→"Pi" error. Error injection to higher V_{TH}-states decreases BER largely after the D.R. because the lateral charge injection is enhanced and Advantage 1-3 become more effective.

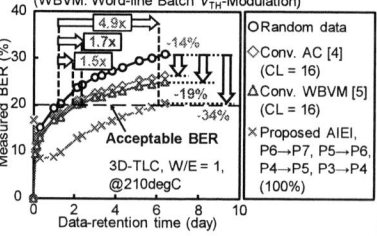

Fig. 13 Measured BER of proposed AIEI with 100% error injection to multiple V_{TH}-states. (a) Error injection of "P6"→"P7", "P5"→"P6", "P4"→"P5" and "P3"→"P4" decreases measured BER by 34% after the D.R. (b) because initial error cells recover to normal cells.

Fig. 14 Comparison of AIEI with conventional Asymmetric Coding (AC) [4] and Word-line Batch V_{TH} Modulation (WBVM) [5]. Measured BER decreases by 14%, 19% and 34%.

Table 1 Summary of this work (measurement).

3D-TLC, W/E = 1, Data-retention time = 6.4 days, CL = 16, @210degC

	Random data	Conv. AC [4]	Conv. WBVM [5]	Proposed AIEI
Measured BER	Baseline	-14%	-19%	**-34%**
Acceptable data-retention time	Baseline	1.5x	1.7x	**4.9x**
Data overhead	-	6.3%	6.3%	**0%**

978-1-7281-9736-4/20 $31.00 © 2020 IEEE   44   2020 IEEE Silicon Nanoelectronics Workshop

A Novel quasi-SLC(qSLC) Program/Erase Scheme in Ultra-Densified Charge-trapping 3D NAND Flash Memory to Enhance System Level Performance

Yuanpeng Li[1, *], Qianwen Wang[1, *], Xuepeng Zhan[1,2], Menghua Jia[1], Rui Cao[1], Jiezhi Chen[#]

[1]School of Information Science and Engineering, Shandong University, Qingdao, P. R. China, [#]email: chen.jiezhi@sdu.edu.cn
[2]State Key Laboratory of High-end Server & Storage Technology, Jinan, P. R. China
[*]These Authors contributed equally to this work

Abstract—In this work, a novel quasi SLC (qSLC) scheme is proposed and verified in TLC 3D NAND flash memory chip. Instead of the standard Program/Erase (P/E) scheme, qSLC performs sequential programming to overwrite the previously programmed data directly. Erase is activated only when the program states reach the highest levels. Since qSLC is a step-up programming scheme between adjacent double levels, the latencies can be well suppressed and the error bits can be largely reduced. In comparison with the standard TLC mode, using the qSLC scheme can obtain ~90% error bits' reduction and at least +75% total workload enhancement.

1. Introduction

Recently, 3D NAND flash has impressing breakthroughs towards ultra-huge data capacities, such as the advanced 3D NAND with semicircular cell structure, QLC and Penta-level-cell (PLC) BiCS flash [1-2]. Besides, the memory layers have been increased from 64 layers to 96 layers and even 128 layers [3-4]. Although there is no doubt that the era of 3D NAND is coming, endurance and the latencies of read/program are still serious concerns as increasing the bit density from SLC/MLC to TLC, QLC and even PLC. In addition, to address the gap between DRAM and NAND flash storage, XL-Flash is newly proposed as storage class memory (SCM) by Toshiba [5], and hybrid controllers with combined SLC and TLC/QLC mix-mode are designed by utilizing SLC with low Program/Read latencies and high endurance and TLC/QLC with large bit densities. However, on the one side, the intrinsic workload and P/E cycling numbers are still limited by P/E related damage to memory cells like the tunneling layer (TNL) [6] and charge storage layers [7]; on the other side, the large capacity of 3D NAND flash memories with high bit densities are still not fully activated in many NAND-based applications (Fig.1).

In this consideration, aiming at 3D NAND flash application in storage-type SCM (S-SCM) [8], it is necessary to optimize P/E sequence to improve the P/E lifetime together with better performance by making fully use of the large data capacity and minimizing the damage to memory cells simultaneously. In this work, a novel approach name as quasi-SLC (qSLC) scheme is proposed, and the total program data capacity in one round P/E cycling is increased by 75% in TLC chip, 87.5% in QLC chip, and 93.8% in PLC chip. By using qSLC scheme in TLC 3D NAND, ~90% error bits can be suppressed during P/E cycling and better endurance properties is obtained.

2. The qSLC Scheme and Characterization Results

Generally, in TLC mode memory chip, the conventional P/E method is to program memory cells from initial erase state to different states ranging from A- to G-level. As shown in Fig.2(a), with repeated P/E cycling, bit error rate (BER) will significantly increase because electrical stress could induce damage to memory cells, which also results in worse retention properties even at room temperature. Unlike the conventional approach, we proposed a novel scheme named as quasi-SLC,

as shown in Fig.2(b). With this approach, we will program 1-bit/cell data step-by-step (two adjacent levels in sequence), and it is not necessary to erase the whole block only when the highest levels are programmed. Taking a TLC chip as an example, erase and A-level are used at the 1st time data programing; then, when those data do not need to be stored anymore and the memory space can be overwritten, we can perform the 2nd time program from erase/A- levels to A- / B-levels without erase, until the 7th time program to the highest F- / G-level. In other words, only one-step erase is necessary after we finish 7 times data programming. Theoretically, the total data capacity can be increased by 75% in TLC chips, and 87.5%/93.8% enhancement can be obtained in QLC/PLC chips, respectively. As summarized in Fig.3, the qSLC approach also benefits chip performance due to less storage levels and reliabilities due to suppressed electric stress damage to memory cells and minimized lateral charge migration. Details of qSLC scheme are shown in Fig.4. Next, with FPGA-based NAND chip tester, we verified the effects of qSLC scheme by using a TLC mode 3D NAND flash chip and characterization results are summarized in Fig.5. Impressively, around 90% error bits are suppressed from the fresh state and even after 2k P/E cycling. Then, to study the endurance of qSLC mode, we compared BER by programming random data with same TLC mode between chips with different P/E cycling schemes, TLC and qSLC (Fig.6(a)). Impressively, it is found that continuous data programming with one-time erase benefits the endurance, indicating that the P/E stress related damage to memory cells could be suppressed in qSLC scheme (Fig.6). Accordingly, it is considered that the qSLC scheme could be effective to expand the applications 3D NAND flash memories in S-SCM.

3. Conclusions

A novel quasi-SLC (qSLC) P/E cycling scheme is proposed in this work to co-optimize the performance and reliability of 3D NAND flash memory systems. By adopting the qSLC scheme in TLC 3D NAND chips, besides larger enhancement to the total workload, obvious error bits' reduction together with no reliability degradation are confirmed. Our results are important for future applications of ultra-densified 3D NAND memories in SCM systems.

Acknowledgment: This work is supported by China Key Research and Development Program (2016YFA0201800), the National Natural Science Foundation of China (91964105, 61874068), the Joint fund for Intelligent Computing of Shandong Natural Science Foundation (ZR2019LZH009), and the Fundamental Research Funds of Shandong University.

References: [1] S. Oshima, *et al.*, Flash memory summit, 2019; [2] N. Shibata, *et al.*, ISSCC 2019, p.209; [3] D. Kang *et al.*, ISSCC 2019, p.215; [4] C. Siau, *et al.*, ISSCC 2019, p.218; [5] S. Ohshima, Flash memory summit 2018; [6] J-D Lee, *et al.*, IRPS 2003, p.497; [7] J. Wu, *et al.*, JPD 2019; [8] R. Kinoshita, *et al.*, Flash memory summit 2019; [9] C. Hu, *et al.*, JAP 1994, p.3695.

6.3

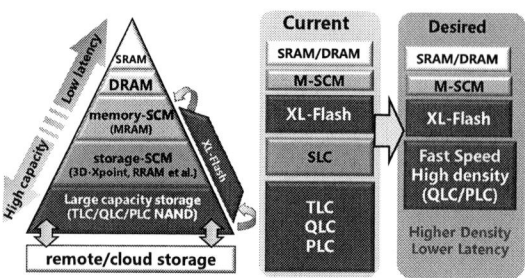

Fig. 1. NAND flash can shorten the gap between DRAM and large capacity storage. It is necessary to construct less process elements to improve the efficiency (energy and speed).

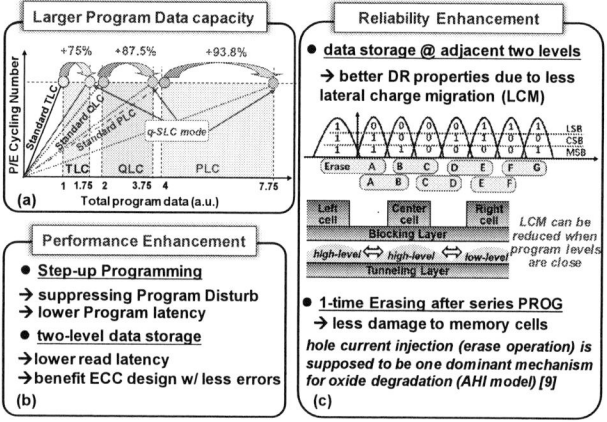

Fig. 3. With the proposed qSLC P/E scheme, we can have (a) larger program data capacities, (b) better performance from less program levels, and (c) better reliabilities, like data retention and endurance due to step-up programming between adjacent levels.

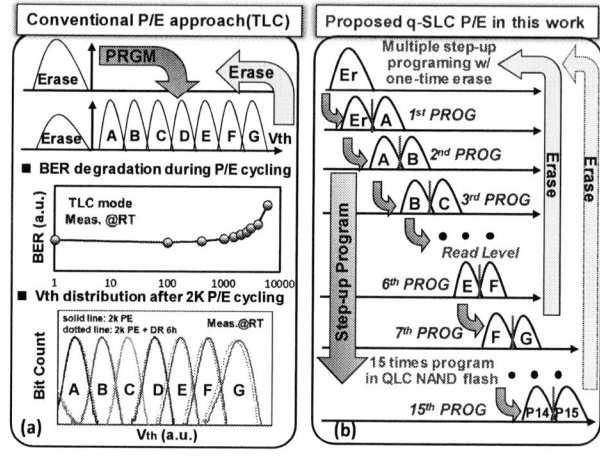

Fig. 2. (a) Conventional P/E approach of TLC 3D NAND flash, BER degradation versus P/E cycling and measured Vth shifts at RT; (b) the proposed qSLC P/E scheme in TLC/QLC NAND flash.

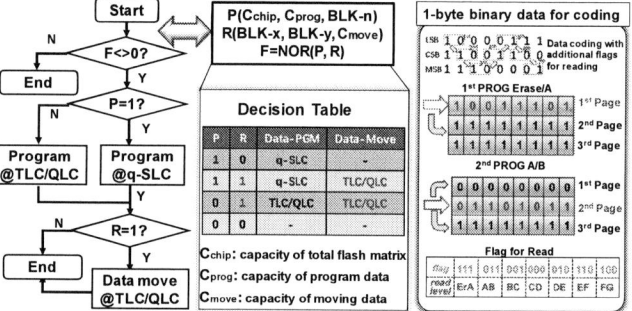

Fig. 4. A flowchart of the proposed qSLC scheme in system level with its decision table for data input and data movement, where the coding method of one-byte binary data programming are demonstrated.

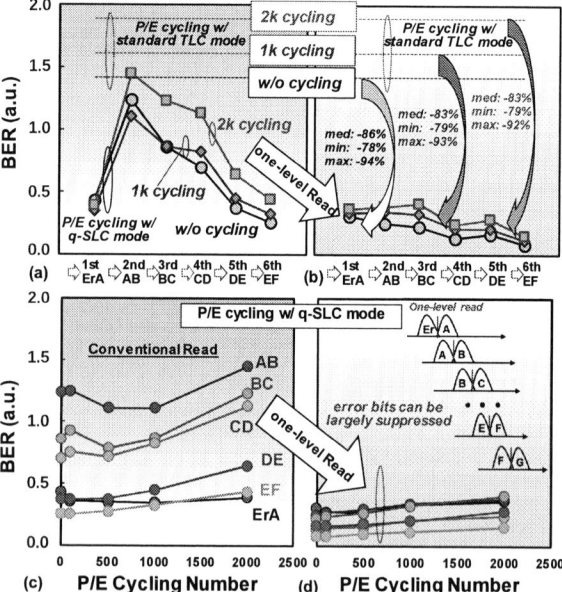

Fig. 5. BER comparisons between standard TLC mode and our proposed qSLC scheme within one round P/E cycling, (a) with and (b) without one-level reading, and BER degradations along with P/E cycling, (c) with and (d) without one-level reading.

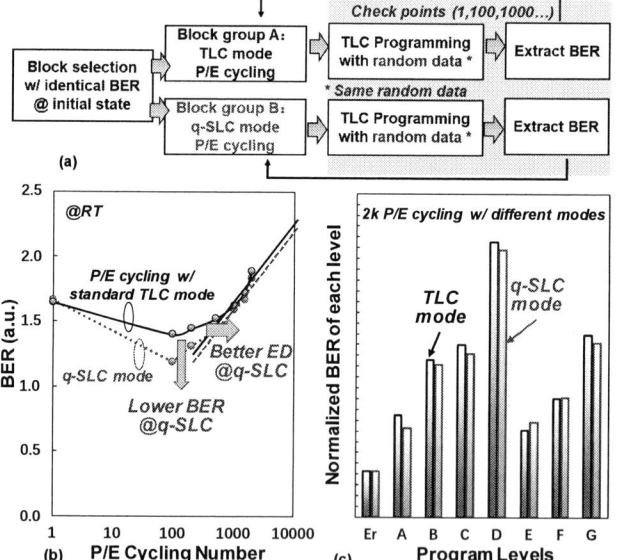

Fig. 6. (a) The characterization scheme to compare BER with different P/E cycling schemes; (b) BER comparison between the standard TLC mode and the proposed qSLC mode, indicating that P/E damage to memory cells are suppressed by using the qSLC scheme; (c) BER of all program levels are compared in detail.

2020 IEEE Silicon Nanoelectronics Workshop

978-1-7281-9736-4/20 $31.00 © 2020 IEEE

3840x Reliability Enhanced Robust NAND flash Optimized to Store Weight Data for Object Detection and Semantic Segmentation of Self-driving Car at High Temperature

Keita Mizushina[1], Shun Suzuki[1], Hiroki Aihara[1] and <u>Ken Takeuchi</u>[2]

[1]Chuo University, [2]The University of Tokyo, E-mail: takeuchi@co-design.t.u-tokyo.ac.jp

Abstract

This paper proposes reliability enhancement techniques in harsh environment to store weight data of machine learning (ML)-based applications, object detection and semantic segmentation for self-driving cars. Proposed techniques consist of robust NAND and Optimized Huffman Coding Compression (OHCC). Proposed robust NAND drastically decreases bit-error rate (BER) in extremely high temperature such as 210degC. Therefore, proposed techniques reduce miss recognition caused by weight data error and contribute to safety self-driving cars. Besides, proposed OHCC modulates weight data of ML-based image recognition applications by utilizing weight data characteristics that concerned around '0'. Consequently, proposed techniques extend data-retention (D. R.) time by 3,840 times for object detection and 2,550 times for semantic segmentation, respectively compared with conventional 3D triple-level-cell (TLC) NAND flash. In addition, proposed techniques achieve to decrease read access time and data-overhead by 39% and 94%, respectively.

Introduction

Self-driving cars require various technologies such as localization/ mapping, perception and prediction, and these are highly controlled by the Engine Control Unit (ECU). ECU needs high reliable storages that endure extremely harsh environment with low cost. Conventional TLC NAND flash do not consider severe conditions such as under hoods where the temperature rises suddenly. The temperature of under hood could reach 175 to 200degC [1] and TLC NAND flash cannot endure such extremely high temperature. Self-driving cars also require large capacity to store important data such as In-Vehicle Infotainment (IVI) and weight data of both object detection and semantic segmentation. As the size of image recognition network increases to improve recognition accuracy, the size of weight data also increases. This paper proposes robust NAND which enables 3D-TLC NAND flash as storage embedded on self-driving cars in harsh environment. To further enhance reliability, this paper also proposes OHCC realized by understanding the characteristics of ML-based application's weight data as shown in Fig. 1. Proposed robust NAND that writes in pseudo multi-level-cell (MLC) on TLC NAND chip and OHCC that utilizes weight data with Huffman coding are shown in Fig. 2. AEC-Q100-Rev-H [2] defined 'Grade 0' ambient operating temperature range is -40degC to 150degC. However, measurement condition in this paper is 210degC.

Machine Learning applications for self-driving cars

Self-driving cars are assumed to use ML-based image recognition applications as shown in Fig. 3. When self-driving cars execute object detection and semantic segmentation, weight data of image recognition need to be stored in embedded storages. In fact, the characteristics of weight data for ML-based image recognition depend on applications such as object detection [3] and semantic segmentation [4]. Because of normalized weight data of almost all kinds of Neural Network, weight data are concentrated near '0' as shown in Fig. 4. In other words, weight data of ML-based image recognition are excessively biased. Data bias is a common feature of both object detection and semantic segmentation. For example, object detection network consists of 3 kinds of convolutional layers. In contrast, semantic segmentation network consists of 3 convolutional/ deconvolutional layers. Central layers have huge parameters, but outer layers have fewer ones. Weight data of ML-based image recognition have error tolerance feature. In other words, weight data can accept a few errors, but storages embedded on self-driving cars must adequately suppress errors to achieve reliable self-driving cars.

Robust NAND flash at extremely High Temperature

Robust NAND technique is proposed to suppress errors caused by extremely high temperature such as 210degC and maintain high precision of ML-based image recognition for self-driving cars. Fig. 5 shows the structure of 3D-TLC NAND flash. TLC NAND flash stores 3 bits/cell, but reliability is reduced due to narrow V_{TH} margin. Generally, 3D-TLC NAND flash uses strong error-correcting code (ECC) [5], but ECC fails to correct errors within a few hours in severe environment. For realizing robust NAND [6], there are various candidates of V_{TH}-state combinations. Previous work [6] realizes pseudo MLC mode by using only lower and middle pages of TLC NAND flash. However, proposed robust NAND considers the best V_{TH}-state combination include upper page. Fig. 6 shows measured V_{TH}-distribution of robust NAND. As shown in Fig. 6, V_{TH}-down error is dominant at extremely high temperature such as 210degC.

Therefore, higher V_{TH}-states require wider V_{TH} margins. BER of robust NAND varies greatly depending on the V_{TH}-state combination. As shown in Fig. 7(a), the best combination is (Er-A-C-G). Fig. 7(b) shows that robust NAND with (Er-A-C-G) combination decreases BER by over 95% after D.R. time. Fig. 8 shows proposed SSD controller.

Optimal Huffman Coding Compression for Robust NAND flash

OHCC technique is proposed to enhance reliability of robust NAND storing weight data of ML-based image recognition. In ML-based image recognition for self-driving cars, exceeding a certain BER causes recognition failure that leads to critical accidents. Therefore, proposed OHCC maximizes robust NAND reliability using Huffman coding. Fig. 9 shows how to build quaternary Huffman coding. Huffman coding counts appearance data pattern and builds Huffman tree step by step. As appearance probability is different, the amount of data stored in each leaf is also different as shown in Fig. 9. Proposed OHCC utilizes features where the amount of data is different. OHCC assigns frequently appearing leaves to reliable states (Fig. 10(a)). Accordingly, large data bias enables high compression rate, and the effect of OHCC becomes large. As shown in Fig. 4, weight data are excessively concentrated near '0'. Hence, proposed OHCC is outstandingly useful for weight data of both object detection and semantic segmentation. The compression ratio of object detection exceeds 60% and semantic segmentation exceeds 47% (Table I). Previous work [7] uses Huffman coding compression decreases BER of 3D-MLC NAND flash by up to 66.2%. In this paper, proposed OHCC decreases BER by 96.5% from conventional 3D-TLC NAND flash and 41% from only robust NAND (Fig. 10(b)) despite harsh environment. Consequently, OHCC technique enhances the reliability of ML-based image recognition on self-driving cars.

Recognition accuracy with robust NAND using OHCC

Fig. 11 shows recognition accuracy of object detection which accepts BER about 3%. Recognition accuracy over 80% maintains for 0.2 hours in case of conventional 3D-TLC NAND flash with LDPC ECC. By using proposed robust NAND, D.R. time is extended to 25 days. Furthermore, D.R. time with robust NAND using OHCC is extended to more than 32 days in harsh environment such as 210degC. As a result, proposed robust NAND using OHCC extends D.R. time by 3,840 times. Weight data of object detection have high error tolerance, and robust NAND using OHCC without strong ECC achieves to maintain recognition accuracy even in extremely high temperature. Fig. 12 shows recognition accuracy of semantic segmentation which accepts BER about 0.7%. Weight data stored in conventional 3D-TLC NAND flash fail semantic segmentation task in a few minutes. Proposed robust NAND with OHCC executes segmentation task more than 17 days and extends D.R. time by 2,550 times. Weight data of semantic segmentation have lower error tolerance compared with object detection's one.

Conclusion

Table II summarizes this work. Proposed robust NAND and OHCC extend D.R. time by 3,840 times in object detection and 2,550 times in semantic segmentation, respectively compared to conventional 3D-TLC NAND flash. These experimental results show that proposal is effective for various ML-based image recognition applications. Because proposed techniques do not use ECC, the read access time is reduced by 39%. Note that proposed OHCC needs Huffman table as data overhead due to lossless compression. However, Huffman table size is much smaller than parity size of LDPC ECC. Hence, proposed techniques achieve to reduce data overhead by 94%. Proposed robust NAND and OHCC could be used to store weight data for various ML-based image recognition tasks, providing low-cost and reliable storage for extremely harsh environment.

Acknowledgement

The authors thank R. Kinoshita for her support. This paper is based on results obtained from a project commissioned by the New Energy and Industrial Technology Development Organization (NEDO).

References

[1] Automotive World Ltd., [Online]. Available: https://www.automotiveworld.com/mbility-magazine/, Mar. 2016. [2] AEC, AEC-Q100. [Online]. Available: http://www.aecou-ncil.com/AECDocuments.html, Sep. 2014. [3] M. Tan *et al.*, arXiv:1905.11946v3 [cs. LG], Nov. 2019. [4] V. Badrinarayanan *et al.*, *PAMI*, pp. 2481-2495, Jan. 2017. [5] S. Tanakamaru *et al.*, *VLSI Circ.*, pp. 126-127, Jun. 2014. [6] S. Tanakamaru *et al.*, *ISSCC*, pp. 336-337, Feb. 2014. [7] H. Watanabe *et al.*, *A-SSCC*, pp. 157-160, Nov. 2017. [8] The Cifar-10 dataset. [Online]. Available: https://www.cs.toronto.edu/~kriz/cifar.html [9] G. Brostow *et al.*, *ECCV*, pp. 44-57, Oct. 2008.

6.4

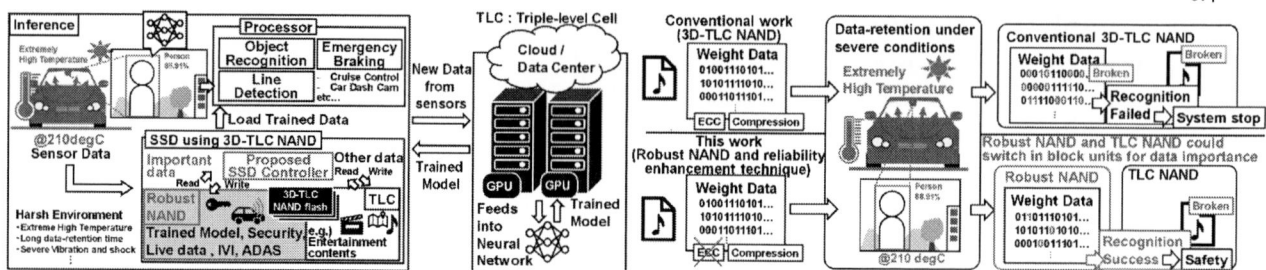

Fig. 1. Concept of this work: Proposed robust NAND and SSD controller. Robust NAND co-works with conventional TLC NAND flash in the same chip. Robust NAND is proposed especially for self-driving cars embedded storages working in harsh environment.

Fig. 2. Overview of this work. Proposed robust NAND mainly stores weight data of machine learning (ML)-based image recognition. Optimal Huffman Coding Compression (OHCC) is proposed for robust NAND reliability enhancement.

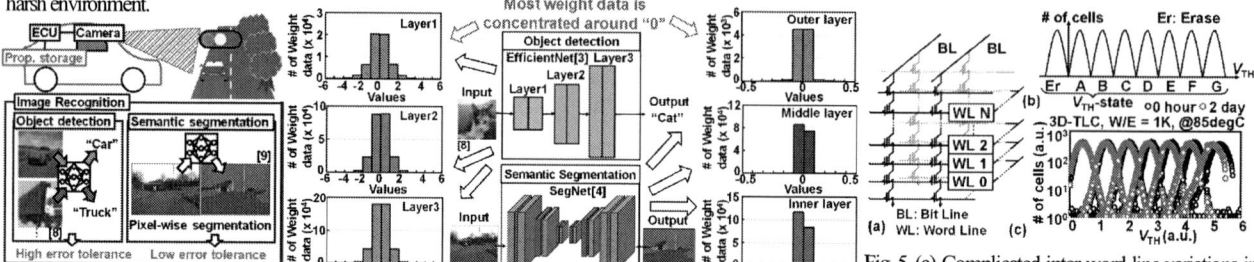

Fig. 3. Object detection and semantic segmentation of image recognition technology [3, 4]. Image recognition contains object detection and semantic segmentation.

Fig. 4. Histogram of weight data for ML-based image recognition. Histogram of the left side is object detection [3] and the right side is semantic segmentation [4]. Most data are excessively concentrated around '0'.

Fig. 5. (a) Complicated inter word-line variations in vertical 3D NAND flash. (b) V_{TH} distribution and (c) measured V_{TH} distribution of 3D-TLC NAND flash.

Fig. 6. Measured number of memory cells V_{TH}-down shift of proposed robust NAND.

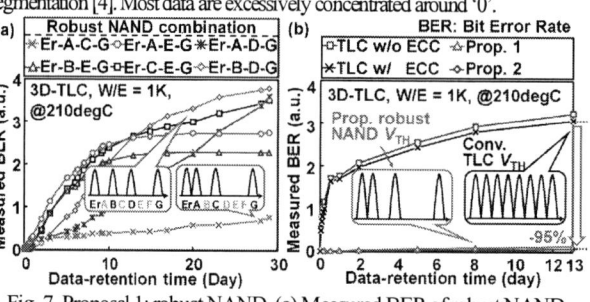

Fig. 7. Proposal 1: robust NAND. (a) Measured BER of robust NAND with various V_{TH}-combinations of states [6] and (b) measured BER of robust NAND and conventional 3D-TLC NAND flash [5].

Fig. 8. Proposed robust compression selector for robust NAND and OHCC in the SSD controller.

Table I Weight data for ML-based image recognition. Accordingly, Huffman coding achieves high compression ratio.

Target	Layer	Compression Rate
Segnet	Conv. 1	47.79%
	Conv. 2	47.73%
	Deconv. 1	47.76%
	Deconv. 2	47.75%
Cifar-10	layer0-0-1	62.50%
	layer0-0-2	62.50%
	layer1-0-1	62.50%
	layer1-0-2	62.50%
	layer2-0-1	62.50%
	layer2-0-2	62.50%

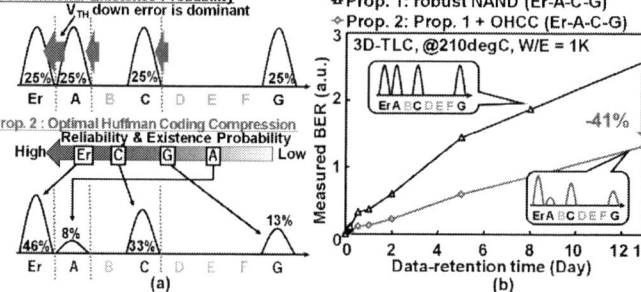

Fig. 9. Quaternary Huffman code. (step. 1) divides data by 4 bits and (step. 2) counts the frequency of appearance. (step. 3) Complete Huffman tree from low frequency leaves [7].

Fig. 10. Proposal 2: OHCC for robust NAND. (a) OHCC adaptively assign high existence probability leaves to high-reliability state in robust NAND. (b) Measured results of proposed robust NAND with OHCC. OHCC achieves to reduce BER by 41% compared with only robust NAND.

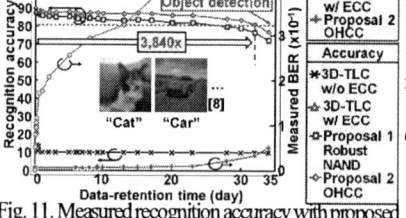

Fig. 11. Measured recognition accuracy with proposed robust NAND using OHCC for object detection. Proposed techniques extend D.R. time by 3,840 times.

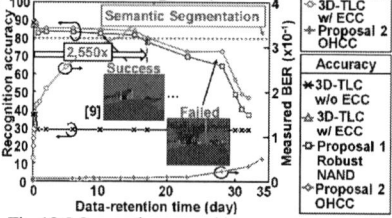

Fig. 12. Measured recognition accuracy with proposed robust NAND using OHCC for Semantic segmentation. Proposed techniques extend D.R. time by 2,550 times.

Table II Summary of measured results of proposed robust NAND and OHCC.

3D-TLC, W/E = 1K, @210degC	Conv. Work (w/o ECC)	Conv. Work (w/ ECC)	Prop. 1 (Robust NAND)	Prop. 2 (Prop. 1 + OHCC)
Read access time [us]	65.6	71.6	44	44
Data overhead [Bytes]	0	1.25 x10⁶	0	80 x10³ (Table size)
Data-retention time (Object detection)	0.2 hour	0.2 hour	25 days	32 days
Data-retention time (Semantic segmentation)	10 minutes	10 minutes	15 days	17 days

2020 IEEE Silicon Nanoelectronics Workshop

6.5

Spatial Color-Perceived Data Control of 3D-TLC NAND Flash for Image Dectection

Chihiro Matsui[1], Shun Suzuki[2], and Ken Takeuchi[1]

[1] Dept. of Electrical Engineering and Information Systems, The University of Tokyo, Japan
[2] Dept. of Electrical, Electronic, and Communication Engineering, Chuo University, Japan
Email: matsui@co-design.t.u-tokyo.ac.jp

Abstract — This paper proposes Spatial Color-Perceived Data Control (SCP-DC) of 3D-TLC NAND flash for image detection with three proposals: Pixel Memory Consistency, Spatial-Perceived Data Modulation and Color-Conscious Data Modulation. By combining three proposals, 3D-TLC flash BER deceases by 82.7% and image recognition accuracy increases by 5.1%.

Keywords: Image detection, spatial perception, color perception, data modulation, 3D-NAND flash

I. INTRODUCTION

Image detection/object detection is a core technology of autonomous driving, robots, and manufacturing machines. Because images generated in such future applications have spatial locality and specific color pattern, this paper proposes **Spatial Color-Perceived Data Control** (SCP-DC) (Fig. 1). In CMOS image sensor, pixel data are selected row by row. The output of the amplifier inside the pixel is read out by the column decoder/latch [1]. Because in the future image detection with machine learning, both data size of pixels and frame rate will become significant increase, this paper assumes that 3D triple-level cell (TLC, 3bits/cell) NAND flash memory [2] stores the image data (Fig. 2). However, 3D-TLC NAND flash has memory cell reliability problems. Fig. 3 shows measured bit-error rate (BER) of memory cells. High Write/Erase (W/E) cycles increase endurance BER of higher state such as P6 and P7. Long data retention time degrades reliability of P7-state. When endurance and data retention errors are combined, BER increases by 15 times.

II. SPATIAL COLOR-PERCEIVED DATA CONTROL OF 3D-TLC NAND FLASH

Fig. 4 shows **Proposal I: Pixel-Memory Consistency**. Image sensor outputs pixel data in the order of pixel row number with Blue (B), Green (G), and Red (R). Because adjacent pixels have similar color patterns, data of the same pixel row have similar color features. In the conventional data assignment, different pixel rows of image sensor are separately stored in Upper/Middle/Lower pages in one cell. Thus, the conventional data assignment is unsuitable for image detection based on the color pattern of the pixels. On the other hand, in the proposed data assignment, buffer memory temporarily stores data from image sensor in the order of pixel row number. Then, the SSD controller re-arrange the data order to write Blue to Upper page, Green to Middle page and Red to Lower page, respectively. As a result, arranged pixel data are written to the word-line of TLC NAND flash. Thus, the same data modulation can be applied to word-lines or pages of NAND flash at once.

Fig. 5 shows **Proposal II & III: Spatial-Perceived & Color-Conscious Data Modulation**. By using Proposal I, the BGR pixel data are written to Upper, Middle and Lower page, respectively. From the previous results [3], among 8bits of image data, the higher-4bits of pixel data, that is, MSB (Most Significant Bit) have large impact on image recognition accuracy. Therefore, the lower-4btis can be eliminated from the pixel bits. **Proposal II: Spatial-Perceived Data Modulation** applies data modulation based on pixel row unit. Conventionally, because three bits of Upper, Middle and Lower pages in one TLC NAND flash cell store data of different pixel rows, it is not possible to detect images based on the color feature. On the other hand, by Proposal I, three bits in the same word-line (Upper, Middle and Lower pages) of TLC NAND flash store the pixel row data of image sensor. Thus, cells of the same word-line of TLC NAND flash stores the data of adjacent pixels. As a result, proposed color-based image detection that selects the optimal data modulation based on pixel row and pages or world-line of NAND flash becomes feasible. Then, **Proposal III: Color-Conscious Data Modulation** chooses optimal data modulation based on color of pixels such as yellow and black. To reduce BER of NAND flash, data modulation such as Asymmetric Coding (AC) [4] is utilized. AC flips the data, "0" or "1," with the predetermined codelength unit to increase the number of cells of low BER such as P0 and P1 states. In this paper, the codelength is 128 that corresponds to the number of pixels in a pixel row multiplied by higher-4bits. If data are flipped, the extra flag becomes "1." If data is NOT flipped, the flag stays at "0." Higher-4bits of black, yellow and red pixel data are originally mapped as P3, P5 and P6 states. In the proposal, to increase the number of cells of highly reliable P0 or P1 state, the optimal modulation among AC-1, 2, ... 8 is selected. In case of black pixel, data in Upper/Middle/Lower pages are flipped with flag (1, 1, 1) that corresponds to AC-1. In case of yellow or red pixels, data are flipped with flags (1, 0, 0) or (1, 1, 0) that corresponds to AC-3 or AC-2, respectively. Because the measured total BER increases with W/E cycles and data retention time (Fig. 3), AC-1 and AC-2 is highly reliable because the lower V_{TH} state (P0 and P1), respectively, cells increase.

III. EVALUATION RESULTS

Fig. 6 shows result of Spatial and Color-Perceived Data Modulation result with CIFAR-10 [5] test_batch data #9246 "Cat" and #23 "Truck." Because #9246 "Cat" image has distinct pixel areas with black or yellow, large number of P3 cells of black is changed to P0 state by Proposal I and II (AC1 for all pixel rows). By applying Proposal I, II and III (Different AC for each pixel row), BER of cells that correspond to yellow pixel area decreases by applying AC-2. Because #23 "Truck" has more complicated color features than #9246 "Cat". In case of such complicated color, AC-1 or AC-2 that increase highly reliable P0 or P1 is selected to minimize BER. The flag cell overhead of AC is 0.78% when the codelength is 128. Proposal I and II reduces BER by 81.6% and 9.5% for images #9246 and #23, respectively. Furthermore, Proposal I, II and III reduces BER by 82.4% and 29.2% for images #9246 and #23, respectively. As a result, the image recognition accuracy increases by 5.1% and 1.6% for #9246 "Cat" and #23 "Truck," respectively.

IV. CONCLUSION

This paper proposed Spatial Color-Perceived Data Control (SCP-DC) of 3D-TLC NAND flash. By co-designing image pixel data and NAND flash data assignment, the proposals improve 3D-TLC NAND flash reliability and image recognition accuracy by 82.7% and 5.1%, respectively.

ACKNOWLEDGMENTS: This paper is based on results obtained from a project commissioned by the New Energy and Industrial Technology Development Organization (NEDO).

REFERENCES: [1] S. Yoshihara et al., *IEEE JSSC*, vol. 41, no. 12, pp. 2998-3006, 2006. [2] H. Tanaka et al., *VLSI Tech.*, 2007, pp. 14-15. [3] Y. Yamaga et al., *VLSI Tech.*, 2018, pp. 109-120. [4] S. Tanakamaru et al., *IEEE ISSCC*, 2011, pp. 204-205. [5] CIFAR-10, https://www.cs.toronto.edu/~kriz/cifar.html.

2020 IEEE Silicon Nanoelectronics Workshop

978-1-7281-9736-4/20 $31.00 © 2020 IEEE

6.5

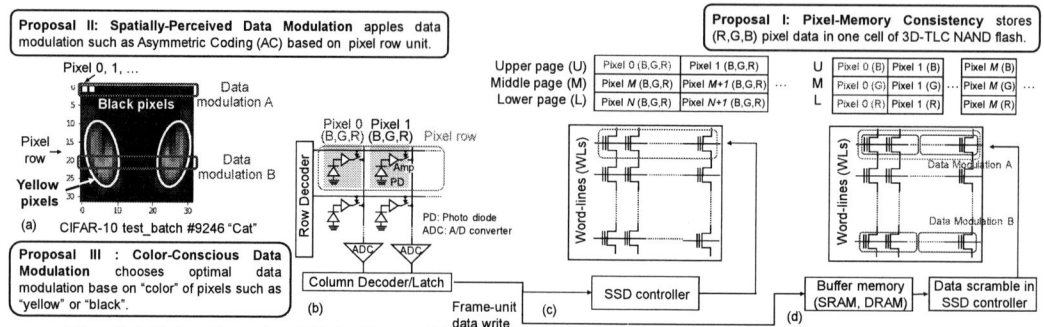

Fig. 1 Proposed **Spatial Color-Perceived Data Control** (SCP-DC) of 3D-TLC NAND flash for image detection. (a) Image from CIFAR-10 [5], (b) CMOS image sensor [1], (c) conventional pixel data assignment, and (d) proposed pixel data assignment. **Proposal I: Pixel Memory Consistency** stores (B, G, R) pixel data in one cell of 3D-TLC NAND flash [2]. **Proposal II: Spatial-Perceived Data Modulation** applies data modulation such as Asymmetric Coding (AC) [4] based on pixel row unit. **Proposal III: Color-Conscious Data Modulation** chooses optimal data modulation based on color of image pixels such as yellow and black.

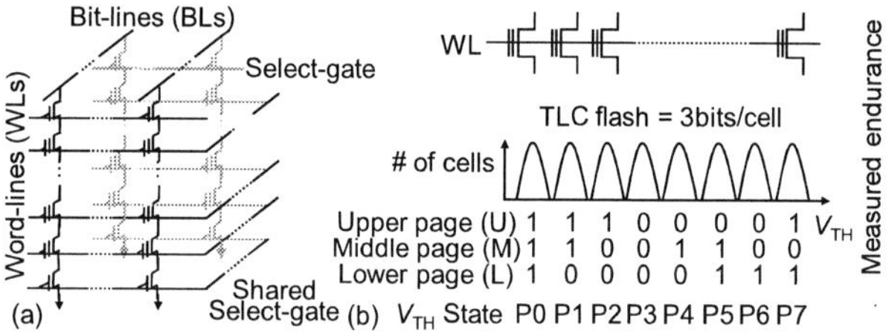

Fig. 2 (a) Architecture [2] and (b) data assignment of 3D-TLC NAND flash

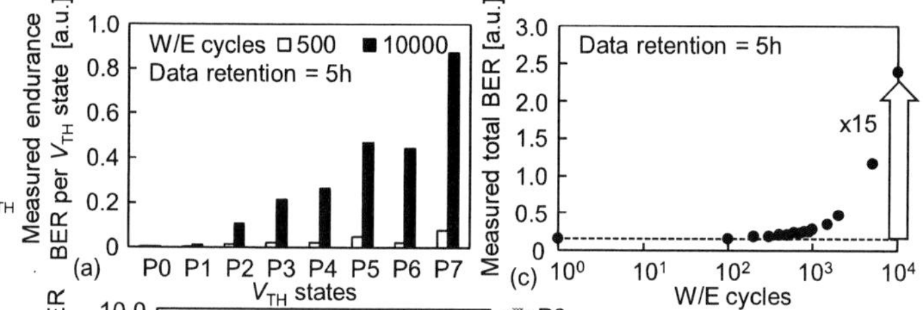

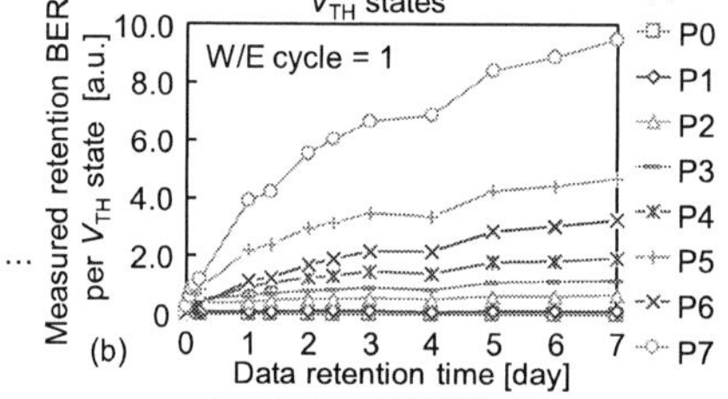

Fig. 3 Measured BER of 3D-TLC NAND flash at 85degC. (a) Endurance BER, (b) data retention BER of each V_{TH} state, and (c) total BER. BER increases with W/E cycles and data retention time.

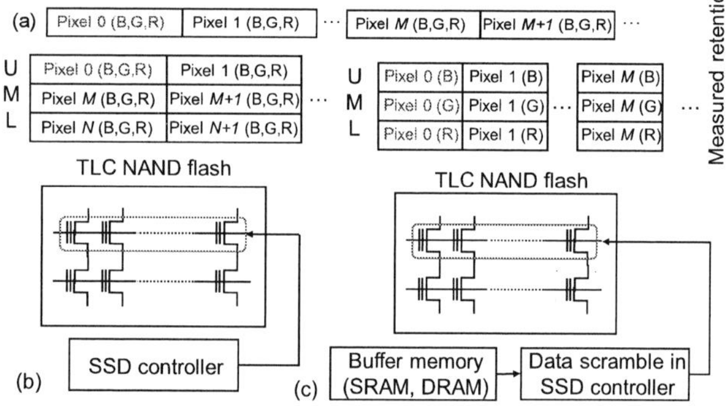

Fig. 4 Proposal I: Proposed Pixel Memory Consistency. (a) Frame-unit data from image sensor. (b) Conventional pixel data assignment and (c) proposed pixel data assignment.

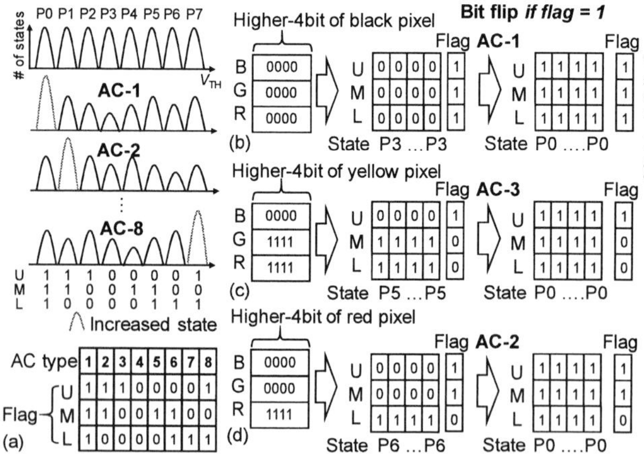

Fig. 5 Proposed II and III: Proposed Spatial-Perceived and Color-Conscious Data Modulation. (a) Data modulation definition by Asymmetric Coding (AC) [4]. Data modulation for (b) black (c) yellow and (d) red pixel rows. The optimal AC such as AC-1 is selected to minimize BER of 3D-TLC NAND flash by considering color of each pixel row.

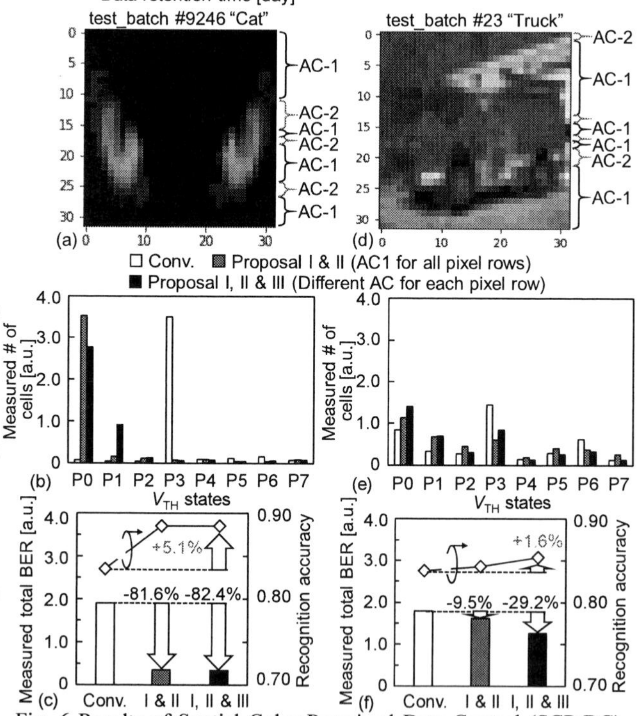

Fig. 6 Results of Spatial Color-Perceived Data Control (SCP-DC). (a)(d) optimal AC that minimizes BER, (b)(e) measured number of cells at each V_{TH} state, and (c)(f) measured total BER and image recognition accuracy of test_batch #9246 "Cat" and #23 "Truck," respectively. By proposed SCP-DC, measured BER decreases by 82.4% and 29.2%, respectively

2020 IEEE Silicon Nanoelectronics Workshop

978-1-7281-9736-4/20 $31.00 © 2020 IEEE

Performance enhancement of BF_2^+ implanted poly-Si junctionless transistors by boron segregation and fluorine effect

Min-Ju Ahn, Takuya Saraya, Masaharu Kobayashi, and Toshiro Hiramoto

Institute of Industrial Science, The University of Tokyo, Japan

Email: mjahn@nano.iis.u-tokyo.ac.jp

Abstract — **We have experimentally investigated the subthreshold characteristics of poly-Si junctionless transistors fabricated by BF_2 implantation and compared with other dopants. Improved subthreshold slopes and small device-to-device variations are obtained and the origins of the improvement are discussed.**

I. INTRODUCTION

Presence of grain boundary (GB) defects in poly-Si significantly degrades the electrical properties of poly-Si based transistors [1]. The ways to suppress these GB defects include various crystallization methods [2-4], high temperature oxygen annealing [5] and hydrogen treatments [6]. In addition, the fluorine (F) treatment effectively improves the quality of poly-Si channel by passivating the GB defects [7].

Recently, much attention has been paid to the junctionless (JL) transistor without source/drain (S/D) junctions due to its simple process and volume conduction. The JL transistor strongly requires thin and highly doped channel for full depletion and high on-current, respectively [8]. However, highly doped channel produces mobility reduction, SS degradation, and severe device-to-device variations [8]. In addition, unlike inversion-mode (IM) transistor, the highly doped JL channel suffers from some damages and defects during heavy ion implantation process depending on doping profiles, implant energy, and dopant types [9]. Therefore, the optimization of channel implantation conditions is essential for understanding the electrical characteristics. However, there are no reports on these issues yet.

In this work, we investigated the dependence of dopant types (P^+, B^+, BF_2^+) on the poly-Si JL transistors and observed superior subthreshold characteristics in case of BF_2 doped channel thanks to B segregation as well as F effects.

II. EXPERIMENTAL

A 120nm thick amorphous Si was deposited on a 200nm thick BOX layer by LPCVD and SPC at 1100°C for 24hours was performed. Then, P^+, B^+, BF_2^+ ions were implanted at 35keV with dose of $3 \times 10^{14} cm^{-2}$, for n-type and p-type channel. The channel region was locally thinned down to 10nm by oxidation at 1100°C, while the S/D regions remained thick. Subsequently, a 10nm dry oxide was thermally grown and a 150nm thick P^+ and N^+poly-Si were deposited for a gate electrode for n-type and p-type channel, respectively. Then, a 300nm SiO_2 was deposited for passivation and H_2 annealing was carried out to terminate defects at 400°C for 30min. Finally, Al metal contacts were formed.

The channel width W, length L, and thickness T are 40μm, 40μm, and 5nm, respectively. For comparison, IM p-type transistors (T=10nm) with P^+poly-Si gate were also fabricated. Fig.1 shows the fabricated four poly-Si transistors (IM, JL_P, JL_B, JL_BF2). In the channel thinning process by local oxidation in JL_B and JL_BF2, the B concentration in the channel regions was reduced due to the segregation of B, resulting in $P^+/P/P^+$ structure. In case of JL_P, on the contrary, the $N^+/N^+/N^+$ structure is formed because P concentration is not reduced during oxidation. The average grain size in the poly-Si is around 230nm extracted from atomic force microscopy (AFM) analysis as shown in Fig.2.

III. RESULTS AND DISCUSSION

Fig.3 plots the I_D-V_G curves and transconductance (g_m) of four transistors at V_D=-0.05V. No hysteresis was observed. Apparently, steep SS and enhanced current drivability are observed in JL_B and JL_BF2 devices compared to others. In addition, threshold voltages $|V_{th}|$ of JL_B and JL_BF2 are higher than that of JL_P. This is due to the reduced channel concentration by B segregation. g_m values of JL_B and JL_BF2 are higher than that of IM because of smaller electric field at on-state owing to the volume conduction [8]. The poor g_m characteristic of JL_P originates from high impurity scattering by high channel concentration. Fig.4 shows the extracted SS as a function of drain current at V_D=-0.05V. SS_{min} and SS_{ave} of JL_BF2 are 97.5mV/dec and 114.4mV/dec, respectively, which are better than that of others. Here, SS_{ave} is average SS in the subthreshold region ($I_D \times L/W = 1 \times 10^{-13}$ to 1×10^{-10}A).

Fig.5 displays cumulative distributions of V_{th}, SS_{min}, and SS_{ave} at V_D=-0.05V. It should be noted that the device-to-device variations of JL_P are large due to high channel concentration ($\sim 1 \times 10^{19} cm^{-3}$) even at large-size planar structure. In contrast, the JL_B and JL_BF2 exhibit very small variations due to low channel concentration.

Table I summarizes key parameters of four devices, clearly proving the quality of poly-Si layer is much better than those of published data of planar poly-Si JL transistors (Table II) [10-13]. The mobility is extracted from g_m values. Poor mobility of IM and JL_P is explained by interface scattering and high impurity scattering, respectively.

Here, JL_BF2 and JL_B are compared. It is noted that JL_BF2 shows better SS and higher mobility. In order to examine the reason, secondary ion mass spectroscopy (SIMS) depth profiles of JL_B and JL_BF2 after channel thinning process were measured and shown in Fig.6. It can be seen that the F concentrations in both top and bottom interfaces are higher in the JL_BF2. In particular, the F concentration of JL_BF2 is over 10 times higher than that of JL_B at the bottom interface, which significantly affect the subthreshold characteristics of JL transistor because the conduction path starts from the bottom interface due to unique volume conduction. It is considered that the F effect enhances carrier transport and improves SS by passivating dangling bonds and GB defects [7]. More details of F effects will be reported elsewhere.

2020 IEEE Silicon Nanoelectronics Workshop

IV. CONCLUSION

Improved subthreshold characteristics and small device-to-device variations of the BF$_2$ doped poly-Si JL transistor with B segregation and F effect have been experimentally characterized and demonstrated. It can be expected that it is more enhanced by applying multi-gate configurations. Thus, it is likely to be useful in future low power applications.

REFERENCES

[1] M. Saitoh *et al.*, VLSI Tech, p.178, 2014. [2] C. J. Su *et al.*, Nanotechnology, 18, p.205, 2007. [3] C. W. Chang *et al.*, EDL, 29, p.474, 2008. [4] H. Yin *et al.*, EDL, 27, p.357, 2006. [5] Y. Morimoto *et al.*, JES, 144, p.2495, 1997. [6] H. N. Chern *et al.*, TED, 41, p.698, 1994. [7] S. K. Kwon *et al.*, JEDS, 6, p.808, 2018. [8] J. P. Colinge *et al.*, Nature Nanotech., 5, p.225, 2010. [9] S. Aziza *et al.*, Mat Sci Semicon Proc, 75, p.43, 2010. [10] H. C. Lin *et al.*, EDL, 33, p.52, 2011. [11] H. B. Chen *et al.*, EDL, 34, p.897, 2013. [12] Y. B. Liu *et al.*, SNW, p.23, 2015. [13] Y. R. Lin *et al.*, TNANO, 17, p.1014, 2018.

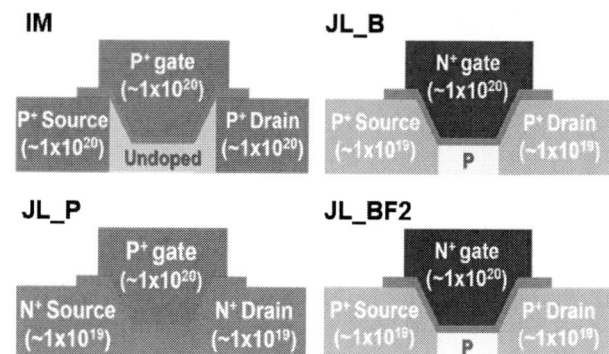

Fig.1: Schematics of fabricated four poly-Si JL transistors (IM, JL_P, JL_B, JL_BF2). P$^+$/P/P$^+$ structure for JL_B and JL_BF2, and N$^+$/N$^+$/N$^+$ structure for JL_P.

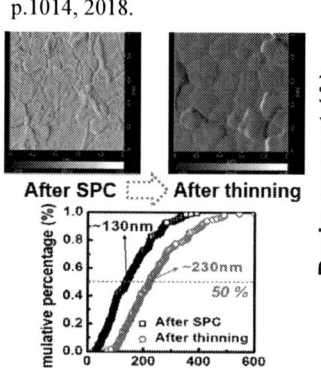

Fig.2: AFM images before and after thinning, and measured average grain size.

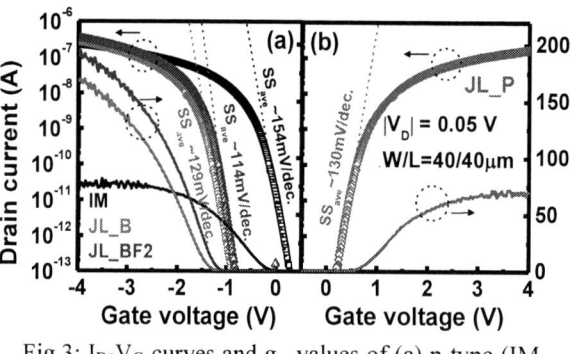

Fig.3: I$_D$-V$_G$ curves and g$_m$ values of (a) p-type (IM, JL_B, JL_BF2) and (b) n-type (JL_P) poly-Si transistors at V$_D$=-0.05V.

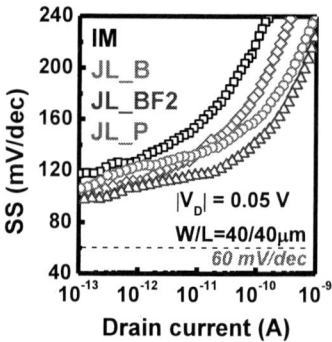

Fig.4: Extracted SS values as a function of I$_D$ at V$_D$=-0.05.

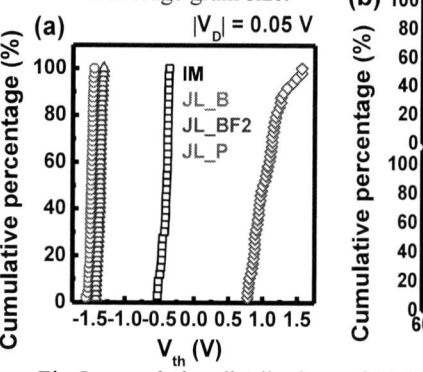

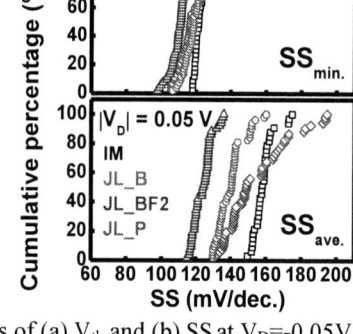

Fig.5: cumulative distributions of (a) V$_{th}$ and (b) SS at V$_D$=-0.05V.

Table I: Key electrical parameters extracted from fabricated four devices.

@V$_D$=50 mV	IM	JL_P	JL_B	JL_BF2
V$_T$ (V)	-0.41	1.05	-1.48	**-1.36**
SS$_{min.}$ (mV/dec.)	117.5	104.3	106.2	**97.5**
SS$_{ave.}$ (mV/dec.)	154.4	130.8	129.9	**114.4**
I$_{on}$/I$_{off}$ (1x10^7)	1.63	1.34	3.32	**3.94**
Mobility (cm^2/V·s)	4.3	4.0	11.6	**12.7**

Table II: Key parameter comparisons in planar poly-Si JL transistors.

	This work	Ref. 10	Ref. 11	Ref. 12	Ref. 13
Structure (Doping type)	Planar (p-type)	Planar (n-type)	Planar (n-type)	Planar (p-type)	Planar (p-type)
Channel thinning	Oxidation	Additional S/D deposition	Dry etching	Dry etching	Dry etching
Crystallization	SPC	*In-situ*	SPC	SPC	SPC
W$_{ch}$xL$_{ch}$ (T$_{ch}$)	40μmx40μm (5nm)	10μmx5μm (10nm)	0.95μmx1μm (15nm)	1μmx0.2μm (25nm)	1μmx1μm (8.6nm)
SS (mV/dec.)	SS$_{min}$:97.5 SS$_{ave}$: 114.4	240	155	184	279
Mobility (cm^2/V·s)	12.7	-	-	-	11.2
I$_{on}$/I$_{off}$ (V$_G$-V$_T$/V$_D$)	3.94x10^7 (-5V/-0.05V)	1.0x10^7 (2.5V/1.0V)	< 1.0x10^6 (2V/0.5V)	1.0x10^5 (-5V/-1.0V)	9.0x10^4 (-5V/-1.0V)

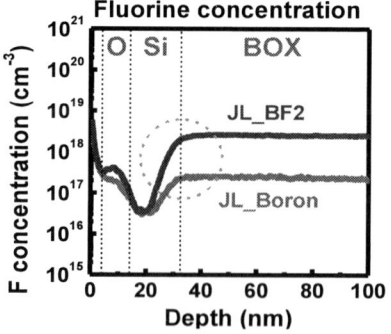

Fig.6: F concentration profiles of JL_BF2 and JL_B after thinning measured by SIMS.

2020 IEEE Silicon Nanoelectronics Workshop

7.2

Crystallinity Effect on Reliability of Sidewall Damascened Nanowire Poly-Si GAA FETs

Chuan-Hui Shen, Wei-Yen Chen, Chun-Chih Chung, Yu-En Huang and Tien-Sheng Chao[*]

Department of Electrophysics, National Chiao Tung University

Email: tschao@mail.nctu.edu.tw

Abstract — **Poly-Si GAA FETs using sidewall damascened method are successfully demonstrated. By manipulating the stress imposed by nitride layer, crystallinity of poly-Si can be modified by changing the thickness of top nitride. Devices with larger grain size and fewer defects lead to superior electrical characteristics. Hot carrier and gate stress reliability of devices were then investigated. With better crystallinity, electrical characteristics degrade less under hot carrier stress due to less electric field enhancement. On the contrary, degradation of gate stress reliability is less sensitive to different crystallinity level. This is owing to the smaller activation energy of hot carrier effect making it more sensitive to crystallinity. With better crystallinity, poly-Si GAA nanowire FETs possess not only better electrical characteristics but also degrade less under stressing.**

Keywords: Crystallinity, GAA, nanowire, poly-Si, reliability.

I. INTRODUCTION

Nowadays, poly-Si was regarded as a potential candidate in future 3D-IC owing to its lower process cost and feasibility of stacking layer on other layers [1]. However, poly-Si is notorious for its defective grain boundaries which lead to lower carrier mobility. In order to boost the performance, the crystallinity of poly-Si channel needs to be improved. It has been reported that stress have impact on poly-Si crystallinity [2]. Thus, we introduced different level of stress to poly-Si by manipulating the nitride layer thickness which poly-Si was in contact with.

Reliability has always been a critical concern for electric devices commercialization. Over decades, hot carrier and gate stress are two main reliability concerns [3]. Here, we investigated the effect of crystallinity on hot carrier and gate stress. In this work, an alternately stacked oxide and nitride structure (NON) was utilized to fabricate poly-Si nanowire channels without using advanced lithography tools. Crystallinity effect on hot carrier stress and gate stress were comprehensively investigated.

II. EXPERIMENT

In our previous work, sidewall damascened nanowire GAA FETs were demonstrated [4]. In this study, same process was used. With different thicknesses of top nitride, different stress then leaded to different level of crystallinity. Hot carrier and gate stress reliability were then investigated comprehensively.

III. RESULTS AND DISCUSSION

Fig. 1 shows the FFT results calculated from TEM images of different top nitride thickness devices. The image of 60-nm top nitride device shows the sharpest diffraction pattern, suggesting that it has the best crystallinity. Fig. 2 shows the I_D-V_{GS} of devices with different top nitride thicknesses. 60-nm top nitride device exhibits the highest I_{on} due to better crystallinity.

Fig. 3 shows the I_D-V_{GS} of devices with different top nitride thicknesses under hot carrier stress. Fig. 4 shows the ΔV_{TH} and I_{on} degradation of devices with different nitride thicknesses. Device with 60-nm top nitride had the least degradation in both

ΔV_{TH} and I_{on} degradation perspective. Additionally, from the exponent of time power-law in Fig. 4 (a), the degradation mechanism in 60-nm top nitride device was closer to oxide trapping charge; while the degradation in 40- and 80-nm top nitride devices were predominated by interface degradation mechanism. Fig. 5 shows the G_m of devices before and after the hot carrier stress. In the case of 40- and 80-nm top nitride devices, not only ΔV_{TH} was observed, but also a G_m degradation occurred referring to severe interface degradation during hot carrier stressing. On the other hand, no G_m peak degradation was observed in 60-nm top nitride device. Only a voltage shift was observed implied that the Ion degradation was mainly owing to ΔV_{TH} which was caused by oxide trapping charge instead of interface degradation. Fig. 6 shows the S.S. degradation of devices with different top nitride thicknesses. Device with 60-nm top nitride shows slight degradation on S.S. indicating no severe interface degradation occurred during the hot carrier stress. On the contrary, devices with 40- and 80-nm top nitride, significant S.S. degradation were observed after hot carrier stress, implying serious interface degradation. This result was consistence with results from time power-law and G_m. Due to high sensitivity of hot carrier effect to electric field, less electric field enhancement by trapped carriers at grain boundaries in 60-nm top nitride device makes it degrade less under hot carrier stress, as shown in Fig. 4 (b).

Fig. 7 shows the I_D-V_{GS} of devices with different top nitride thicknesses under gate bias stress. Fig. 8 shows the ΔV_{TH} and I_{on} degradation of devices with different top nitride thicknesses under gate bias stress. No distinct difference was observed. According to the exponent of power-law in Fig. 8 (a), the degradation mechanism of devices was likely to be oxide charge trapping. Fig. 9 shows the G_m before and after gate bias stress. Voltage shift were observed in all cases. Based on the results of G_m, it could be suggested that the degradation of the devices were dominated by oxide charge trapping which was consistent with the result from exponent of power-law. The sensitivity of gate stress to crystallinity was much lower than that of hot carrier stress might be ascribed to the activation energy difference in hot carrier effect and F-N tunneling. Since the activation energy of hot carrier is 1.1 eV, much lower than that of F-N tunneling 3.1 eV, hot carrier stress was more sensitive to crystallinity.

IV. CONCLUSIONS

NON sidewall damascened poly-Si GAA nanowire FETs were successfully fabricated. By modifying top nitride thickness, different level of crystallinity can be obtained. With better crystallinity, electrical performance of devices can be improved. Owing to activation energy difference, sensitivity to crystallinity of degradation in hot carrier and gate stress are different.

2020 IEEE Silicon Nanoelectronics Workshop

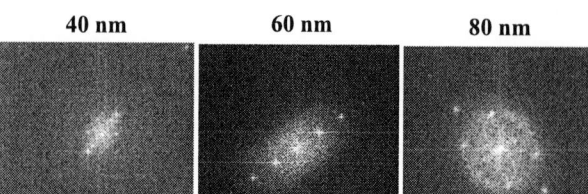

Fig. 1 shows the Fast Fourier Transform (FFT) images of NON sidewall damascene GAA nanowire poly-Si device with 40-, 60- and 80-nm top nitride. 60-nm device exhibit the sharpest diffraction points, implying the best crystallinity.

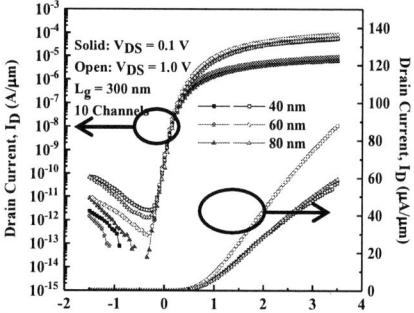

Fig. 2 shows the I_D-V_{GS} of devices. Device with 60-nm top nitride shows the highest on-state current. This is attributed to its better crystallinity of poly-Si channels.

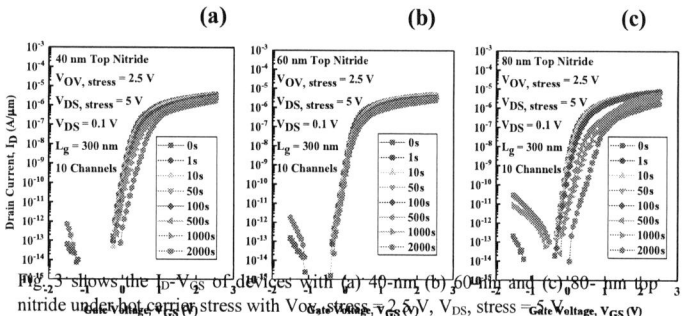

Fig. 3 shows the I_D-V_{GS} of devices with (a) 40-nm (b) 60-nm and (c) 80-nm top nitride under hot carrier stress with $V_{OV, stress}$ = 2.5 V, $V_{DS, stress}$ = 5 V.

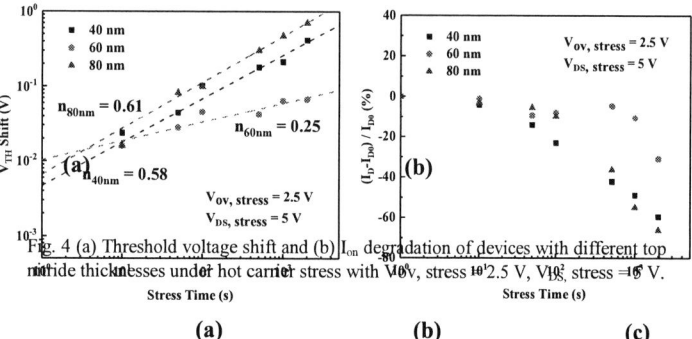

Fig. 4 (a) Threshold voltage shift and (b) I_{on} degradation of devices with different top nitride thicknesses under hot carrier stress with $V_{ov, stress}$ = 2.5 V, $V_{DS, stress}$ = 5 V.

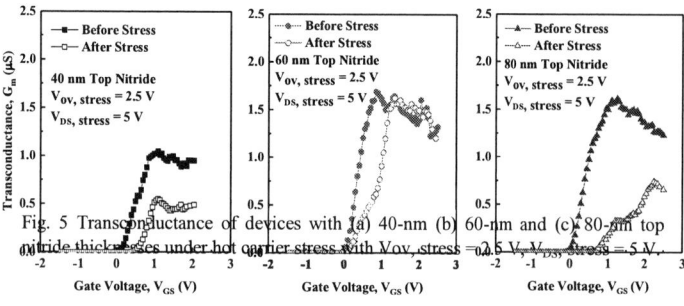

Fig. 5 Transconductance of devices with (a) 40-nm (b) 60-nm and (c) 80-nm top nitride thicknesses under hot carrier stress with $V_{ov, stress}$ = 2.5 V, $V_{DS, stress}$ = 5 V.

7.2

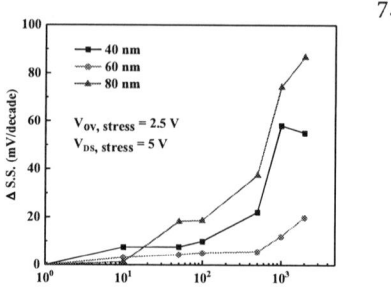

Fig. 6 Subthreshold swing degradation of devices with different top nitride thicknesses under hot carrier stress with $V_{ov, stress}$ = 2.5 V, $V_{DS, stress}$ = 5 V.

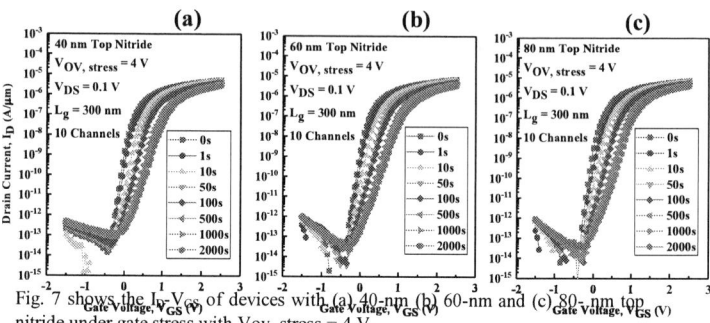

Fig. 7 shows the I_D-V_{GS} of devices with (a) 40-nm (b) 60-nm and (c) 80-nm top nitride under gate stress with $V_{ov, stress}$ = 4 V.

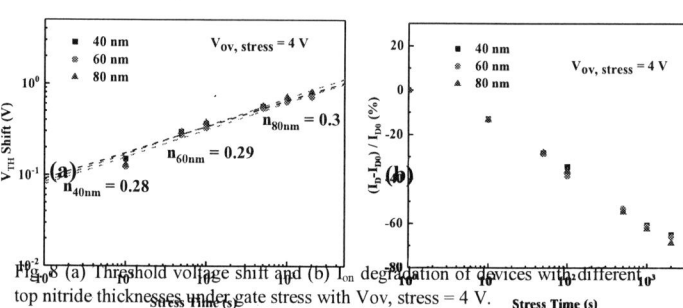

Fig. 8 (a) Threshold voltage shift and (b) I_{on} degradation of devices with different top nitride thicknesses under gate stress with $V_{ov, stress}$ = 4 V.

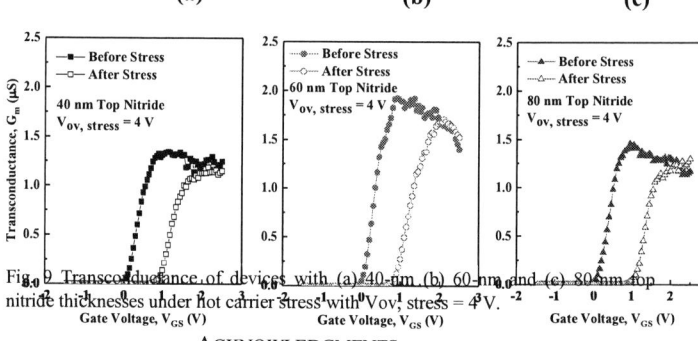

Fig. 9 Transconductance of devices with (a) 40-nm (b) 60-nm and (c) 80-nm top nitride thicknesses under hot carrier stress with $V_{ov, stress}$ = 4 V.

ACKNOWLEDGMENTS

This work was supported by the Ministry of Science and Technology, Taiwan, under contract MOST-106-2221-E-009-151-MY3 and MOST-107-2633-E-009-0093, and by Taiwan Semiconductor Research Institute, Taiwan, under Contract JDP 108-Y1-039.

REFERENCES

[1] F. Hsueh *et al.*, *IEDM*, pp. 2.3.1-2.3.4, 2016
[2] Y. Kimura *et al*, *Journal of Applied Physics,* vol. 86, pp. 2278-2280, 1999
[3] M. Cho *et al.*, *IEEE Transactions on Electron Devices,* vol. 60, pp. 4002-4007, 2013
[4] C. Shen *et al*, *IEEE Transactions on Nanotechnology,* pp. 1-1, 2020

2020 IEEE Silicon Nanoelectronics Workshop

978-1-7281-9736-4/20 $31.00 © 2020 IEEE

Superior subthreshold slope of gate-all-around (GAA) p-type poly-Si junctionless nanowire transistors with highly suppressed grain boundary defects

Min-Ju Ahn, Takuya Saraya, Masaharu Kobayashi, and Toshiro Hiramoto

Institute of Industrial Science, The University of Tokyo, Japan

Email: mjahn@nano.iis.u-tokyo.ac.jp

Abstract — GAA p-type poly-Si junctionless nanowire transistor have been fabricated and evaluated. It exhibits excellent subthreshold characteristics close to ideal subthreshold slope (60mV/dec.) as well as high on/off current ratio (~1.2x10^8) and low off-current (<10^{-13}A).

I. INTRODUCTION

Polycrystalline silicon (poly-Si) transistors have been considered as the key building elements for 3D integration. However, unavoidable defects at grain boundaries (GBs) in poly-Si channel undermine the carrier transport and switching properties [1]. The adoption of gate-all-around (GAA) nanowire (NW) as the channel structure has demonstrated superior electrical characteristics due to its small-size volume with reduced GB defects as well as enhanced gate controllability [2].

On the other hand, the concept of junctionless (JL) scheme without S/D junctions is promising for nano-scaled poly-Si transistors due to its unique volume conduction which is influenced much less by interface scattering. The thickness of JL channel must be thin enough because the channel should be entirely depleted, which has a strong influence on key factors for low power applications such as subthreshold slope (SS), threshold voltage (V_T) and off-current (I_{off}) [3]. Meanwhile, many studies have focused on the n-type, but few have addressed the p-type poly-Si JL NW transistors.

In this work, p-type GAA JL poly-Si NW transistors were fabricated by improved process and superior SS is successfully demonstrated.

II. EXPERIMENTAL

Fig.1 shows fabrication processes and a 3D schematic. A 120nm amorphous Si was deposited on the 200nm BOX layer by LPCVD, and SPC at 1100°C for 24 hours was performed [4]. Then, BF_2^+ ions, instead of B^+ ions, was implanted at 35keV with dose of 3x10^{14}cm^{-2} to form a p$^+$ poly-Si layer. The active region was locally thinned down to 10nm by oxidation, by which the B concentration of thinned active region was reduced due to the segregation, while the S/D regions remained thick and high concentration. The NW pattern was defined by E-beam lithography and RIE, followed by wet-etching for suspending NW from the BOX layer. Fig.2 displays the top-view scanning electron microscopic (SEM) image of the defined NW just after RIE process. The length (L_{NW}) and width (W) of NW are ~250 and ~20nm, respectively. A 10nm dry gate oxide was thermally grown and a 150nm in-situ doped N$^+$poly-Si was deposited for a gate electrode. Then, a 300nm SiO_2 was deposited for passivation and H_2 annealing was carried out to terminate defects at 400°C for 30min. Finally, Al metal contacts were formed. The effects of the F ions and segregation will be reported in [5].

Fig.3 shows a 2D schematic of device structure. NW is connected to large S/D and extension regions. The L_{NW} and W of NW range from 250 to 1050nm and 5 to 20nm, respectively. The effective width (W_{eff}) is a perimeter of NW, and ranging from 20 to 50nm. The gate length L_G is 30μm and NW channel thickness T_{ch} is <5nm. For comparison, a planar (PL) JL transistor was also fabricated. The average grain size in the poly-Si channel is ~230nm estimated from AFM analysis (data not shown).

III. RESULTS AND DISCUSSION

Fig.4 compares the normalized I_D-V_G curves of PL and NW transistors (W_{eff}=20nm, L_{NW}=250nm) at V_D=-0.05 and -1.2V. No hysteresis was observed. It should be noted that the PL transistor shows steep SS of ~100mV/dec, indicating the quality of poly-Si layer is much better than those of published data of PL poly-Si JL transistors [6-8]. This is due to the effects of F ions and segregation in our improved process [5]. The NW transistor shows even steeper SS and enhanced current drivability thanks to GAA structure. Extracted SS as a function of drain current is shown in the Fig.5. SS$_{min}$ of PL and NW transistors are 97mV/dec and 63mV/dec, respectively, at V_D=-0.05V. SS$_{ave}$ of PL and NW transistors are 114mV/dec and 74mV/dec, respectively, at V_D=-0.05V, where, SS$_{ave}$ is average SS in the subthreshold region from $I_D×L/W = 1×10^{-13}$ to $1×10^{-10}$A. The NW transistor shows excellent SS in a wide range of I_D, and the SS values are superior to the published data [8-11].

Fig.6 shows dependence of SS$_{ave}$ on W_{eff} and L_{NW}. It can be clearly observed that SS$_{ave}$ decreases with decreasing W_{eff}, which is reasonable because the total GB defects decrease in narrow NW. However, SS slightly increases with decreasing L_{NW}. Since the gate length is long enough in our devices, the short channel effect is negligible. The origin of L_{NW} dependence is not clear at present.

Fig.7 shows I_D versus gate overdrive voltage (V_{OV}=V_G-V_T) of a p-type GAA JL poly-Si NW transistor with V_{sub} of 0V and -25V, where V_{sub} is the substrate bias. The inset shows I_D-V_D curves. Notably, improved SS$_{ave}$ and higher I_{on} with low V_{OV} can be achieved by applying V_{sub}. The reasons for this improvement will be reported elsewhere.

From these results, it is demonstrated that GAA p-type poly-Si JL NW transistors by improved process show superior subthreshold characteristics.

IV. CONCLUSION

GAA p-type poly-Si JL NW transistors with superior SS, low I_{off} and high I_{on}/I_{off} were fabricated and demonstrated, which are very promising for future low power applications.

REFERENCES

2020 IEEE Silicon Nanoelectronics Workshop

[1] M. Saitoh *et al.*, VLSI Tech, p.178, 2014. [2] T. Y. Liu *et al.*, IEEE EDL, 34, p.523, 2014. [3] J. P. Colinge *et al.*, Nature Nanotech., 5, p.225, 2010. [4] K. H. Jang *et al.*, JJAP, 59, 021004, 2020. [5] M. J. Ahn et al., submitted to SNW, 2020. [6] H. B. Chen *et al.*, IEEE EDL, 34, p.897, 2013. [7] Y. B. Liu *et al.*, SNW, p.23, 2015. [8] Y. R. Lin *et al.*, TNANO, 17, p.1014, 2018. [9] L. C. Chen *et al.*, IEEE JEDS, 4, p.50, 2016. [10] D. R. Hsieh *et al.*, IEEE JEDS, 7, p.282, 2019. [11] Y. R. Lin *et al.*, IEEE JEDS, 6, p.1187, 2018.

- ● SPC poly-Si
- ● Ion implantation
- ● Channel thinning
- ● Define NW
- ● Gate oxidation
- ● Define N⁺poly-Si gate
- ● Passivation
- ● Al contacts

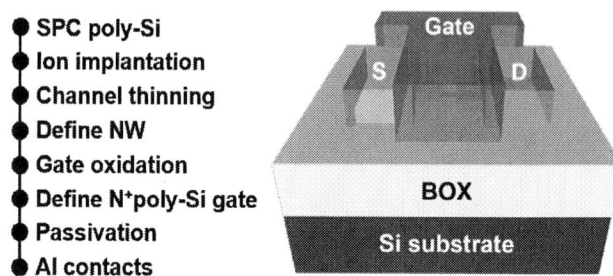

Fig.1: Key fabrication process flows and 3D schematic of fabricated GAA poly-Si JL NW transistor.

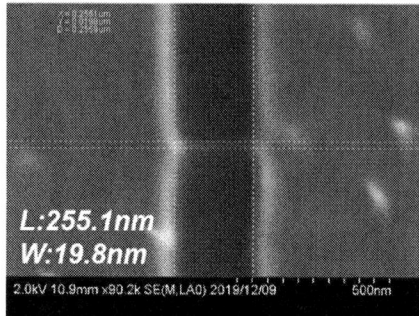

Fig.2: A top view of SEM image of the defined nanowire just after RIE. The width is ~20nm at this stage and is reduced down to ~5nm after gate oxidation and cleaning processes.

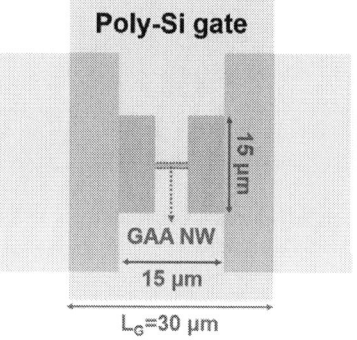

Fig.3: A top schematic view of the device structure. NW is connected to large S/D and extension regions. L_G= 30μm, where short channel effects are negligible.

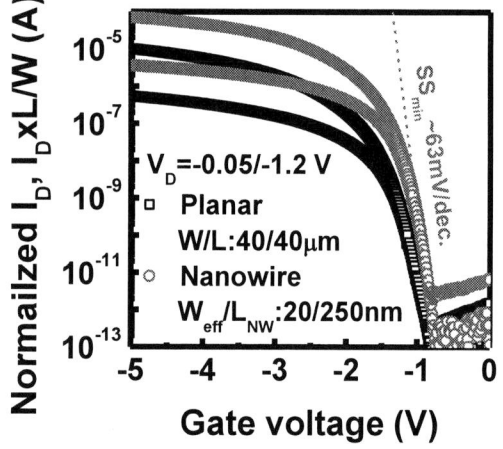

Fig.4: Normalized I_D-V_G curves of PL and NW transistors at V_D=-0.05 and -1.2 V.

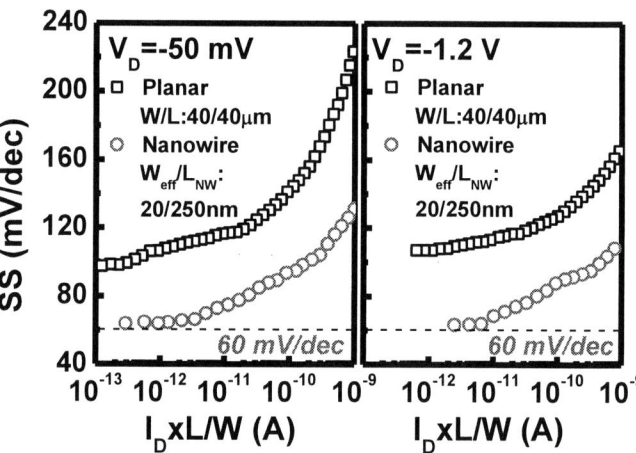

Fig.5: Extracted SS values as a function of I_D at V_D=-0.05 and -1.2 V. W_{eff}=20nm and L_{NW}=250nm in the NW transistor.

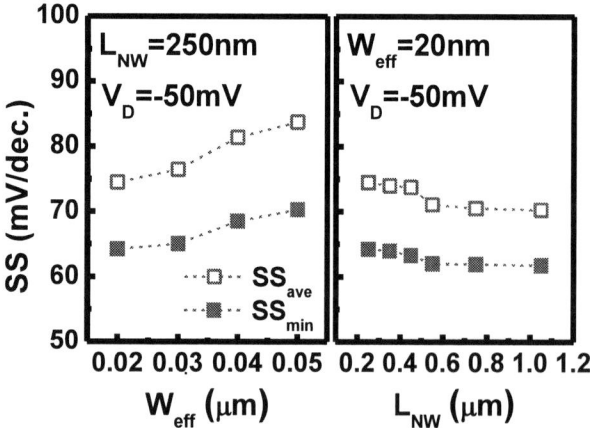

Fig.6: Dependence of SS_{min} and SS_{ave} on W_{eff} and L_{NW} in NW transistors.

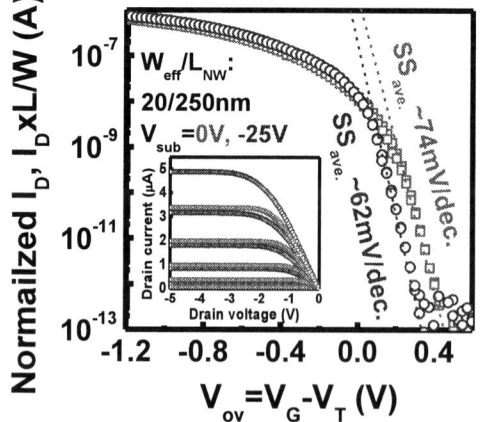

Fig.7: I_D-V_{OV} curves of a NW transistor with V_{sub}=0V and V_{sub}=-25V at V_D=-0.05V. The inset shows I_D-V_D curves.

2020 IEEE Silicon Nanoelectronics Workshop

978-1-7281-9736-4/20 $31.00 © 2020 IEEE

Characteristics of Dual-gated Poly-Si Junctionless Nanowire Transistors with Asymmetrical Source/drain Offsets

You-Tai Chang[1], Ruei-Jen Wu[1], Kang-Ping Peng[1], Chun-Jung Su[2], Pei-Wen Li[1], and Horng-Chih Lin[1*]

[1] Institute of Electronics, National Chiao Tung University, Hsinchu, Taiwan

[2] Taiwan Semiconductor Research Institute, 26 Prosperity Road I, Hsinchu Science Park, Taiwan

[*] hclin@faculty.nctu.edu.tw

Abstract — **In this paper, a novel gate-all-around (GAA) junctionless (JL) nanowire (NW) transistor with dual gate was proposed, fabricated and characterized. The fabricated transistors exhibit well-behaved performance with on/off current ratio of ~10^6 and subthreshold swing of 76 mV/decade. An important finding of notes is that when drain bias is applied to the end of the NW with a longer channel offset, the drain current is lower than that applied to the shorter end.**

I. INTRODUCTION

The GAA NW FET has been considered one of suitable candidates for next-generation technology due to its excellent gate controllability and immunity to short channel effects (SCEs) [1]. It's interesting to investigate these two gates controlling the devices. The main gate could control the switching of the devices and the sub-gate could enhance current through the channel. Also, both the main gate and sub-gate can be logic inputs that enable to design more efficiently on XOR implement [2].

II. DEVICES STRUCTURE AND FABRICATION

Figure 1 shows the schematic diagrams of process flow for fabrication of dual-gate NW transistors. Firstly, a 300-nm-thick wet oxide was grown on the p-type substrate as a block layer, followed by a LPCVD of 200-nm-thick Si_3N_4 as the sacrificial layer. Next, a 100-nm-thick *in-situ* n^+-doped poly-Si was deposited and patterned by I-line lithography as source and drain (S/D) studs. Then, a 180-nm-thick *in-situ* n^+-doped poly-Si layer was further deposited, followed by dry etching to form S/D spacers (Fig. 1(a)) aimed to shorten the effective channel length ($L_{channel}$) which was defined as the distance between S/D studs. Afterwards, an 20-nm-thick *in-situ* n^+-doped poly-Si layer was deposited to serve as the JL channel layer. Then, a CVD SiO_2 was deposited and subsequently patterned by I-line lithography and etching to form a dummy structure (Fig. 1(b)). Next, a nitride layer was deposited and etched back to form nitride spacer abutting the dummy oxide and serve as hard mask for subsequent NW formation (Fig. 1(c)). After removing the dummy oxide by using buffered oxide etchant (BOE), the poly-Si NW channels were defined by dry etching with the spacer layers of Si_3N_4 as an etching hardmask (Fig. (d)). Afterwards,

nitride spacer and the nitride sacrificial layer under channel layer were removed by hot phosphoric acid (H_3PO_4) to suspend the NW channel. Next, gate oxide was deposited and annealed at 800°C in N_2O ambient for 30 seconds. After depositing a n^+-doped poly-Si layer, the main gate was defined by e-beam lithography and dry etching (Fig. 1(e)). A 25-nm-thick TEOS oxide layer was deposited by LPCVD to block main gate from the sub-gate. Finally, a n^+-poly-Si layer was deposited and patterned by I-line lithography as sub-gate (Fig. 1(f)).

III. RESULTS AND DISCUSSION

Figure 2 shows the plan-view SEM image of the fabricated dual-gate transistors. The main-gate with a channel length of L_{MG} governs the middle, gated-NW region for switching operation of the transistors, whereas the resistance of offset NW regions between main-gate and source/drain (S/D) with lengths of $L_{offset,1}$ and $L_{offset,2}$, respectively, are modulated by the sub-gate.

We measured the transfer characteristics by sweeping sub-gate voltage (V_{G2}) from 0 to 2 V on two transistors with a given L_{MG} of in combination with different L_{offset} of 4600 and 620 nm, respectively, as shown in Figs. 3(a) and (b). Increasing V_{G2} significantly decreases the series resistance of the offset regions between the main-gated NW and S/D studs, effectively enhancing on-state current (I_{on}). For the devices with longer L_{offset}, the improvement of I_{on} from applying V_{G2} would be more significant. The normal mode marked in Fig. 4 denotes that the source pad was grounded and the drain pad was biased with applied drain voltage ($V_{DS,N}$). In contrast, bias conditions of source and drain were exchanged for the reverse mode, that is, a positive "drain voltage" ($V_{DS,R}$) is applied to the source pad while keeping the drain pad being grounded. Figure 4 shows that the value of I_D measured under normal mode is higher than that in reverse mode, and such current improvement is significantly enhanced when applied V_{DS} gets larger. Increasing V_{G2} effectively reduces the series resistance of the offset NW regions so that the difference in the driving current of transistors operating in normal- and reverse-modes becomes gets smaller. Figure 5(a) illustrates the TCAD simulated structure (only the core NW channel and the surrounding gate oxide are shown) with assumed misalignment of $L_{offset,1}$ = 70 nm and $L_{offset,2}$ = 30 nm. Figures 5(b) and (c) show

2020 IEEE Silicon Nanoelectronics Workshop

the simulated electrostatic potential profiles along the NW channel with V_{G2} of 0 and 2 V, $V_{G1} = 1$ V, and $V_{DS} = 1.5$ V under normal mode and reverse mode, respectively. The voltage drop, denoted as $V_{OS,D}$, across this offset region extracted from the simulation results, accordingly affects the effective drain voltage that practically drives the main channel region gated by the main-gate.

IV. CONCLUSION

The sub-gate shows sufficient gate controllability for reducing the resistance of the NW offset regions and accordingly enhances the on current. When the drain bias is applied to the end of NW with a longer channel offset, drain current is lower than the case of drain bias

to the shorter end. TCAD simulations point out that a longer the offset with applied drain bias would lead to a higher voltage drop wherein and, accordingly, reducing the effective bias for driving the channel region gated by the main-gate.

Acknowledgement This work was supported in part by the Ministry of Science and Technology, Taiwan, under Grant MOST108-2221-E-009-011-MY3 and MOST109-2639-E-009-001.

REFERENCES

[1] H.-C Lin *et al*, *IEEE Trans. Electron Devices*, vol. 53, no. 10, p. 2471 (2006).

[2] M. D. Marcki *et al*, *IEEE Trans. Nanotechnology*, vol. 13, no. 6, p. 1029 (2011).

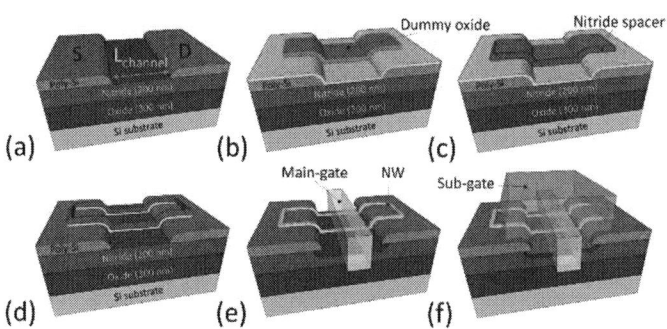

Fig. 1 Process flow of the device fabrication.

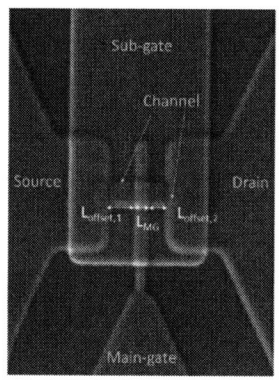

Fig. 2 Top-view SEM image of a fabricated dual-gate poly-Si NW device.

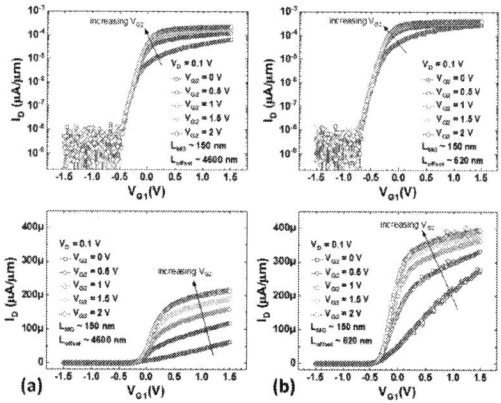

Fig. 3 Influence of V_{G2} on the transfer characteristics of devices with $L_{MG} = 150$ nm and $L_{offset} =$ (a) 4600 nm and (b) 620 nm.

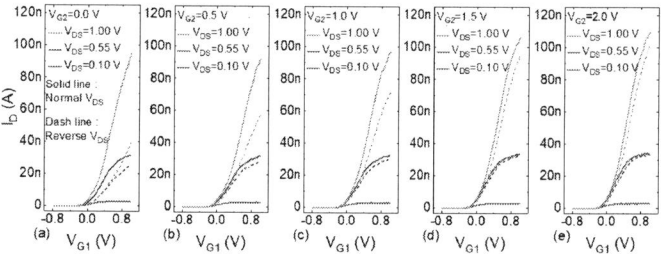

Fig. 4 Transfer characteristics of a device with $L_{MG} = 100$ nm, $L_{offset,1} \sim 70$ nm (source side), and $L_{offset,2} \sim 30$ nm (drain side) under normal mode and reverse mode and $V_{G2} =$ (a) 0, (b) 0.5, (c) 1, (d) 1.5 and (e) 2 V.

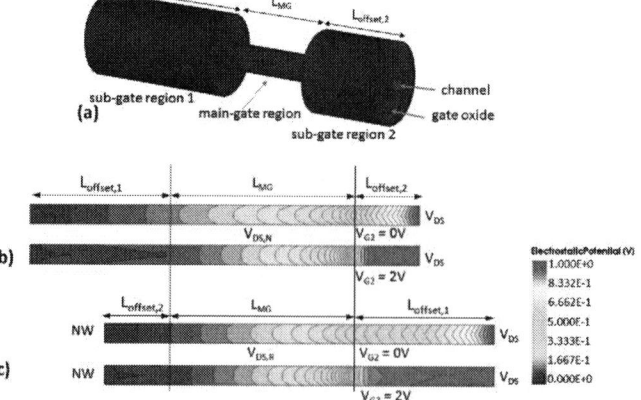

Fig. 5 (a) The simulated structure with $L_{MG} = 100$ nm, $L_{offset,1} \sim 70$ nm, and $L_{offset,2} \sim 30$ nm (only core NW and gate oxide are shown, with the gate oxide thickness of 5 nm in the main-gated region, and 25 nm in the offset regions). (b) Simulated potential in the NW channel $V_{G2} = 0$ and 2 V and $V_D = 1$ V under normal mode. (c) Simulated potential in the NW channel $V_{G2} = 0$ and 2 V and $V_D = 1$ V under reverse mode.

2020 IEEE Silicon Nanoelectronics Workshop

Ge and GeSn-based Nano-electronic and Photonic Devices

Xiao Gong, Ying Wu, Dian Lei, Shengqiang Xu, and Kaizhen Han

Department of Electrical and Computer Engineering, National University of Singapore, 117576

Email: elegong@nus.edu.sg

Abstract — **In this talk, I will review our recent progress of various nano-electronic and photonic devices employing Ge and GeSn, including Ge/GeSn-based multi-gate transistors, GeSn as S/D materials to achieve ultra-low contact resistivity, and photodetectors for wavelength detection at ~2 μm.**

Keywords: GeSn, transistors, photodetectors.

I. Introduction

Germanium-Tin ($Ge_{1-x}Sn_x$) alloy has many attractive properties that make it promising for many applications in nano-electronic and photonic devices and systems. GeSn has both higher hole and electron mobilities than Ge and Si to possibly enable scaling of supply voltage for future high performance field-effect transistors (FETs). In addition, the tunable bandgap of GeSn by changing Sn compositions helps to extend photo detection range beyond what can be achieved by Ge, covering wavelength larger than 1.55 μm.

In this paper, we discuss our research and development of utilizing GeSn for various applications in nano-electronic and photonic devices. This includes multi-gate GeSn p-channel FETs (pFETs) on GeSn-on-insulator (GeSnOI) substrates, realization of ultra-low contact resistivity by incorporating Sn into Ge for the metal/GeSn S/D contact, and GeSn photodetectors for applications at 2 μm wavelength range.

II. GeSn pFinFETs with Sub-10 nm Fin Width on GeSnOI Substrates

A large-area GeSnOI substrate was fabricated using direct wafer bonding (DWB) and layer transfer techniques at a 200 mm wafer scale, as shown in Fig. 1 (a) [1]. Very clear and similar peak positions and shapes were observed in the XRD curves for all the locations, indicating good uniformity of the substrate. This GeSnOI substrate, together with the multi-gate architecture, enabled the realization of high performance $Ge_{1-x}Sn_x$ p-FETs with extremely scaled channel length down to 50 nm [2, 3]. An optimized dry etch process was developed to fabricate GeSn FinFETs with fin width less than 10 nm, as illustrated in Fig. 1 (b) and (c). Excellent electrical characteristics were achieved with subthreshold swing (*SS*) as small as 63 mV/decade and intrinsic transconductance of 900 μS/μm at V_{DS} of -0.5 V (Fig. 2 and Fig. 3). High-field effective mobility is ~275 cm²/V·s,

III. Metal/GeSn p-type Contact with Ultra-low Contact Resistivity

GeSn shows a smaller hole Schottky barrier and a hole effective mass (m_h) as compared to Ge. These advantages make GeSn attractive for low-resistance p-type contacts. *In-situ* Ga doping was introduced during the epitaxial growth of GeSn using molecular-beam epitaxy (MBE) and XRD curves in Fig. 4 show high-crystalline quality GeSn films with well-defined GeSn peaks and clear thickness fringes [4]. Using these materials, an average ρ_c of 6.5×10^{-10} Ω·cm² for Ti/Seg. p+-GeSn contact was extracted from 14 sets of Nano-TLM structures, as shown in Fig. 5. The benefits of Sn incorporation and Ga segregation can be clearly seen in Fig. 6 [5]. The ultra-low ρ_c of Ti/p$^+$-GeSn is retained after anneal annealing at 420 °C for 1 hour, showing adequate thermal stability for BEOL process in current CMOS technology [6].

IV. GeSn Photodetectors for Applications at 2 μm Wavelength Range

The absorption coefficient and wavelength can be extended by increasing the Sn composition in GeSn (Fig. 7). High-speed photo detection was demonstrated beyond traditional telecommunication bands and reaching 2 μm band, realized by GeSn/Ge multiple-quantum-well (MQW) photodiode on Si substrate (Fig. 8) [7]. GeSn/Ge MQW structure was employed to increase the critical thickness (h_c) of epitaxial GeSn on Ge virtual substrate. A decent responsivity of 15 mA/W was obtained at 2 μm. A low leakage current density of 44 mA/cm² was achieved at reverse bias of 1 V, which is among the lowest reported values for all GeSn photodiodes. Beyond traditional O to U bands all up to the new 2 μm window with a fixed input power, photocurrent is reduced at longer wavelength while the detector shows clear response to cover beyond 2 μm (Fig. 9). A 3-dB bandwidth ($f_{3\text{-dB}}$) larger than 10 GHz was experimentally demonstrated directly at 2 μm as shown in Fig. 10.

Acknowledgment

We would like to acknowledge the funding support from National Research Foundation (NRF) Singapore, under its Quantum Engineering Programme (Award QEP-P3) and Singapore Ministry of Education Tier 2 grants (R-263-000-D45-112 and R-263-000-D77-112).

References

[1] D. Lei *et al.*, *Appl. Phys. Lett.*, 109, 022106, 2016.

[2] D. Lei *et al.*, *VLSI Symposia*, 2017, 198-199.

[3] D. Lei *et al.*, *VLSI Symposia*, 2018, 197-198.

[4] Y. Wu *et al.*, *J. Appl. Phys.*, 122, 224503, 2017.

[5] Y. Wu *et al.*, *VLSI Symposia*, 2018, 77-78.

[6] Y. Wu *et al.*, *IEEE Electron Device Lett.*, 40, 1575-1578, 2019.

[7] S. Xu *et al.*, *IEDM*, 2018, 544-547.

8.1

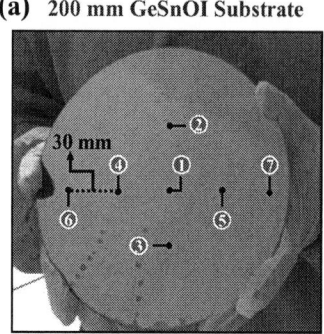

(a) 200 mm GeSnOI Substrate

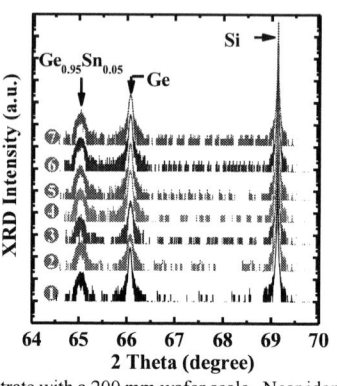

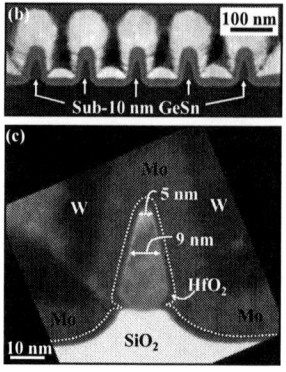

Fig. 1. (a) The photograph of the formed GeSnOI substrate with a 200 mm wafer scale. Near identical XRD curves were observed at different locations. (b) TEM image of GeSn p-channel FinFETs with 5 fins in parallel. (c) Zoom-in TEM image of one fin.

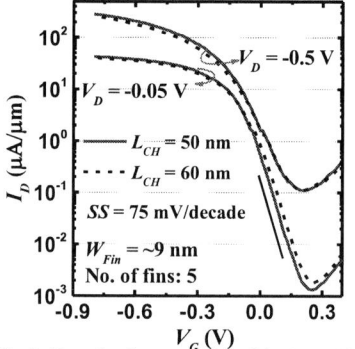

Fig. 2. Transfer characteristics of devices with channel length of 50 and 60 nm and fin width of 9 nm.

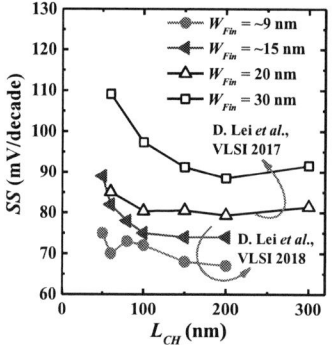

Fig. 3. SS of less than 80 mV/Dec. is maintained by the 9 nm fin. The best device has SS of 63 mV/decade.

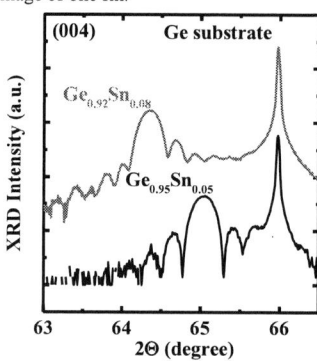

Fig. 4. The XRD shows high-crystalline quality GeSn films with well-defined GeSn peaks and clear thickness fringes.

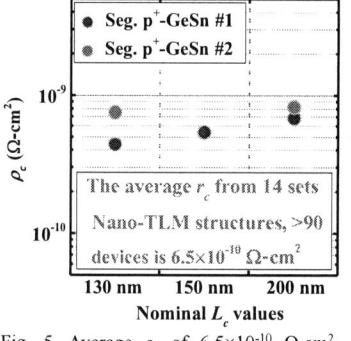

Fig. 5. Average ρ_c of 6.5×10^{-10} $\Omega\cdot$cm^2 (lowest ρ_c of 4.4×10^{-10} $\Omega\cdot$cm^2) for Ti/Seg. p$^+$-GeSn measured from 14 sets of Nano-TLM structures.

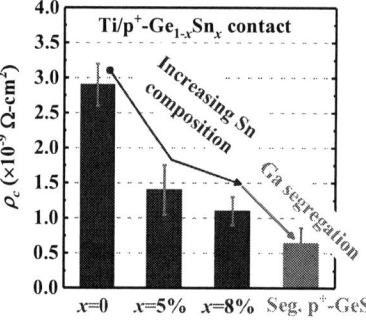

Fig. 6. Reduction of specific contact resistivity ρ_c with increasing Sn composition. Ga surface segregation leads to ρ_c in the sub-10^{-9} $\Omega\cdot$cm^2 regime.

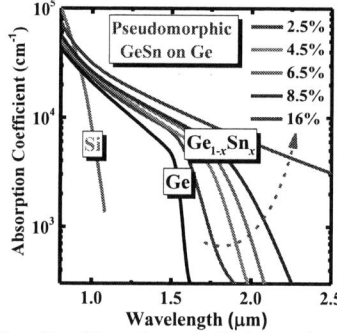

Fig. 7. Absorption coefficient of our pseudomorphic GeSn grown on Ge substrate with different Sn compositions ranging from 2.5% to 16%.

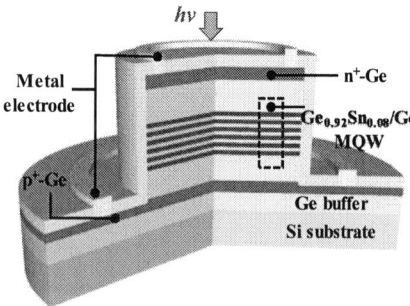

Fig. 8. (a) Three-dimensional internal schematic of the high speed photodiode. Ge$_{0.92}$Sn$_{0.08}$/Ge MQW structure was employed.

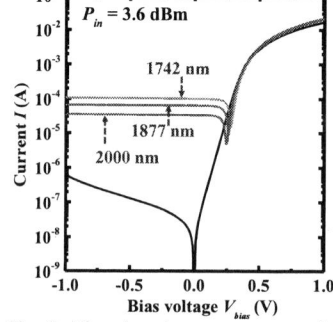

Fig. 9. The photodetector response from bands up to 2 um window with a fixed input power.

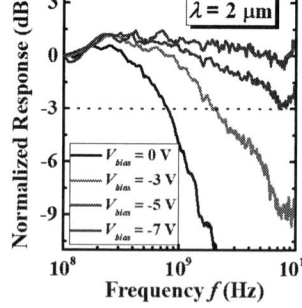

Fig. 10. Normalized small signal frequency response of the 20-μm-diameter photodiode at 2 μm.

978-1-7281-9736-4/20 $31.00 © 2020 IEEE

2020 IEEE Silicon Nanoelectronics Workshop

Gap in pagination due to formatting issues.

Pages 61-62

Polarization Dependence of Incident Angle Sensitivity in SOI Photodiode with 2D Hole Array Grating

Anitharaj Nagarajan[1,2], Shusuke Hara[3], Hiroaki Satoh[3],
Aruna Priya Panchanathan[2] and Hiroshi Inokawa[1,3]

[1]Graduate School of Science and Technology, Shizuoka University, Japan
[2]Department of Electronics and Communication Engineering, SRM Institute of Science and Technology, India
[3]Research Institute of Electronics, Shizuoka University, Japan,
Email: inokawa.hiroshi@shizuoka.ac.jp

Abstract — **The plenoptic imaging technique requires angle sensitive pixel (ASP) to capture the direction (θ, ϕ) of light. In this work, we experimentally demonstrate the polarization dependence of incident angle sensitivity in the silicon-on-insulator (SOI) photodiode (PD) stacked with 2D hole array grating. Measured results can be explained based on phase matching condition between the diffracted light from the grating and the SOI waveguiding mode. The theoretical prediction and the measured results show that the incident light with specific incident angle shift in the contours of the peak angle when the grating period is varied. The proposed device has potential application as image sensor pixels in lensless imaging, refocusing and depth mapping.**

I. INTRODUCTION

Light field camera or plenoptic camera forms image by capturing the light field information reflected from the object. i.e., the direction along with the intensity of light. In contrast, conventional camera records only the intensity. The plenoptic function of a light ray could be represented by the 7D function, $P(\theta, \phi, \lambda, t, x, y, z)$, where ($\theta$, ϕ) denotes the direction information, λ is wavelength, t is time, and (x, y, z) are spatial information. Microlens arrays are used to capture the light field effectively but making the plenoptic camera bulky and complex. We have proposed a miniaturized angle sensitive pixel based on SOI photodiode with line and space gold grating which is compatible with the CMOS technology, and investigated the capability of full azimuth-elevation angle detection and its effect of grating period [1, 2].

In this work, we demonstrate the polarization dependence of the 2D angle (θ, ϕ) sensitivity of SOI PD stacked with the gold grating composed of 2D hole array in contrast to the previous work with 1D grating. The measured results are presented as a spatial pattern of the external quantum efficiency in polar coordinates for various polarization angles, and compared with the theoretical prediction by the phase matching condition between the diffracted light from grating and SOI waveguiding mode.

II. FABRICATED DEVICE AND EXPERIMENTAL CONDITION

The top view of the fabricated SOI PD with 2D hole array grating inscribed over the light sensitive area by electron beam lithography is shown in Fig. 1. Figure 2 shows the bird's eye view of the device structure and the definition for polarization (ϕ_{pol}) and incident angle with azimuth (ϕ) and elevation (θ) angles.

III. RESULTS AND DISCUSSION

The theoretical concept behind the incident angle detection by the proposed device is shown in Fig. 3. When the phase matching condition is satisfied between the diffracted light from the grating and SOI waveguiding mode, high light absorption in SOI layer can be obtained. Figure 4 is the spatial pattern which represents theoretical optical information for the full azimuth and elevation direction (azimuth angle and elevation angle as polar and radial axes, respectively) of the SOI PD with 2D hole array grating of 105-nm-thick SOI with different grating period, $p = 286$, 306 and 326 nm, especially for the polarization angle of $\phi_{pol} = 45°$. As the period size decreases the four-fold symmetrical polar plot shift towards the center, because the phase matching condition is satisfied for normal incidence when $p = 273$ nm. The experimental demonstrations for the fabricated SOI PD with the thickness of 105 nm and the fixed grating period at $p = 286$ nm are done by using a laser with the wavelength of 685 nm. The polarization dependence is shown in Fig. 5. For $\phi_{pol} = 0$ and 90°, the patterns with two-fold symmetry in vertical and horizontal directions, respectively, and for $\phi_{pol} = 45°$, the superposition of the patterns of $\phi_{pol} = 0$ and 90° are observed. The peak position in the experimental spatial pattern is justified by comparing with the theoretical spatial pattern of $p = 286$ nm in Fig. 4. The proposed device is capable of detecting the information for not only 2D incident angle but also polarization.

IV. CONCLUSION

The polarization dependence of incident angle sensitivity in the proposed SOI PD is demonstrated experimentally and interpreted theoretically. Since the spatial pattern clearly depends on the incident polarization, the proposed device can detect incident angle and polarization simultaneously, and may contribute to the development of a new ASP for plenoptic camera that enables image refocusing, depth mapping and lensless imaging.

REFERENCES

[1] H. Satoh, et al., "Surface plasmon antenna with gold line and space grating for enhanced visible light detection by a silicon-on-insulator metal–oxide–semiconductor photodiode," *IEEE Trans. Nanotechnol.*, vol. 11, no. 2, pp. 346-351, Nov 2011.

[2] A. Nagarajan, et al., "Directivity of SOI Photodiode with Gold Surface plasmon Antenna," *2019 Silicon Nanoelectronics Workshop (SNW)* pp. 1-2 IEEE.

8.3

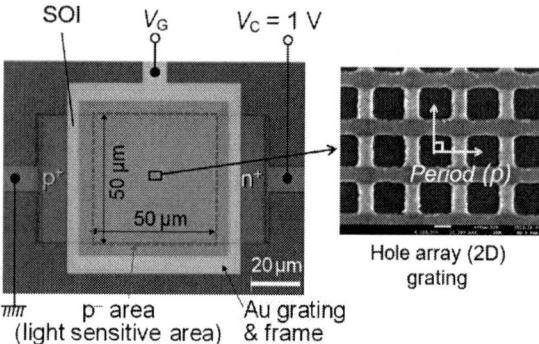

Fig. 1. Top view of the fabricated SOI PD with 2D hole array grating.

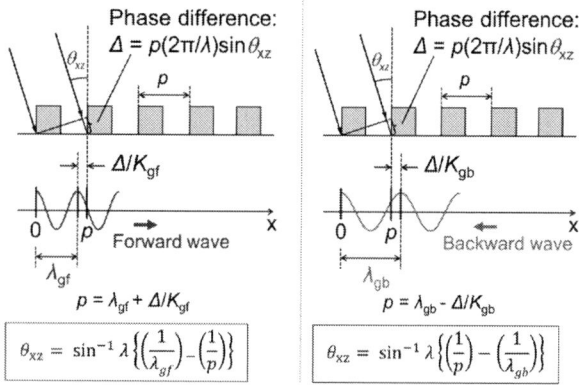

Fig. 3. Phase matching conditions between diffracted light from antenna and the waveguide mode in SOI layer for forward and backward waves. λ_{gf} and λ_{gb} are the forward and backward propagating wavelength respectively and the wavenumbers for forward and backward waves are $k_{gf} = 2\pi/\lambda_{gf}$ and $k_{gb} = 2\pi/\lambda_{gb}$ respectively. For the waves propagating in y direction, θ_{yz} has to be considered instead of θ_{xz}.

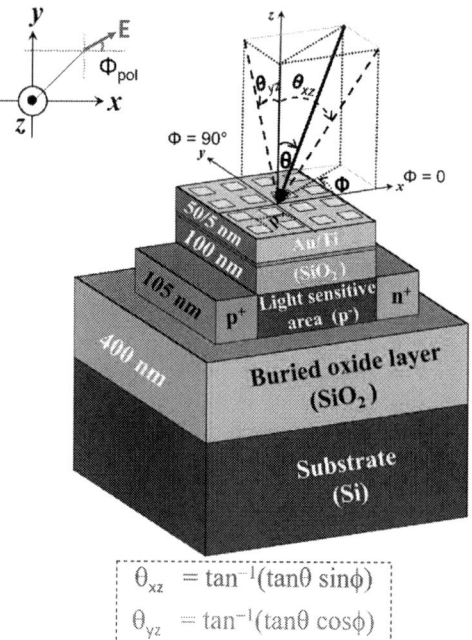

$$\theta_{xz} = \tan^{-1}(\tan\theta \sin\phi)$$
$$\theta_{yz} = \tan^{-1}(\tan\theta \cos\phi)$$

Fig. 2. Bird's eye view of the SOI PD with 2D hole array grating, and the definition of polarization angle (ϕ_{pol}), and azimuth (ϕ) and elevation (θ) angles.

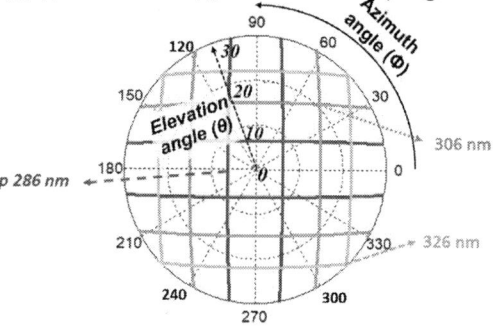

Fig. 4. Theoretical spatial pattern calculated for fundamental TM mode in devices with SOI thickness of 105 nm and various grating periods.

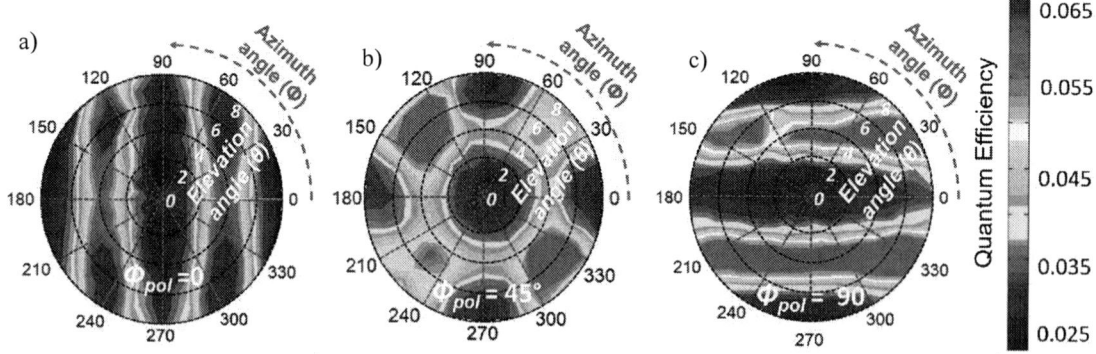

Fig. 5. Measured spatial pattern of SOI PD with grating period p = 286 nm and SOI thickness of 105 nm for various polarizations, a) $\phi_{pol} = 0$, b) 45° and c) 90°.

978-1-7281-9736-4/20 $31.00 © 2020 IEEE

2020 IEEE Silicon Nanoelectronics Workshop

8.4

Photon-number Statistics Observed by SOI MOSFET Single-Photon Detector with Real-time Signal Processing

Revathi Manivannan[1], Hiroaki Satoh[2], and Hiroshi Inokawa[1,2]

[1]Graduate School of Science and Technology, Shizuoka University, Japan,
[2]Research Institute of Electronics, Shizuoka University, Japan
Email: inokawa.hiroshi@shizuoka.ac.jp

Abstract — **In this work, an FPGA-based signal processor is developed specifically for the SOI MOSFET single-photon detector and other photodetectors based on single-charge counting. We take statistics of the photon arrival (photo-generation of holes) for different light intensities, and successfully verify that they follow the Poisson distribution.**

Keywords: SOI MOSFET single-photon detector, FPGA, Signal processing, Poisson distribution

I. INTRODUCTION

Single-photon detectors are used in wide range of applications, such as fluorescence detection for biomedical researches, particle sizing, confocal microscope, LiDAR, quantum cryptography, etc. [1,2]. Especially, SOI MOSFET single-photon detector [3], which counts photo-generated holes one by one without multiplication, features small dark counts, low-voltage operation and photon-number resolution. However, the output of this type of single-photon detectors based on single-charge counting shows complex waveforms including multilevel steps corresponding to generation and recombination of holes (electrons), and requires dedicated signal processing. In order to resolve this issue, we have developed the signal processing algorithm, implemented in FPGA, and verified its operation.

II. DEVICE STRUCTURE AND METHOD OF SIGNAL PROCESSING

Fig. 1 shows the structure of single-photon detector. Since negative and positive voltages are applied to the gate and substrate respectively, photo-generated holes are stored under the gate, and the presence of the holes is detected as the change in electron current flowing through the back-side channel.

The photon counting method to verify the photon arrival statistics is shown in Fig. 2. The Zynq ZC702 FPGA board by Xilinx is utilized to process the input signal in real time. In this particular experiment, the recorded signals from the detector are reproduced by an arbitrary waveform generator (AWG). The input is digitized by analog-to-digital converter (ADC), and then processed based on the signal processing algorithm, where the differentiation, integration and discrimination are sequentially performed. Fig. 3 shows typical waveforms at different stages of the signal processing. The events of hole generation by photon incidence is detected, and number of holes involved in each event is indicated by red symbol. Single holes are properly counted even if the multiple photons enter at the same time. In order

to generate the histogram of photon number arriving in a unit observation time, counters corresponding to photon numbers from 0 to 31 are prepared.

III. RESULTS AND DISCUSSION

Fig. 4 shows the obtained histograms of photon numbers that arrive in a unit observation time T of 33.28 ms for different light intensities. Since the total observation time T_{total} is 133.12 s, 4,000 ($=T_{total}/T$) events are involved. It can be observed that the counts in larger photon numbers increase as the light intensity increases. Fig. 4 also includes the theoretical histograms based on Poisson distribution for average number of photons λ arriving in T. The error bar is the square root of the theoretical counts multiplied by the square root of N, considering that the AWG repeats the same recorded data N=200 times. The absence of experimental data (red bar) indicates the counts in a particular bin is less than N.

The experimental histogram matches the theoretical one within the error bar for all the levels of light intensity, suggesting that the observed hole generation follows the Poisson statistics as expected, and the signal is processed properly. It should be noted that the developed signal processor can also be applied to other photon detectors based on single-charge counting, such as quantum dot (QD) detector, charge sensitive infrared phototransistor (CSIP) [4], and single-electron transistor (SET)-based photodetector [5], etc.

IV. CONCLUSION

A digital signal processor for the output SOI MOSFET single-photon detector has been developed, which shows complex multilevel pulses reflecting photo-generation and recombination of holes, and its operation was verified by taking photon number statistics for various light intensities. The obtained histograms matched the Poisson distribution within the margin of error, suggesting the proper operation of the processor. The developed signal processor is expected to expand the real-time usage of various photodetectors based on single-charge counting.

ACKNOWLEDGMENTS

This work was supported by the Cooperative Research Project Program of RIEC, Tohoku University.

REFERENCES

[1] W. Becker, A. Bergmann, M. A. Hink, K. Konig, K. Benndorf, and C. Biskup, "Fluorescence lifetime imaging by time-

correlated single-photon counting," *Microscopy research and technique,* Wiley periodicals, 2004, pp. 58-63

[2] H. Kosaka, A. Tomita, Y. Nambu, T. Kimura, and K. Nakamura, "Single-photon interference experiment over 100 km for quantum cryptography system using balanced gated-mode photon detector," *Electronic Letters,* vol. 39, no. 16, pp. 1199-1201, 2003.

[3] W. Du, H. Inokawa, H. Satoh, and A. Ono, "SOI metal-oxide-semiconductor field-effect transistor photon detector based on single-hole counting," *Optics letters,* vol. 36, no. 15, pp. 2800-2802, August 2011.

[4] S. Komiyama, "Single-Photon Detectors in the Terahertz Range," *IEEE J. Sel. Top. Quantum Electron.*, vol. 17, no. 1, pp. 54-66, January/February 2011.

[5] A. N. Cleland, D. Esteve, C. Urbina, and M. H. Devoret, "Very low noise photodetector based on the single electron transistor," *Appl. Phys. Lett.*, vol. 61, no. 23, pp. 2820-2822, December 1992.

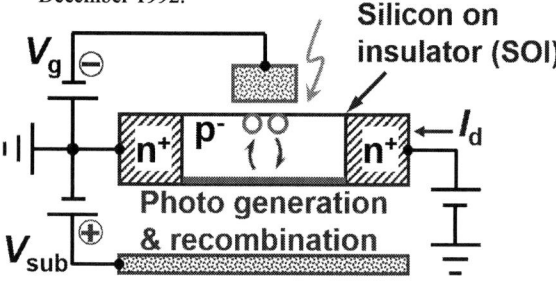

Fig. 1. Structure of the SOI MOSFET single-photon detector. In this study, a device with L=65 nm and W=105 nm is used.

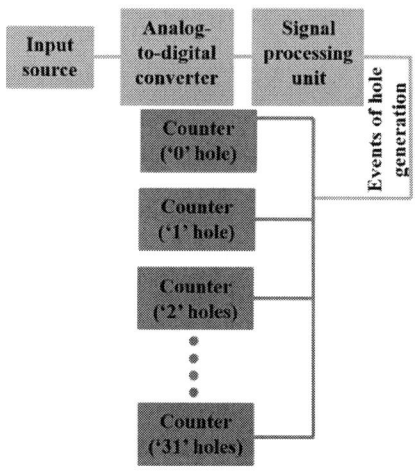

Fig. 2. Signal flow to obtain photon-number statistics.

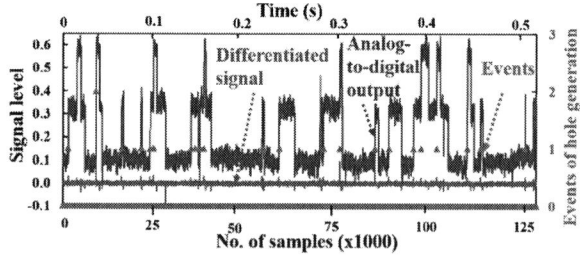

Fig. 3. Signal waveforms at different stages of the processing for the signal obtained under the illumination at the wavelength of 550 nm and the light intensity of 4.5 x 10⁻⁶ W/cm².

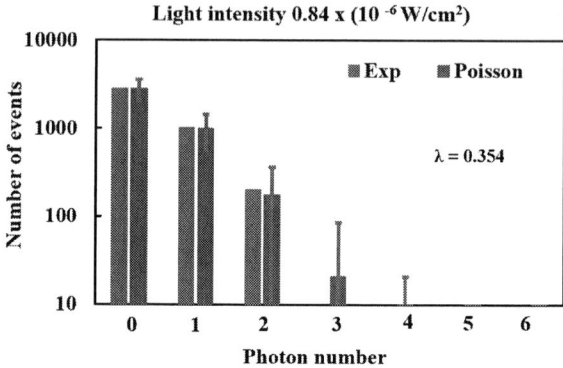

(a)

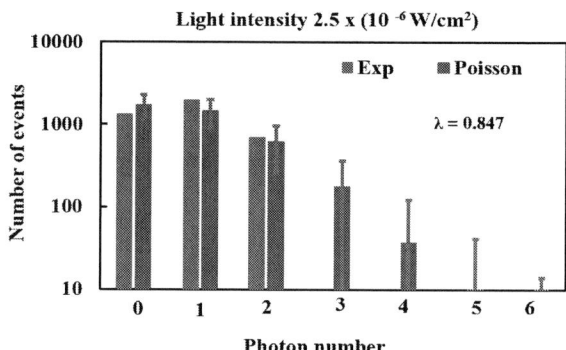

(b)

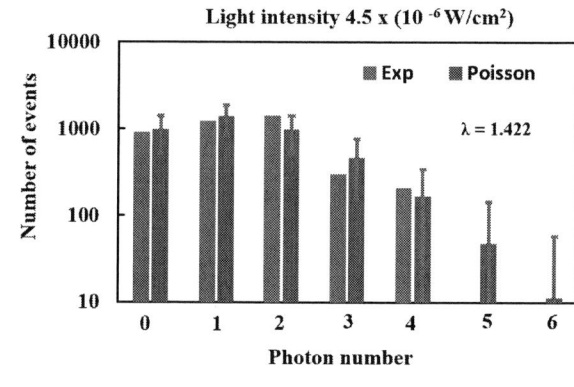

(c)

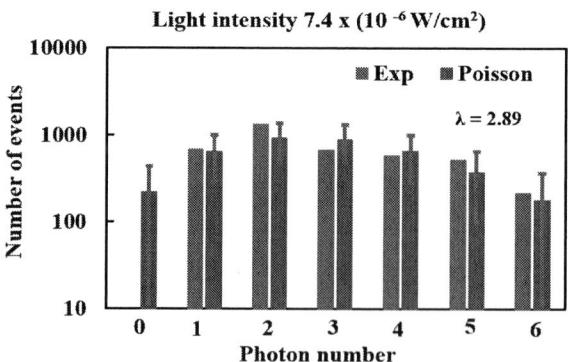

(d)

Fig. 4. Histograms of photon counts for different light intensities.

2020 IEEE Silicon Nanoelectronics Workshop

Scaled transistors with 2D materials from the 300mm fab

I. Asselberghs, T. Schram, Q. Smets, B. Groven, S. Brems, A. Phommahaxay, D. Cott, E. Dupuy, D. Radisic, J-F de Marneffe,
A. Thiam, W. Li, K. Devriendt, A. Gaur, D. Verreck, T. Maurice, D. Lin, P. Morin, I.P. Radu

imec, Leuven, Belgium
inge.asselberghs@imec.be

Abstract

Integration of 2D-materials brings a new set of challenges to a 300 mm Si CMOS fab. We have opted for double gated WS_2 transistors as test vehicle with transition metal dichalcogenides as channel. Moreover, we explore two different routes within a modified industry standard flow, differentiating in channel deposition method either by direct deposition or via layer transfer. The integration flow mitigates the constraints of high surface sensitivity and low adhesion for these materials. Device performance of I_{on}/I_{off} up to 10^7 is achieved and WIW mapping is obtained opening the route for further process understanding.

Introduction

The atomic thickness, intrinsic flatness and Van der Waals bonding of layered 2D- materials bring a new set of integration challenges. From the entire class of 2D-materials, we focus on Transition Metal Dichalcogenides (MX_2) for integration screening in a Si CMOS fab. Relying on the improved electrostatic control and reduced short channel effects, MX_2 carries the promise for increased performance in scaled transistors [1-3]. Today, the best experimental demonstrators are mostly build using natural or synthetic flakes as channel and traditional lab techniques such as electron beam lithography and lift-off are employed to fabricate the test structures [4-8]. Here, we selected an integration route close to an industry standard flow. Pursuing this approach over the last few years, we gradually have increased transistor performance and yield by stepwise improving the MX_2-deposition method in combination with integration flow tuning. Figure1, shows the performance evolution of back gated IV-characteristics over time.

Integration route

Theoretical predictions [9] identify WS_2 among the best candidate material with high potential towards performance. The expected stability and the minimal additional safety measures to be taken for production further motivate the continued to focus on this material. Elaborating on our findings [3,10], two different integration routes (Fig.2), differentiated by the deposition method for the 2D-layer, are pursued. In the case of direct deposition, restrictions are imposed on the process parameters for growth to preserve wafer and interface integrity while maintaining material quality. With the transfer process requiring more steps in a flow, benefit arises from the release of constraints imposed on the thermal budget.

A. Stack deposition

We opted for a $W(CO)_6$ and H_2S based MOCVD process using a modified epi reactor for growth [11]. The operation temperature of 750°C is selected to have minimal process impact on the back-gate dielectric. For the transfer route, layers are grown at 950°C on sacrificial wafers. The change in temperature budget is noted by difference in grain size (Fig.3), while in both cases the overgrowth of monocrystalline larger crystals is still present. Gate stack deposition on layered van der Waals materials [12] is extremely challenging. While various promising approaches are reported in literature, here, we have opted for the use of a sub 1 nm Si-seed deposit by MBD which is air oxidized followed by a conventional HfO_2 process where nucleation starts at the SiO_2-seeds and layer closure at 10 nm HfO_2 is achieved (Fig. 4). The complete 300 mm transfer of only a 7Å thick WS_2 layer between substrates is achieved by a combination of temporary W2W bonding and laser debonding techniques [13]. Figure 5 shows the wafer level defect inspection after gate stack deposition for as deposit (a) and transferred (b) layers, indicating complete and uniform transfer with only some minor materials loss at the wafer rim. To complete the removal of the temporary bonding material an organic wet strip is combined with a soft remote H-plasma clean [14].

B. Integration challenges

The limited adhesion of the transferred layers compared to the as deposit case, forced us towards the implementation of a novel active patterning recipe (fig.6). Complete delamination occurs after deposition of the typical SiO_2 hard mask. The novel recipe is based upon Spin on Carbon/Spin on Glass (SoC/SoG) approach which shows no delamination.

A damascene process is used for both the Ti-based side contacts, and high-k first, metal last top gate [10]. The extra 3 nm of HfO2, as seen in the cross-section TEM (Fig. 7a) of a typical double gated transistor, is added during the top gate trench fill to isolate the channel and the gate in the overlap region (fig 7) . A HR-TEM (Fig. 7b) made at the gate level, shows the presence of the individual WS_2 layer. It reveals also the non-uniformity induced by the SiO_2 seed layer deposition.

C. Electrical read-out

Fig.8a shows typical I_d-V_{bg} curves for direct grown WS_2 devices with only a back gate, which controls the contacts and the entire channel. No aging of the device behavior is observed, indicating that the device encapsulation protects the channel from the environment (e.g. oxidation). The devices are ambipolar, indicating the Fermi level is unpinned, which is promising for CMOS. Fig.8 b-c show that no systematic WIW

2020 IEEE Silicon Nanoelectronics Workshop

variability is observed for I_{on}, while a radial pattern is observed for the V_t, likely related to WIW process variation.

Summary

We report on the integration of WS_2 double gated transistors by tweaking an industry standard flow in a 300 mm CMOS fab. This can be achieved for both the direct deposition as the 300 mm transfer route. Transferred wafers are more susceptible for adhesion loss compared to direct deposit wafers. This opens the route for detailed WIW variation mapping allowing for building further process understanding.

References

[1] T. Agarwal et al., IEDM (2017); [2] R. Chau, IEDM (2019); [3] C. Huyghebaert et al., IEDM (2018); [4] Q. Smets et al., IEDM (2019); [5] KKH. Smithe et al., ACS Nano, **11.8**, p.8456 (2017); [6] C-C. Cheng et al., VLSI (2019); [7] D. Ovchinnikov et al., ACS Nano, **8.8**, p.8174 (2014); [8] Y.Y. Gong, 2D Materials, 3.2, p.021008 (2016); [9] A. Afzalian et al, SISPAD (2019); [10] T. Schram et al., ESSDERC (2017); [11] M. Caymax et al., SSDM (2019); [12] V. Kaushik et al., Solid State Tech. (2015); [13] A. Phommahaxay et al., IWLPC (2019); [14] D. Marinov et al., PESM (2019).

Acknowledgements

This work was done under the imec IIAP core CMOS programs. The authors acknowledge support by the EC under the Graphene flagship (contract no. CNECT-ICT-604391).

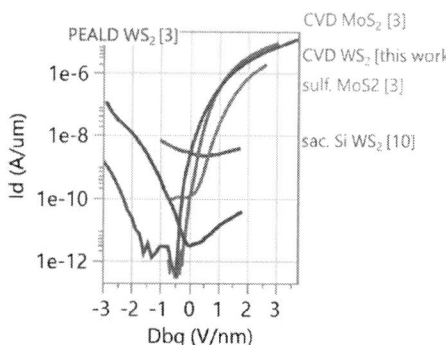

Fig.1: Comparison graph of IV-characteristics of different generations of MX_2 layers integrated in imec's 300 mm CMOS fab. [back gated transistor with Lg = 220 nm].

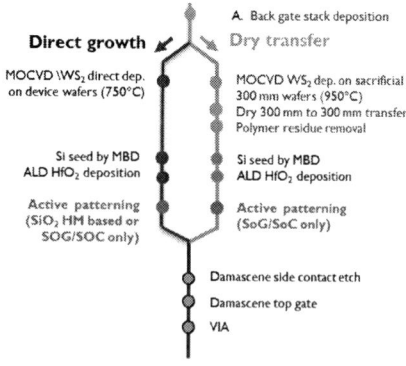

Fig.2: Process flow for the fabrication of 300 mm wafers with double gated.

WS_2 devices, for both direct grown (left branch) and 300-to-300 mm transferred WS2 (right branch).

Fig.3: AFM image of WS_2 after CVD-growth on SiO_2 coated wafers at (a) 750C and (b) 950C, respectively.

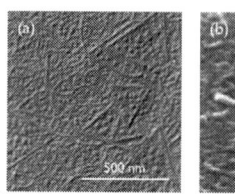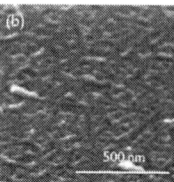

Fig. 4: SEM of 950°C grown MOCVD WS_2 after MBD Si-seed deposition (a) and the subsequent HfO_2 growth (b).

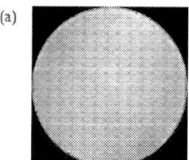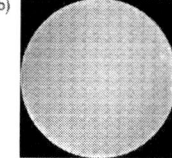

Fig. 5: wafer level defect inspection after gatestack deposition for as deposit (a) and transferred WS_2 (b).

HfO_2 top gate deposition	40 nm oxide HM deposition	SoG/SoC coating
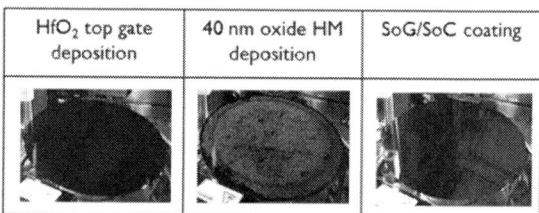		

Fig. 6: 300 mm wafer with transferred WS_2 (a) after HfO_2 dep, (b) HfO_2+40 nm SiO_2 HM, and (c) HfO_2+SoC/SoG coating.

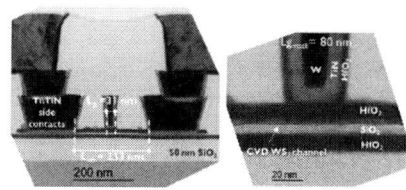

Fig. 7: (a) TEM of a double gated device with direct grown MOCVD WS_2 and a 10 nm HfO_2/ALD TiN top gate. (b) HR-TEM of the channel region. 1 to 2 monolayers of WS_2 and 10 nm HfO_2 overgrowth on SiO_2 seeds are observed.

direct grown WS_2 & back gate only

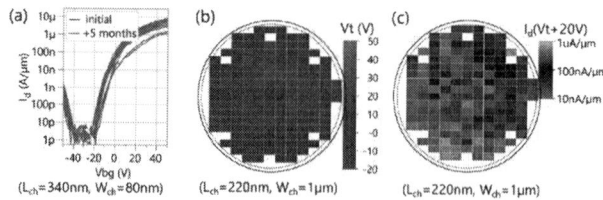

Fig. 8: (a): I-V characteristics measured immediately after device fabrication and 5 months later. No aging is observed Wafer maps of direct grown WS2 devices (back gate only) with (b) Vt taken at 10nA/um, and (c) Id at fixed overdrive. High yield is observed, but with center-to-edge variation in Vt. The n-type on-current shows a less pronounced pattern over the wafer.

2020 IEEE Silicon Nanoelectronics Workshop

Half-meshed and fully-meshed suspended graphene for transport gap engineering

Fayong Liu[1], Manoharan Muruganathan[1], Shinichi Ogawa[2], Yukinori Morita[2], Marek Schmidt[1], Hiroshi Mizuta[1,3]

[1]School of Material Science, Japan Advanced Institute of Science and Technology, Japan
[2]National Institute of Advanced Industrial Science and Technology, Japan, [3]Hitachi Cambridge Laboratory, U.K.
Email: fayong@jaist.ac.jp

Abstract — **We report on successful fabrication of large-area suspended graphene nanomesh (GNM) by using the helium ion beam milling (HIBM) technique. By positioning the meshed area on the suspended graphene, a heterostructure composed of graphene nanoribbon and GNM is demonstrated. Increased driving current with a wider transport gap is observed for the heterostructure devices in contrast to a fully meshed devices.**

Keywords: Graphene nanomesh, focused helium ion microscope, thermal activation energy

I. INTRODUCTION

Graphene nanoribbons (GNRs) are of interest to bandgap engineering due to its geometrically confined transport gap opening [1]. By separating the GNR with its surrounding materials, suspended GNR provides an idea 2D platform towards the applications of electronic and mechanical devices [2-3]. However, after scaling the GNR into the sub-10 nm regime, the fabrication of suspended GNR is quite challenging in the state of the art. Besides, the small driving current and fragile structure increase the difficulty in the conventional gate modulation measurement to characterize the transport gap opening. Graphene nanomesh (GNM) has been reported to have a larger driving current and observe a transport gap opening [4-5], which also considers as ultra-scaled GNR arrays. In this study, we introduce the suspended GNM patterned by helium ion beam milling (HIBM) and demonstrate a heterostructure device with GNR and GNM. We investigate the transport gap opening by extracting the thermal activation energy (E_A) from the Arrhenius plot for both GNMs and heterostructure devices.

II. DEVICE FABRICATION

The device fabrication started with transferring chemical-vapor-deposition (CVD) graphene on to a heavily doped Si substrate with a 285-nm-thick thermal SiO_2 layer. The first electrode layer (5/80 nm Cr/Au) was to achieve good adhesion with the substrate by removing the graphene underneath via electron beam lithography (EBL), shown in Fig. 1a. The second electrode layer (5/70 nm Cr/Au)was to make good contact between the graphene and the first electrode layer via EBL patterning (Fig. 1b). Subsequently, the GNRs were patterned with hydrogen silsesquioxane (HSQ) by EBL and plasma etching (Fig. 1c). After dipping into the buffered hydrofluoric acid (BHF) and drying with a CO_2 critical point dryer (Fig. 1d), 500 nm wide and 1.2 μm long suspended GNRs were achieved. We patterned the nanopores on the suspended GNRs by HIBM with different pitches (center to center) [6]. By controlling the dwell time of the beam, the nanopore diameter was fixed at 6 nm approximately [6]. To get the heterostructure, we verified the meshed area to achieve the fully-meshed and half-meshed devices, illustrated in Fig. 2. By zooming on the meshed area, the nanopores were observed with the helium ion microscope in Fig. 3b, 4b.

III. MEASUREMENTS RESULTS

Two reference devices including both the fully-meshed and half-meshed devices were measured to confirm the conventional gate modulation characterize by using Si substrate as the backgate, shown in Fig. 5-6. The half-meshed device shows multiple charge-neutral-points (CNPs) characterization between 0 V and 10 V. We have observed the similar multiple CNPs phenomena in the n-type and p-type doped graphene junction device [7]. It implies that a heterostructure with two different electrical properties of materials is formed in the half-meshed device. Moreover, the half-meshed device shows a larger drain current, higher on-off ratio, and the wider transport gap characterizations in contrast to the fully-meshed device. In order to evaluate the transport properties of the half-meshed and fully-meshed devices, we extracted the E_A from the Arrhenius plot. The $I_D - V_D$ characteristics (Fig. 7) for both two kinds of devices with different pitches were measured from 300K to 10K (Fig. 8-9). The conductance at a certain temperature for each device was extracted by the linear fitting. As the half-meshed device includes a GNR part, it suppresses the heavy conductance decrement during scaling the pitch of the meshed area, which relates to the high driving current. By replotting the conductance in the Arrhenius plot shown in Fig. 10-11, the E_A for each device was extracted from 300K to 100K region, shown in Fig. 12 [8]. The E_A of half-meshed devices shows a similar tuneability with the fully-meshed devices by controlling the pitch of the nanopores. It not only proves that the GNM is tuned to different material from pristine GNR, but also illustrates that the recombination of different band structures between GNR and GNM forms the multiple CNPs and extends the transport gap. From these results, it proposes a new approach for the transport gap engineering by manipulating the combination of GNR and pitch-well-controlled GNM beyond the GNR method.

VI. SUMMARY

We successfully fabricated the half-meshed and fully-meshed suspended graphene devices by HIBM technique. The heterostructure was achieved in the half-meshed device, which shows a better performance for transport gap engineering. The E_A shows a good tuneability in both two kinds of devices by the well-controlled pitch of nanopores.

ACKNOWLEDGMENTS

The authors acknowledge T. Iijima and H. Ota for the usage of the HIBM at the AIST SCR station for the helium ion irradiations. This work was supported by the Grant-in-Aid for Scientific Research No. 18H03861, 19H05520 from the Japan Society for the Promotion of Science (JSPS).

REFERENCES

[1] M.Y. Han *et al.*, *Phys. Rev. Lett.*, 98, 206805, 2007.
[2] J.C. Meyer *et al.*, *Nature*, 446, 60–63, 2007.
[3] J. Sun *et al.*, *Appl. Phys. Lett.*, 105, 033103, 2014.
[4] S. Berrada *et al.*, *Appl. Phys. Lett.*, 103, 183509, 2013.
[5] F. Ouyang *et al.*, *ACS Nano*, 5, 4023–4030, 2011.
[6] F. Liu *et al.*, *Micromachines*, 11, 387, 2020.
[7] Z. Wang *et al.*, *ACS Nano*, 13, 7, 7502-7507. 2019.
[8] M.Y. Han *et al.*, *Phys. Rev. Lett.*, 104, 056801, 2010.

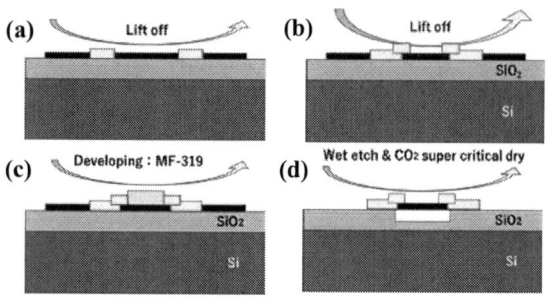

Fig. 1: Main steps of the fabrication processes. (a) patterning the first layer of electrodes. (b) patterning the second layer of electrodes. (c) patterning the GNR. (d) suspending the GNR by wet etching.

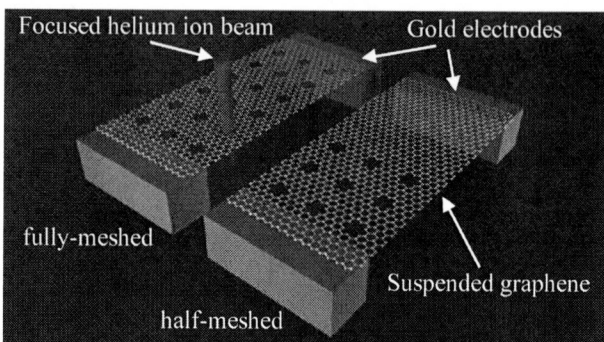

Fig. 2: Schematic illustration of the fully-meshed and half-meshed suspended graphene.

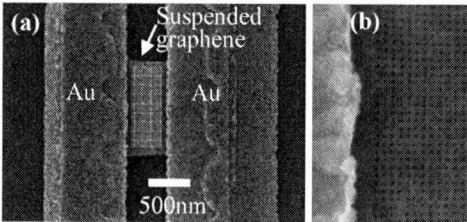

Fig. 3: Helium ion beam microscopy images for fully-meshed device (a) before HIBM, white dots shows the milling area. (b) the nanopores on the suspended graphene after HIBM.

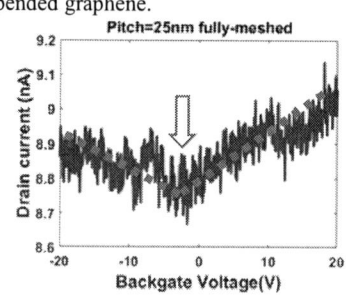

Fig. 5: Backgate characteristic of the fully-meshed device after HIBM with 25 nm pitch at 300K. The Drain voltage is 5 mV. One main CNP was found around 0 V.

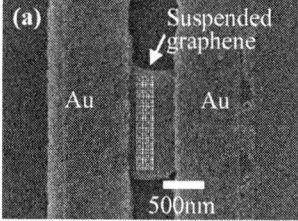

Fig. 4: Helium ion beam microscopy images for half-meshed device (a) before HIBM, white dots shows the milling area. (b) the nanopores on the suspended graphene after HIBM.

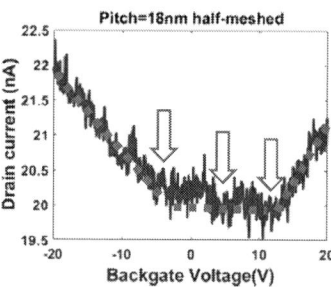

Fig. 6: Backgate characteristic of the half-meshed device after HIBM with 18 nm pitch at 300K. The Drain voltage is 5 mV. Multiple CNPs were found between 0 V to 10 V.

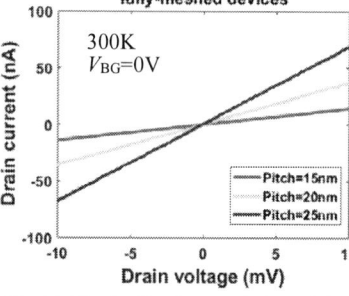

Fig. 7: The I_D-V_D characteristics of the fully-meshed devices with different pitches

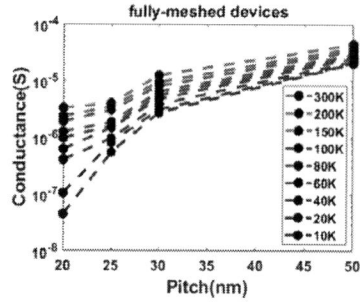

Fig. 8: Conductance variation of the fully-meshed devices with different pitches

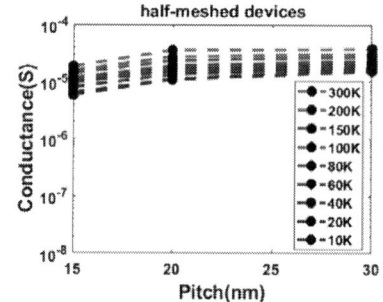

Fig. 9: Conductance variation of the half-meshed devices with different pitches

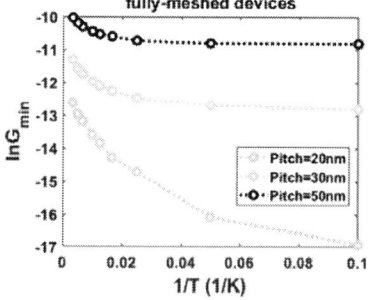

Fig. 10: Arrhenius plot for the fully-meshed devices with different pitches.

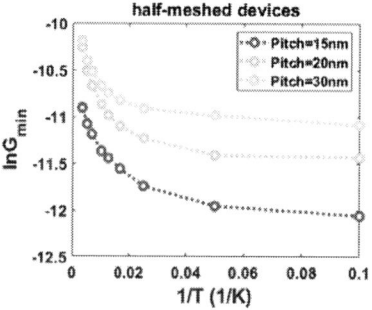

Fig. 11: Arrhenius plot for the half-meshed devices with different pitches.

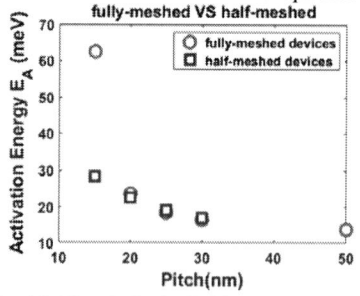

Fig. 12: The E_A for both fully-meshed and half-meshed devices with different pitches. (fully-meshed device with pitch 15 nm is a reference from difference chip with the same process)

2020 IEEE Silicon Nanoelectronics Workshop

A Computational Performance Evaluation of Negative-Capacitance MOSFETs based on Ultra-thin body Silicon and Monolayer MoS₂

Sheng Luo*, Xiaoyi Zhang, and Gengchiau Liang

Department of Electrical and Computer Engineering, National University of Singapore (NUS), 117583 Singapore

Email: elelshe@nus.edu.sg

Abstract

A computational study at atomic level was performed for negative-capacitance MOSFETs (NCFETs) with ultrathin body silicon (UTB-Si) and monolayer MoS₂ (1L-MoS₂) as channel materials and HfO₂ as ferroelectric layer material. ~25mV/dec average subthreshold slope (SS) for at least 4-order of drain current was observed in 1L-MoS₂ NCFET and shows steeper slope comparing to the other simulated UTB-Si cases. It was found that both lower tunneling current and improved capacitance matching enabled the steep-slope feature in 1L-MoS₂ NCFET.

I. Introduction

The negative capacitance effect has been found in ferroelectric (FE) materials, which enables the break of Boltzmann limits (60mV/dec) for steeper subthreshold slope (SS), and hence offers a solution for the aggressive channel scaling under the Moore's law [1-2]. Several recent experiments based on various thickness of MoS₂ NC-FETs [3-4], differing from the commonly used silicon channel, demonstrated SS with less than 50mV/dec. Nevertheless, the device physics of the MoS₂ NC-FET and the performance benchmark with its ultra-thin body counterparts, UTB-Si, were still not conducted. In this work, we investigated the device performance of the ferroelectric based NC-FETs using UTB-Si and monolayer MoS₂ as channel materials. The SS properties of both cases were analyzed by comparing different current components and projected operation path on P-E relations.

II. Simulation Methodology

The schematic cross-sectional view of a double gate FEFETs (DG-NCFET) was shown in **Fig. 1(a)**. The channel length for all simulated cases were 10nm. A V_{DD} of 0.6V and 1nm SiO₂ was used in this work. The atomic structure of the simulated materials was shown in **Fig. 1(b)**. Various thickness of UTB-Si from ~2.2nm (~16 atomic layer) to ~5Å (~4 atomic layer) was implemented and its transport direction is in [100] direction and $sp^3d^5s^*$ tight-binding model [5] was used for their Hamiltonians. To derive the MoS₂'s Hamiltonian, *ab initio* calculation under DFT framework using PAW method was carried out, which was enabled by VASP with HSE06 calculation [6-9]. The tight-binding model was then derived from MLWFs [10] methods. For the ferroelectric material, the P-E relation of HfO₂ was implemented with parameters extracted from experiments [11]. The simulation scheme was shown in **Fig. 2**. The positioned-dependent Gibbs free energy calculation [12] was included in the Poisson Solver to cover the non-uniform polarization along the channel. Also, by judging the states in which had the minimum Gibbs free energy, the operation region in P-E relation was determined.

III. Results and Discussions

The $I_{DS} - V_{GS}$ of the double-gate NC-FETs based on each simulated channel material were shown in **Fig. 3(a)**. It is worth noting that the MoS₂ demonstrated steeper slope comparing to the UTB-Si's counterparts. **Fig. 3(b)** also demonstrated lower point SS within several decades of drain current, featuring an average SS ~25mV/dec for the at least 4 decade drain current. On the other hand, the 4-layer Si had the best subthreshold property among the simulated UTB-Si cases with an average SS ~36mV/dec. Performance summary of all simulated cases were shown in **Fig. 4**. In order to investigate the impact of the negative capacitance, both the reverse body factor ($\frac{d\psi_s}{dV_g}$) and the operating path on P-E relations were shown in **Fig.5**. The reverse body factor [**Fig.5(a)**] reflected the band profile shifting on each step gate bias, directly reflected the voltage amplification induced by negative capacitance. The results corresponded to the SS results as 1L-MoS₂ DG-NCFET demonstrated the highest ($\frac{d\psi_s}{dV_g}$) ratio, indicating better capacitance matching comparing to the other simulated channel materials and device configuration. The detailed operation path on the P-E [**Fig.5(b)**] of both 4L-Si and 1L-MoS₂ DG-NCFET showed both cases had FE polarized at Vg=0V, ensuring the negative capacitance were induced in the subthreshold region. The FE polarization at Vg=0V could be attributed to the non-flat band condition induced in ultra-scaled channel device. Furthermore, comparison of the tunneling current within the operation window [**Fig.6**] showed the impact of the tunneling current. Both 1L-MoS₂ and 4L-Si had tunneling current suppressed due to the large effective mass (m*/m₀=~0.45 and m*/m₀=~0.38, respectively), while the 16L-Si (m*/m₀=~0.25) NCFET current at the subthreshold region was dominated by tunneling current. The suppression of the tunneling enabled the 1L-MoS₂ DG-NCFET to achieve steep SS in short channel devices.

In conclusion, we have investigated the device performance of various UTB-Si and 1L-MoS₂ based NCFETs. Steep slope with SS ~25mV/dec was observed in 1L-MoS₂ DG-NCFET, attributing to both lower tunneling current and capacitance matching. Both features enabled the 1L-MoS₂ FEFETs to be the candidate for LP applications in next-generation FETs.

Acknowledgement

We acknowledge support from the Singapore National Research Foundation (NRF) through a Competitive Research Program (Grant No: NRF-CRP6-2010-4) and the Ministry of Education under MOE2017-T2-1-114. The project is also supported by NUS Hybrid-Integrated Flexible (Stretchable) Electronic Systems Program.

References

[1] S. Salahuddin *et al.,* Nano Lett., vol. 8, no. 2, pp. 405-410, 2008.
[2] A. I. Khan *et al.,* ature Materials, vol. 14, pp 182-186, Dec. 2015.[3] Si, Mengwei, *et al.,* Nature nanotechnology 13.1 (2018): 24.[4] Yu, Zhihao, *et al.,* Electron Devices Meeting (IEDM), 2017 [5] S. Datta, Quantum Transport: Atom to Transistor, Cambridge 2005[6] A. Szab *et al.,* Phys. Rev. B, vol. 92, no. 3, p. 035435, 2015[7] W.

Kohn *et al.*, J. Phys. Rev., vol. 140, p. A1133, Nov. 1965.[8] G. Kresse *et al.*, Phys. Rev. B, vol. 59, p. 1758, Jan. 1999.[9] J. Heyd, *et al.*, J. Chem. Phys., vol. 118, no. 18, p. 8207, 2003[10] A. A. Mostofi, *et al.*, Comput. Phys. Commun., vol. 185, no. 9, p. 685–699, 2014.[11] D.

Zhou, *et al.*, Acta Materialia, vol. 99, pp. 240-246, 2015.[12] Zhang, Xiaoyi *et al.*, Semiconductor Science and Technology (2018) [13] International Roadmap for Devices and Systems. Available: https://irds.ieee.org/

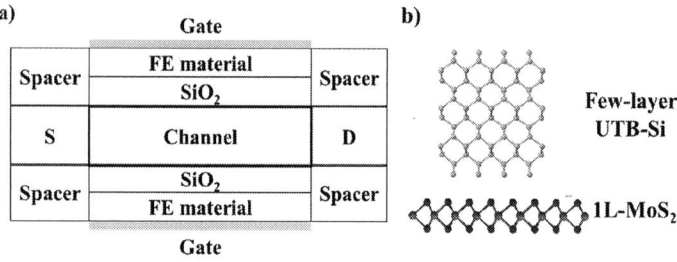

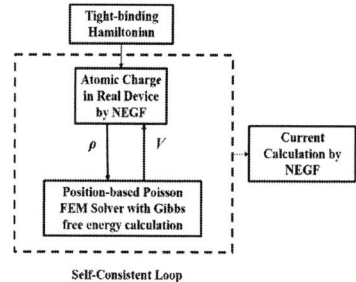

Fig. 1. a) Cross-sectional view of the device investigated in this work. (b) Atomic structures of the channel materials used in simulation.

Fig. 2. Procedures of the self-consistently solved atomic device performance simulation. Gibbs free energy calculation is involved in the Poisson Solver.

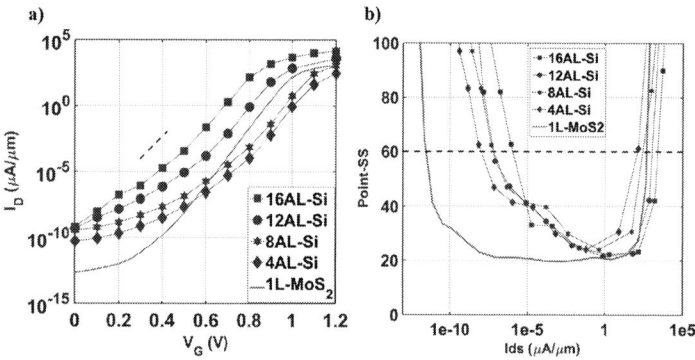

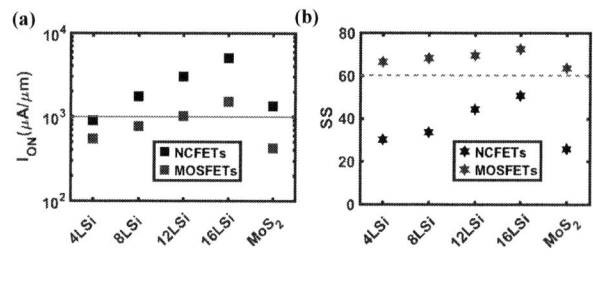

Fig. 3. a) I-V characteristics comparison between multiple thickness UTB-Si and 1L-MoS2 b) Point Subthreshold Swing (SS) at different I_{ds} for each simulation case.

Fig. 4. (a) Summarized SS (point and average) values for each simulation cases. (b) Extracted drive current benchmarked with the IRDS2027's projection with the inclusion of R_{SD}=131Ω.μm.

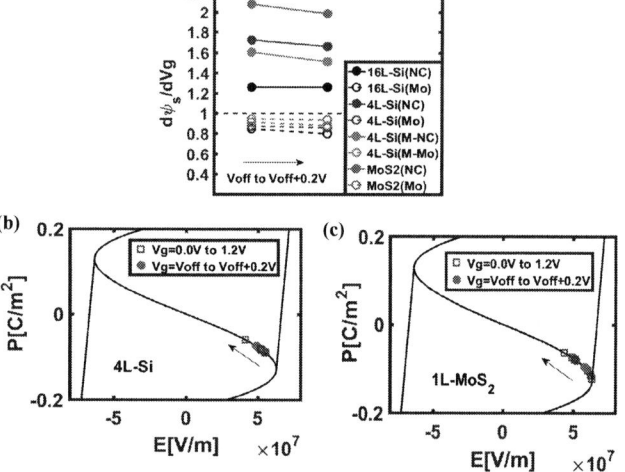

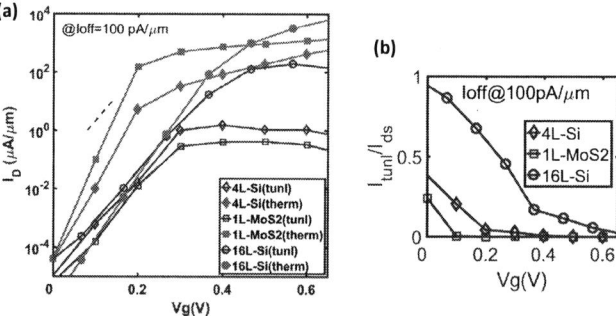

Fig. 5. (a): Reverse body factor ($\frac{d\psi_s}{dVg}$) of each swept at the step of 0.1 V in each simulated case. ψ_s was extracted at each top-of-barrier (ToB) of each conduction band (Ec) profile. 4L-Si(M-NC/Mo) represented the 4L-Si NCFET and MOSFET with MoS2 dielectric constant. (b)-(c): Highlighted operation region for both cases when devices operated at sub-threshold region. Blue rectangles represented the operation path of Vg from 0V to 1.2 V, while the red dots were highlighted was the subthreshold region from V_{off} to V_{off}+0.2 V

Fig. 6. (a) and (b): (a) Thermionic emission and tunneling current comparison between 1L-MoS2, 4L-Si and 16L-Si at LP operation window. The off-state current defined as 100 pA/μm (LP). (b) Ratio of tunneling current I_{tunl} against the total drain current I_{ds}.

2020 IEEE Silicon Nanoelectronics Workshop

Deconvolution of Hot Carrier and Cold Carrier Injection in ZnO TFTs

P.Bolshakov, R.A. Rodriguez-Davila, M. Quevedo-Lopez, and C.D. Young

Materials Science and Engineering Dept., University of Texas at Dallas, 800 W. Campbell Rd., Richardson, TX, USA

Email: chadwin.young@utdallas.edu

Abstract—Positive bias instability stress (PBI) and hot carrier injection stress (HCI) was done on ZnO thin-film transistors (TFTs) with 100ºC Al_2O_3. The threshold voltage (V_T), transconductance (g_m), and subthreshold slope (SS) were monitored. HCI stress with two intermittent sense measurements where the first I_{DS}-V_{GS} is measured at the drain contact and the second is measured at the source contact to separate the contribution of the hot carrier and cold carrier injection on the V_T shift. PBI stress was done to determine the viability of the carrier injection separation using only HCI.

Keywords - ZnO, TFTs, Al_2O_3, V_T, PBI, HCI

I. INTRODUCTION

Large area/flexible electronics may rely on oxide-based semiconductors that are desirable because of their compatibility with low-temperature fabrication required for large-area/flex-compatible technologies. ZnO is an oxide-based candidate to be used as an active layer in thin-film transistors (TFT) circuitry due to inexpensive processing and noteworthy electrical performance [1-3] and possible uses in flexible circuits [1]. For flex compatibility, deposition of high-k gate dielectrics at these low temperatures will be required as well. With all the low-temperature processing, thin-film transistor reliability must be evaluated due to threshold voltage (V_t) instability experienced by TFTs with high-k dielectrics [4-6]. In this work, TFTs are constant voltage stressed while monitoring critical parameters to assess the reliability of ZnO-based TFTs.

II. DEVICE AND STRESS PROCEDURE DESCRIPTION

Zinc-oxide TFTs are fabricated using traditional photolithography to pattern staggered-bottom-gate and top-contacts, as previously reported (**Fig. 1**) [7] with the final device in **Fig. 2**. The devices are fabricated on a glass substrate with patterned 135 nm of indium tin oxide (ITO) to serve as the gate electrode. Then, a 15 nm Al_2O_3 gate dielectric is deposited by atomic layer deposition (ALD) at 100ºC followed by 45 nm of zinc oxide as the semiconductor channel deposited by pulsed laser deposition (PLD). Stress testing was performed by applying stress voltages of either 5.5 or 6.0 V, with intermittent I_D-V_G sense measurements.

III. DATA/RESULTS AND DISCUSSION

Low-temperature high-k dielectrics are essential for compatibility with large-area/flex electronics. A deconvolution of the contribution of hot carrier (HC) injection and cold carrier (CC) injection to the ΔV_T using only HCI stress measurements could allow for a reduction in measurements time and devices under test. To demonstrate consistent trends across devices, multiple TFTs were measured for each sense scheme at two different voltages. All devices had dimension of W/L = 160/20 μm, as shown in the plan-view picture of a device in **Fig. 2**. The **Fig. 3**. and **Fig. 5** illustrate the evolution of I_D-V_G degradation for devices stressed at 5.5 V and 6.0 V, respectively, with four different sense schemes. The first scheme in **Fig. 3a** is the conventional PBI stress measurement. The HCI stress in **Fig. 3b** and **Fig. 3c** is on the same TFT where **(b)** is the 1st sense measurement at the drain contact and **(c)** is the 2nd sense

measurement at the source contact. Presumably the 2nd sense does not see the HC injection unlike the 1st sense measurement. The HCI in **Fig. 3d** is HCI with only a single sense measurement at the source contact. The same exact setup is done for **Fig. 5** at stress voltage of 6.0 V. For both stress voltage, the HCI at drain contact (**Fig. 3b** and **Fig. 5b**) appears to cause an increase in OFF current, suggesting degradation near the drain contact. One can observe V_T shifts with little to no degradation in either g_m or SS in all schemes. It should also be noted that the Δg_m in **Fig. 4** for stress at 5.5 V shows negligible degradation but the Δg_m in **Fig. 6** for stress at 6.0 V shows degradation at longer stress times. This suggests that near interface trap generation requires larger voltage stress and longer stress times to emerge. Extraction of the ΔV_T across all four sets of TFTs in **Fig. 4** and **Fig. 6** yields a more comprehensive understanding of the behavior of the V_T with stress time. At stress voltage of 5.5 V and 6.0 V, HCI with sense at drain contact demonstrates the larger relative ΔV_T. Presumably due to the contribution of both HC and CC injection. The other 3 sense schemes all show close agreement with each other in the ΔV_T. This is expected as the ΔV_T trends for the PBI and the two HCI sense at the source contact can be attributed to CC injection. The agreement between PBI and HCI at source also allows for separation of the HC injection and CC injection in terms of their contribution to ΔV_T. This can be done by subtracting the ΔV_T of HCI at the source contact from the ΔV_T of HCI at the drain contact as shown in **Fig. 7**. For both stress voltages, the red represents the HC + CC from HCI at drain contact while the blue and green, which are in close agreement, represent the CC from HCI at source contact and PBI, respectively. Furthermore, **Fig. 8** compares the ΔV_T after 1000 sec stress for both stress voltages. For stress of 5.5 V (red), the HC injection (Drain – Source) contribution is ~25% but for 6.0 V stress (blue), the HC injection contribution is ~15%. This is attributed to a larger increase in the contribution of CC injection compared to HC injection at the higher stress voltage. This suggests that with increasing voltage stress during HCI measurements, the component of CC injection is increasing at a higher rate than the component of HC injection. This can be explained by the larger gate area compared to the drain contact area resulting in a greater contribution from the field generated at the gate versus the field generated at the drain contact as shown in **Fig. 9**. Future work would involve smaller channel dimensions to determine their impact of HC and CC injection deconvolution using HCI.

IV. CONCLUSIONS

To determine the viability of using only HCI to separate the contribution of hot carrier injection and cold carrier injection in ZnO TFTs, multiple HCI sense schemes and PBI measurements are done. The results show that HCI with two intermittent sense measurements at the drain contact and the source contact can be used to attribute the percentage of contribution of hot carrier injection and cold carrier injection to the threshold voltage shift. Furthermore, the larger increase in cold carrier injection contribution compared to hot carrier injection contribution at higher voltage stress suggests that the contribution of each changes at different rates.

2020 IEEE Silicon Nanoelectronics Workshop

978-1-7281-9736-4/20 $31.00 © 2020 IEEE

9.4

REFERENCES

[1] G. Gutierrez-Heredia *et al.*, TSF, vol. 545, 2013.
[2] Y. Kawamura *et al.*, JDT, vol. 9, no. 9, 2013.
[3] M. S. Oh, *et al.*, APL, vol. 93, no. 3, 2008.
[4] R. A. Chapman, *et al.*, TED, vol. 63, no. 10, 2016.
[5] D. Siddharth, *et al.*, WoDiM, 2014.
[6] D. Siddharth, *et al.*, IIRW, 2014.
[7] R. A. Rodriguez-Davila *et al.*, IPFA, 2018.

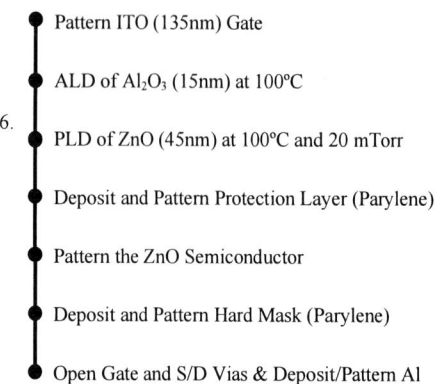

- Pattern ITO (135nm) Gate
- ALD of Al_2O_3 (15nm) at 100°C
- PLD of ZnO (45nm) at 100°C and 20 mTorr
- Deposit and Pattern Protection Layer (Parylene)
- Pattern the ZnO Semiconductor
- Deposit and Pattern Hard Mask (Parylene)
- Open Gate and S/D Vias & Deposit/Pattern Al

Fig. 1. Process flow for ZnO TFTs with both Al_2O_3 and ZnO deposited at 100°C.

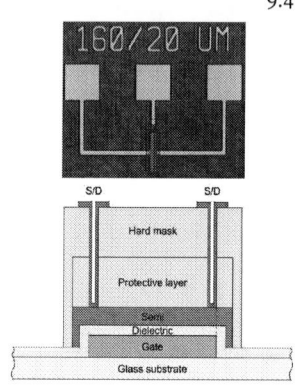

Fig. 2. Plan-view and cross-section schematic of the ZnO TFT structure with a channel width of 160 μm and a channel length of 20 μm.

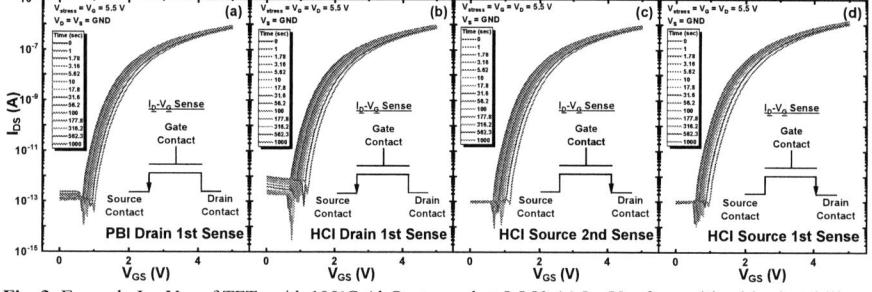

Fig. 3. Example I_{DS}-V_{GS} of TFTs with 100°C Al_2O_3 stressed at 5.5 V. (a) I_{Ds}-V_{GS} for positive bias instability (PBI) stress with intermittent sense measurements during gate only stress. Hot carrier injection (HCI) stress with two intermittent sense measurements with the (b) 1st sense measurement at the drain contact where the stress voltage was applied followed by the (c) 2nd sense measurement at the source contact where no stress was applied. (d) HCI stress with the sense measurement done at the source contact where no stress was applied. For HCI with sense at drain contact (b), the degradation appears to increase the OFF current.

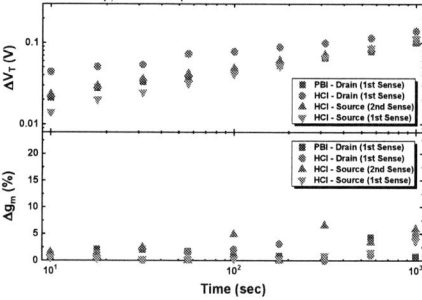

Fig. 4. The change in threshold voltage (V_T) and change in transconductance (g_m) for all four types of measurements at stress voltage of 5.5 V. The small change in Δg_m for all sense schemes suggests no interface state generation is contributing to the ΔV_T.

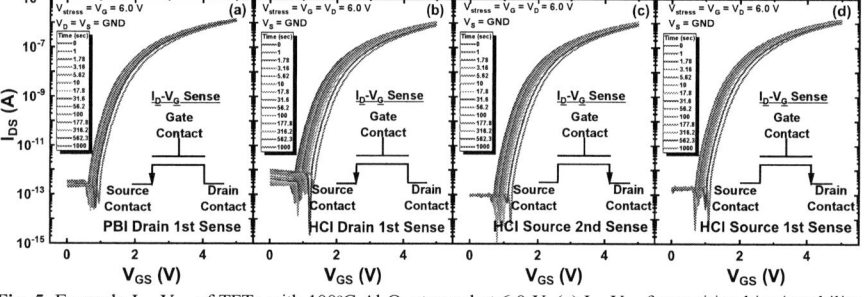

Fig. 5. Example I_{DS}-V_{GS} of TFTs with 100°C Al_2O_3 stressed at 6.0 V. (a) I_{Ds}-V_{GS} for positive bias instability (PBI) stress with intermittent sense measurements during gate only stress. Hot carrier injection (HCI) stress with two intermittent sense measurements with the (b) 1st sense measurement at the drain contact where the stress voltage was applied followed by the (c) 2nd sense measurement at the source contact where no stress was applied. (d) HCI stress with the sense measurement done at the source contact where no stress was applied. For HCI with sense at drain contact (b), the degradation appears to increase the OFF current.

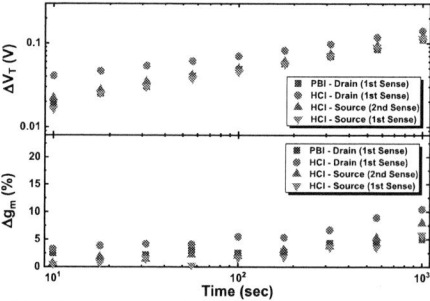

Fig. 6. The change in threshold voltage (V_T) and change in transconductance (g_m) for all four types of measurements at stress voltage of 6.0 V. The Δg_m appears to trend exponentially at higher stress voltage indicating interface state generation.

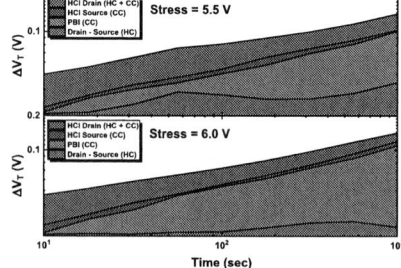

Fig. 7. Comparison of ΔV_T for the PBI (green) and HCI sense at drain (red) and HCI sense at source (blue). There is a close agreement between the PBI and HCI with sense at source suggesting cold carrier (CC) injection can be separated from hot carrier (HC) injection only using HCI measurements.

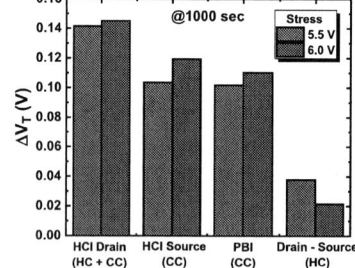

Fig. 8. Comparison of ΔV_T at a fixed time between the TFTs with stress at 5.5 V and 6.0 V. At a stress voltage of 5.5 V about 25% of the ΔV_T is attributed to HC injection while at 6.0 V about 15% of the ΔV_T is attributed to HC injection. This suggests that at higher stress voltages the contribution of HC injection to ΔV_T may become negligible compared to CC injection.

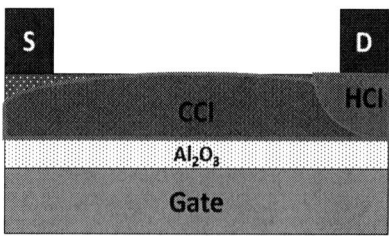

Fig. 9. Graphic demonstrating the fields generated from stress at the gate and stress at the drain. There is an overlap of the vertical field where CCI occurs and the lateral field where the HCI occurs. The relatively long channel length suggests the source side of the device would mostly be dominated by the vertical field and CCI.

2020 IEEE Silicon Nanoelectronics Workshop

978-1-7281-9736-4/20 $31.00 © 2020 IEEE

Realization of Diverse Spike-timing-dependent Plasticity with Nanosecond Timescale Based on Metal Oxide Resistive Switching Memory

Ruiyi Li, Peng Huang[*], Yulin Feng, Zheng Zhou, Xiangyu Wang, Wensheng Shen, Xiangxiang Ding, Lifeng Liu[#], Xiaoyan Liu, Jinfeng Kang

Institute of Microelectronics Peking University, Beijing 100871, China

Email: *phwang@pku.edu.cn, #lfliu@pku.edu.cn

Abstract —The spike-timing-dependent-plasticity (STDP) is considered to be the basic function for the synapse device to simulate biological activities of the brain. In this paper, we demonstrate four standard STDP forms with nanosecond timescale based on the analog property of RRAM. An operation scheme to achieve diverse STDP rules is proposed according to the relationship between conductance change and the applied voltage of pulses. The diverse ultra-fast STDP rules are realized experimentally in fabricated Al$_2$O$_3$/ HfO$_2$-based RRAM. The maximum weight change is up to 300% under nanosecond timescale pulses. Moreover, the nonvolatile characteristic of the device enables long-term potentiation and depression.

I. INTRODUCTION

In the coming era of big data, the generation rate of the data is quite faster than the processing and analysis capabilities. The conventional computers based on Von Neumann architecture are becoming more and more inefficient.[1] Inspired by human brain, neural networks with the features of massive parallelism and distributed storage are proposed to enhance information analysis capabilities.[1] The resistive random access memory (RRAM) is considered to be one of the most promising candidates as the electronic synapse to construct neural networks, due to its scalability, low power consumption and the capability to simulate biological behaviors.[2] The spike-timing-dependent-plasticity (STDP) is the foundation of learning and memory in biological neural system, so it is the primary function that the electronic synapses should obtain.[1] In this paper, the electrical property of TiN/HfO$_2$/Al$_2$O$_3$/Pt RRAM device has been investigated, focusing on the gradual resistance change behavior. A method to generate different STDP rules is proposed along with a logical design of the applied pulse waveform. We demonstrate four standard forms of STDP from the same intermedia resistive state by tuning the pulse waveforms. The maximum conductance change is up to 300% and the timescale within the order of nanoseconds is obtained, which is 10^6 times faster than the same function in the human brain.

II. EXPERIMENTS

The TiN/HfO$_2$/Al$_2$O$_3$/Pt device was fabricated. A 100-nm Pt was deposited by magnetron sputtering. Two cycles of Al$_2$O$_3$ layer and five cycles of HfO$_2$ layer were deposited by atomic layer deposition at 300 °C, which was repeated for 10 cycles. The total thicknesses of Al$_2$O$_3$ and HfO$_2$ were 2nm and 3nm, respectively. 100-nm TiN was grown by reactive sputtering in an Ar:N$_2$ (18:2) atmosphere followed by lithography to form electrode patterns. DC and AC measurements were performed by Agilent B1500 semiconductor parameter analyzer and Agilent 81160A pulse generator.

III. RESULT AND DISCUSSION

Fig. 1 shows the multilevel characteristics in the device by controlling compliance currents and RESET voltages. The gradual resistance modulation is exhibited in Fig.2, which shows the feasibility of achieving the analogue synaptic weight change of the STDP rules. Fig. 3 displays the modulation of conductance from HGS to LGS by applying different pulse amplitudes or widths to investigate the proper operation scheme for STDP rules. To make it more efficient for applications, the pulse-amplitude modulation is chosen. Fig. 4 shows the change of conductance by various pulse amplitudes from an intermediate initial state (0.4mS~0.5mS). As can be seen in Fig. 4, the change of conductance gradually

increases from 0.5mS to 3.2mS by increasing the pulse amplitude from 1.3V to 1.6V, and decreases from -0.16mS to -0.45mS with the decrease of the pulse amplitude from -1.6 V to -2.2 V. It is found that the conductance state of 0.4mS~0.5mS can undergo both increase and decrease so it is a suitable initial state to realize the STDP rules. There is a positive correlation between the pulse voltage and the conductance change, which can be fitted as:

$$\Delta G = G_0^+ + k^+ \exp(V_{net} / V_{th}^+) \quad V_{net} > 0$$
$$\Delta G = G_0^- + k^- \exp(V_{net} / V_{th}^-) \quad V_{net} < 0 \tag{1}$$

where ΔG is the conductance change, V_{net} is the applied net voltage, k^+, k^-, G_0^+, G_0^- are the fitting parameters, V_{th}^+ and V_{th}^- represent the threshold voltages. Based on equation (1), the operation scheme for diverse STDP rules can be designed. Take the asymmetric Hebbian learning rule as an example. The effect of this STDP rule can be quantified by fitting the biological data as[6]:

$$\Delta G = A \exp(-\Delta t / \tau) \tag{2}$$

where A and τ refer to the scaling factor and time constant of the STDP function, Δt is the pre-/post-spike interval. Combining equation (1) with (2), the net voltage that can achieve the synaptic weight modulation as described in equation (2) can be deduced as:

$$V_{net} = V_{th}^+ \ln[A \exp(-\Delta t / \tau) - G_0^+] - V_{th}^+ \ln k^+ \quad \Delta t > 0$$
$$V_{net} = V_{th}^- \ln[A \exp(-\Delta t / \tau) - G_0^-] - V_{th}^- \ln k^- \quad \Delta t < 0 \tag{3}$$

However, not all pulses of which the net voltage satisfies the relationship in equation (3) can induce the STDP characteristics. There are two basic principles that should be obeyed, including the threshold and superposition principle. The threshold principle refers that no single pulse can tune the conductance. In this work, the threshold is 1.1V and -1.4V for SET and RESET processes, corresponding to the threshold voltage in equation (1). The superposition principle states that the conductance modulation results from the net voltage defined as the voltage differential between the pre- and post-pulse. One reasonable operation scheme is exhibited in Fig. 5(a) and the calculated net voltage with different Δt is shown in Fig. 5(b). The experimental result is shown in Fig. 6, which indicates that the maximum weight change is close to 300% and the timescale is on the order of nanoseconds.

Fig. 7 shows the realization of other three standard STDP rules. It should be noticed that the designed pulse waveforms in this work (inset in Fig. 7) is only an illustration. Other operation schemes that obtain the same net voltage can also be used. Fig. 8 shows the retention of the conductance, verifying the non-volatility.

IV. CONCLUSION

The HfO$_2$/Al$_2$O$_3$-based RRAM with the property of gradual resistance change has been proved the potential as an electronic synapse device. Based on the relationship between conductance change and pulse amplitude, the operation schemes to realize four standard STDP rules is proposed. Experimentally, the ultra-fast diverse STDP rules are realized. The maximum weight change is about 300%, and the timescale is on the order of nanoseconds.

ACKNOWLEDGMENTS

This work was supported by National Key Research and Development (2019YFB2205102, 2018YFE0203801) and the National Natural Science Foundation of China under Grant 61841404 and 61874006.

REFERENCES

[1] S. Yu, et al., Springer 2017. [2] P. H, et al., ACS Appl. Electron. Mater 2019. [3] S. Yu, et al., TED 2011. [4] Panwar N, et al., EDL 2017. [5] Erika C, et al., Front. Neurosci. 2016. [6] Bi G.Q, et al., J Neurosci 1998.

2020 IEEE Silicon Nanoelectronics Workshop

10.1

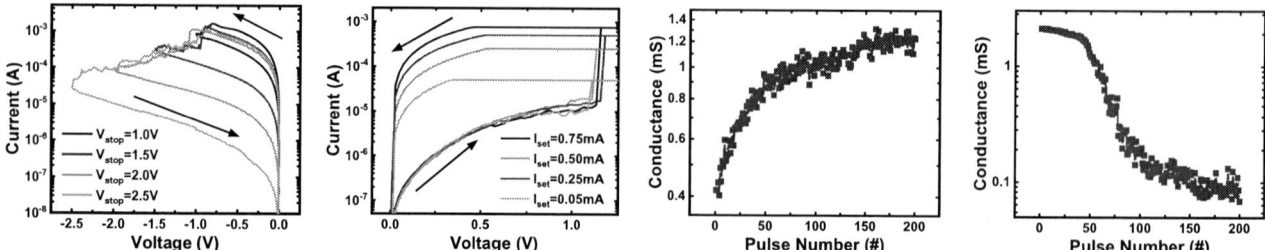

Fig.1 The multilevel characteristics in the device by controlling compliance currents and RESET voltages.

Fig.2 (a) The gradual set process by applying 200 pulses with 1.2V voltage amplitude and 15ns pulse width. (b) The gradual reset process by applying 200 pulses with -1.6V voltage amplitude and 15ns pulse width.

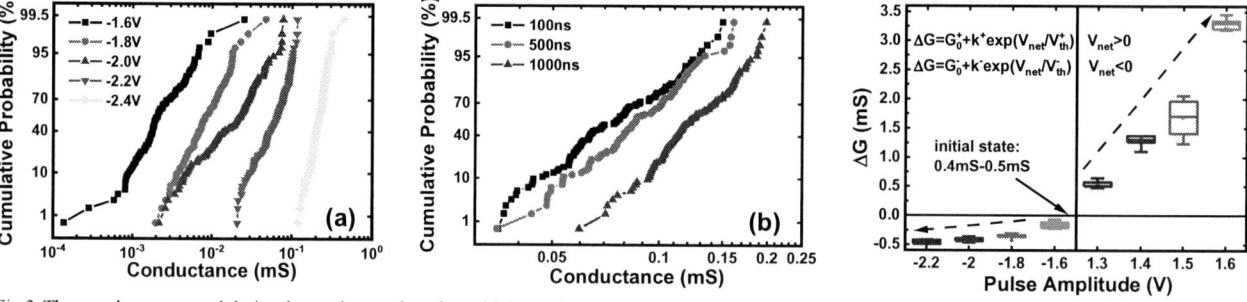

Fig.3 The conductance modulation by various pulses from high conductance state (HGS) (a) with the same width (50ns) and different amplitudes (from -1.6V to -2.4V) or (b) with the same voltage (1.6V) and different widths (100ns, 500ns, 1000ns).

Fig.4 The conductance modulation by various pulse amplitudes with the same width (15ns) from intermediate initial state (0.4mS~0.5mS)

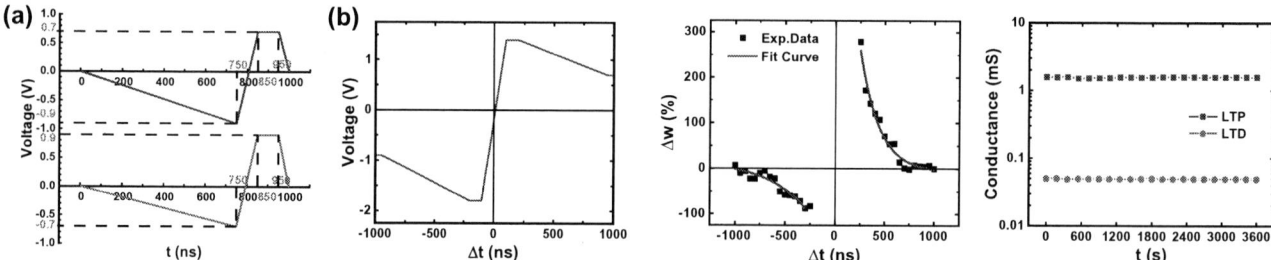

Fig.5 (a) The design of the pulse waveforms for the asymmetric Hebbian learning rule. (b) The net voltage applied across the electronic synapse.

Fig.6 The realization of the asymmetric Hebbian learning rule.

Fig.8 The nonvolatile characteristic of the device.

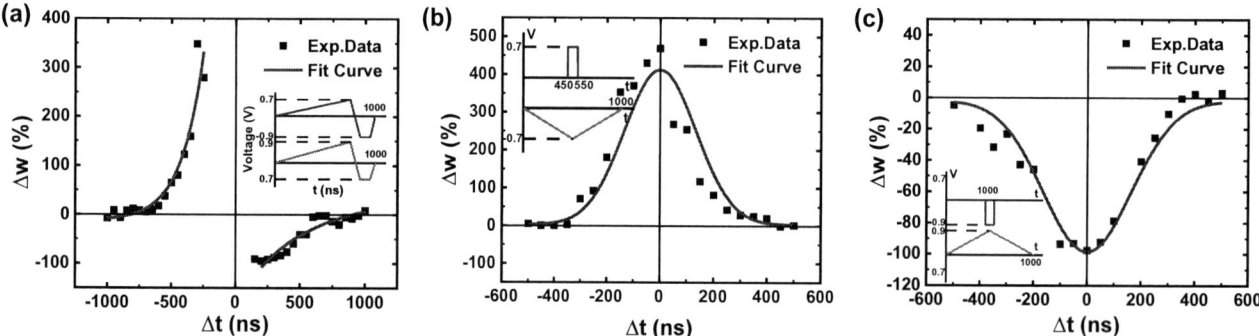

Fig.7 The implementation of other three standard STDP learning rules and the design of the corresponding applied pulse waveforms (inset). (a) The asymmetric anti-Hebbian learning rule. (b) The symmetric Hebbian learning rule. (c) The symmetric anti-Hebbian learning rule.

2020 IEEE Silicon Nanoelectronics Workshop

978-1-7281-9736-4/20 $31.00 © 2020 IEEE

Unsupervised Learning Architecture Based on Spike-Timing-Dependent Plasticity Using Flash Memory Synaptic Devices

Won-Mook Kang[1], Soochang Lee[1], Jangsaeng Kim[1], Byung-Gook Park[1], and Jong-Ho Lee[1]

[1]Department of EECS and ISRC, Seoul National University, Seoul 151-742, Korea,

Email: jhl@snu.ac.kr

Abstract — In this paper, we propose a hardware-based neural network architecture to implement pulse scheme required for unsupervised learning using flash memory synaptic devices. We perform circuit simulation to verify the systematic operation of multiple neuron system, and calculate energy consumption during synaptic weight changes. Also, the recognition accuracy is obtained in MNIST handwritten digit pattern learning by applying spike-timing-dependent plasticity rule through system level simulation.

I. INTRODUCTION

A spiking neural network (SNN) with electronic synaptic devices and integrated neuron circuits has emerged as a candidate for efficient hardware-based computing architecture. Among various neural network, SNN using spike-timing-dependent plasticity (STDP) rule to change synaptic weights is an intuitive, simple, and efficient way to enable unsupervised learning [1]. Recently, various electronic synaptic devices and pulse schemes for implementing synaptic plasticity by applying the STDP rule have been proposed [2]. However, research on circuits and architectures to implement pulse schemes suitable for electronic synaptic devices has not been emphasized. In this paper, we propose an appropriate pulse schemes to utilize AND type flash memory synaptic device array for STDP learning rule, and design an efficient computing architecture composed of circuit blocks. The implementation of the proposed pulse scheme and systematic operation of the pulse generating circuits are confirmed through circuit simulation. System level simulations are also performed based on the characteristics of fabricated electronic synaptic devices and the operation of the designed computing architecture.

II. RESULTS AND DISCUSSION

Fig. 1 (a) show a schematic cross-sectional view of a fabricated thin-film transistor type flash memory synaptic device. A SiO_2 / Si_3N_4 / SiO_2 gate insulator stack is formed between the control gate and the channel polysilicon, and used to store the synaptic weight. Fig. 1 (b) shows the long-term potentiation (LTP) and long-term depression (LTD) characteristics of the fabricated electronic synaptic device obtained by applying identical erase (ERS) and program (PGM) pulses. Fig. 2 (a) represents the proposed pulse scheme that enables selective LTP/LTD of electronic synaptic devices using STDP learning rule in AND-type crossbar array. The proposed pulse scheme does not require additional inhibition pulses for selective memory operation in the array. The systematic operation of the input and output neurons in charge of implementing this pulse scheme is achieved through the designed architecture shown in Fig. 2 (b). Neurons within a single neuron layer share a global pulse generator, which is responsible for generating the various pulses required for neurons. Global pulse generators use signals generated by the neurons, eliminating the need for an additional external controller in the architecture. The proposed input and output neuron circuits, which serve to provide the appropriate pulses for electronic synaptic devices depending on the operation of the neurons, are shown in Fig. 3 (a) and (b), respectively. V_{DDX} represents a positive voltage that can be obtained by a charge pumping circuit, and V_X determines the output pulse width of the pulse extension module. The values of capacitors included in the pulse generator module should be determined according to the number of electronic synaptic devices connected to the neuron. Fig. 4 shows the results of circuit simulation to verify the operation of the input and output neurons. The simulation results indicate the proposed pulse scheme is successfully implemented by the designed architecture. During one period in which the LTP/LTD operation of the electronic synaptic device occurs, the input and output neurons consume 0.474 nJ and 2.81 nJ, respectively. To evaluate the proposed architecture, the MNIST handwritten digit pattern learning simulation was performed on a fully connected 2-layer network including 784 input and 200 output neurons, and 88.2 % of recognition accuracy was obtained after 20 epochs training process.

III. CONCLUSION

We have proposed and investigated computing architectures that support the proposed pulse scheme available for AND type array of flash memory synaptic devices and STDP-based unsupervised learning rule. In a fully connected 2-layer neural network, an accuracy of 88.2% was obtained for the MNIST data set.

ACKNOWLEDGMENT

This work was supported by the MOTIE (10080583) and KSRC support program, the Brain Korea 21 Plus Project in 2020.

REFERENCES

[1] G. K. Chen et al., "A 4096-Neuron 1M-Synapse 3.8-pJ/SOP Spiking Neural Network With On-Chip STDP Learning and Sparse Weights in 10-nm FinFET CMOS," in IEEE Journal of Solid-State Circuits, vol. 54, no. 4, pp. 992-1002, April 2019.

[2] V. Milo et al., "Demonstration of hybrid CMOS/RRAM neural networks with spike time/rate-dependent plasticity," 2016 IEEE International Electron Devices Meeting (IEDM), San Francisco, CA, 2016, pp. 16.8.1-16.8.4.

10.2

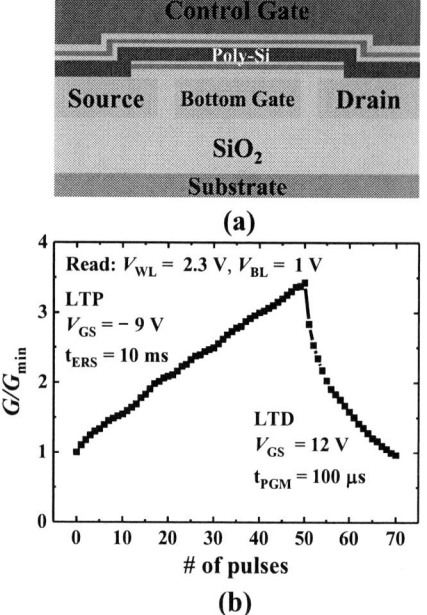

(a)

(b)

Fig. 1 (a) Cross-sectional view of fabricated flash memory synaptic device. (b) Measured LTP//LTD characteristics of the fabricated device by applying identical ERS and PGM pulses.

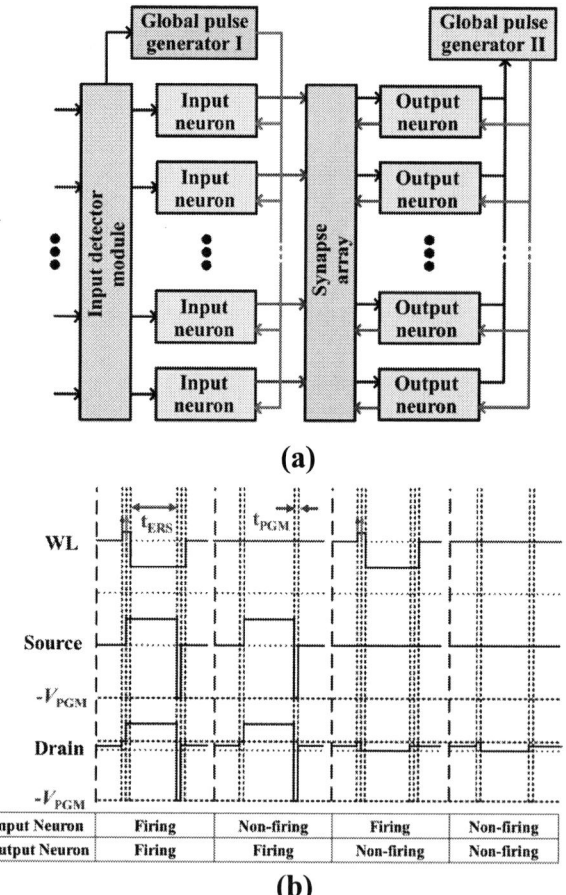

(a)

(b)

Fig. 2 (a) A conceptual diagram of designed SNN architecture. (b) A pulse scheme that enables selective LTP/LTD of electronic synaptic devices using STDP learning rule in AND-type crossbar array.

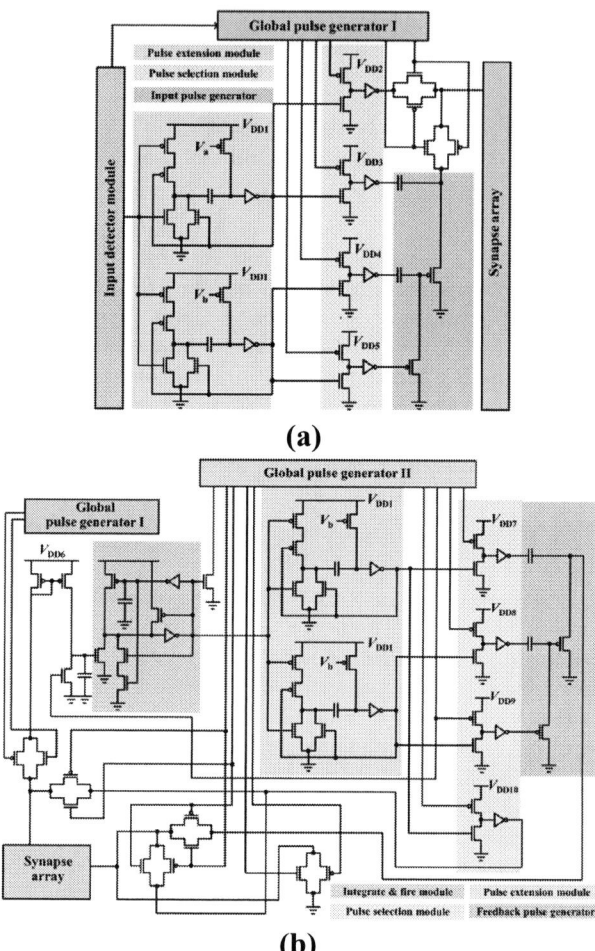

(a)

(b)

Fig. 3 Proposed (a) input and (b) output neuron circuits.

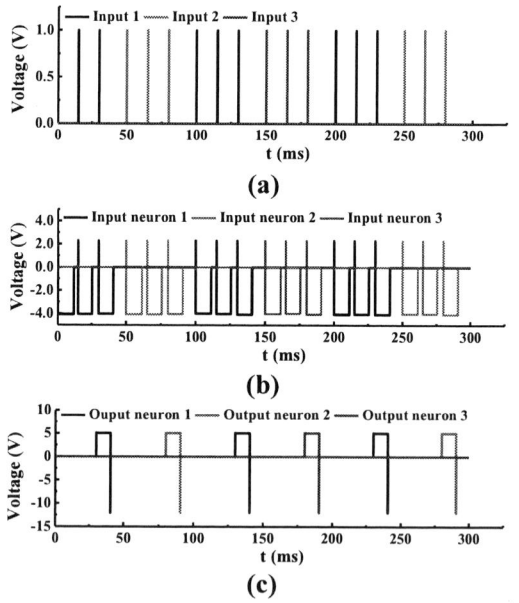

(a)

(b)

(c)

Fig. 4 (a) Input signals received by representative input neurons. (b) Input pulses generated by the input neuron circuit in each representative input neuron. (c) Feedback pulses generated by the output neuron circuit in each representative output neuron.

2020 IEEE Silicon Nanoelectronics Workshop

978-1-7281-9736-4/20 $31.00 © 2020 IEEE   78

Effects of Thermal Annealing on Ta_2O_5 based CMOS compatible RRAM

Somsubhra Chakrabarti[1], Jia Min Ang[1,2], Jia Rui Thong[1,2], Kunqi Hou[1], Mun Yin Chee[1], Putu Andhita Dananjaya[1], Desmond Loy Jia Jun[1,2], , Yong Chiang Ee[1] and Wen Siang Lew[1]

[1] Nanyang Technological University, 21 Nanyang Link, Singapore 637371, Singapore
[2] Globalfoundries, 60 Woodlands Industrial Park D Street 2, Singapore 738406, Singapore
Email: *s.chakrabarti@ntu.edu.sg*

Abstract — **In this paper, study on the thermal annealing effect is done on nanoscale $W/TaO_x/Pt$ resistive random access memory (RRAM) structure. Electrical characterization shows that device performance is improved after undergoing thermal annealing, with the cycle-to-cycle and device-to-device variability showing less variation with lower operating voltage. The device also undergoes CMOS BEoL compatible thermal budget.**
Keywords: TaO_x, CMOS compatible, thermal annealing

I. INTRODUCTION

RRAM is an emerging class of NVM due to its excellent scalability, CMOS compatibility, high speed operation, high endurance and longer retention time [1]. Ta_2O_5 is one of the promising material which is widely used as switching layer [2]. In this paper the effect of thermal annealing on $W/TaO_x/Pt$ RRAM structure is investigated through electrical as well as physical characterization. The structure satisfies the thermal budget, which is essential criterion for CMOS BEoL process. The structure also shows stable endurance and retention.

II. EXPERIMENTAL

The $W/TaO_x/Pt$ RRAM structure with cell size of 200nm was patterned through the electron-beam lithography. DC sputtering deposition of a 30 nm-thick W layer was done on the SiN substrate to form the BE. This is followed by a RF sputtering deposition of a 7 nm-thick TaO_x layer to form the switching layer of the device. A final DC sputtering deposition of a 30 nm-thick Pt layer was done to complete structure. Thermal annealing of the devices was done at 400°C in vacuum condition for 3 hours.

III. RESULT & DISCUSSION

The consecutive 20 cycles of current-voltage (I-V) characteristics of as-deposited and annealed samples are showed in **Fig. 1 (a)** and **(b)** respectively. The cumulative probability of set/reset voltage for as-deposited as well as annealed samples are plotted in **Fig. 1. (c)**. The σ/μ of set voltage of as-deposited samples is 0.16 (C-to-C) and 0.29 (D-to-D). Where the σ/μ of set voltage of annealed samples is 0.13 (C-to-C) and 0.22 (D-to-D). The average (μ) value of set voltage is also decreased after annealing (from 0.96 V

to 0.73 V). There is no significant change in reset voltages after annealing. The cumulative probability of LRS/HRS for as-deposited as well as annealed samples are plotted in **Fig. 1. (d)**. The σ/μ of LRS of as-deposited samples is 0.98 (C-to-C) and 0.96 (D-to-D). Where the σ/μ of LRS of annealed samples is 0.35 (C-to-C) and 0.36 (D-to-D). There is no significant change in HRS after annealing. **Fig. 2 (a)** shows the cross-sectional TEM of the $W/TaO_x/Pt$ structure after thermal annealing, where **Fig. 2 (b)** shows the HRTEM of the same. There is no sign of diffusion of metal into 7 nm-thick TaO_x film. To investigate further EDX line profile is done and there is no sign of metal diffusion (**Fig. 2 (c, d, e)**). **Fig. 3** indicates the color map of chemical analysis of the RRAM structure. The XPS characteristics of annealed TaO_x switching layer is shown in **Fig. 4**. Doublet $Ta4f_{7/2}$ and $Ta4f_{5/2}$ are observed at 25.4 eV and 27.2 eV respectively [3]. The O1s peak is observed at 530.3 eV [4]. Both as-deposited and annealed samples shows Ohmic conduction in both LRS and HRS (**Fig. 5 (a, b)**). The structure also shows long retention of $>10^4$ s, stable read endurance (read at 0.1 V with PW of 10 μs) of $>10^6$ cycles and stable program/erase endurance of $>10^4$ cycles with 200 ns write/erase pulse (**Fig. 6 (a, b, c)**).

IV. CONCLUSION

In summary, annealed $W/TaO_x/Pt$ RRAM structure shows less variability and less operating voltage than the same of as-deposited one. The structure is also stable after going through BEoL compatible thermal budget.

ACKNOWLEDGMENTS

The work was supported by the RIE2020 ASTAR AME IMF-ICP Grant (No. I1801E0030).

REFERENCES

[1] F. Pan et. al. "Recent progress in resistive random access memories: Materials, switching mechanisms, and performance" Sci. Eng. Rep. 83(2014) 1-59.
[2] S. R. Lee et. al. "Multi-level switching of triple-layered taox rram with excellent reliability for storage class memory" VLSI sym. 2012.
[3] J. G. S. Moo et. al. "An XPS depth-profile study on electrochemically deposited TaOx" J. Sol. State Electrochemistry. 17 (2013) 3115.
[4] E. Atanassova et. al. "XPS study of N2 annealing effect on thermal Ta2O5 layers on Si" Appl. Surf. Sci. 225 (2004) 86.

2020 IEEE Silicon Nanoelectronics Workshop

10.3

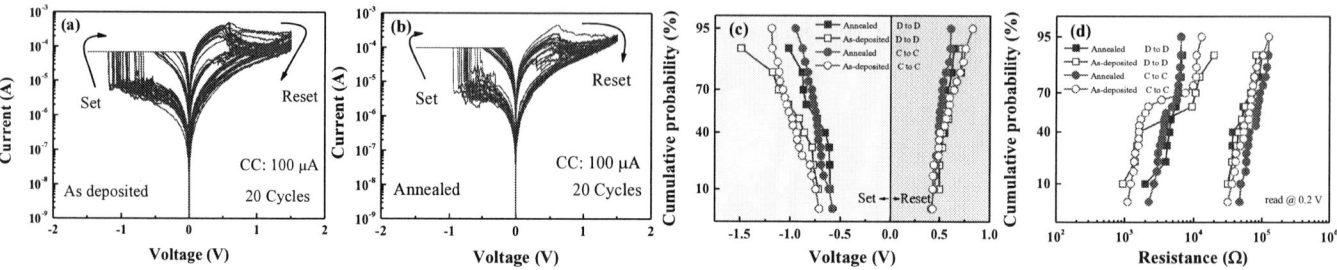

Fig. 1. Consecutive 20 cycles of current-voltage characteristic of **(a)** as-deposited and **(b)** annealed $W/TaO_x/Pt$ RRAM device. **(c)** Cumulative probability plot of set/reset voltage for both cycle-to-cycle and device-to-device of as-deposited as well as annealed samples. **(d)** Cumulative probability plot of LRS/HRS resistance for both cycle-to-cycle and device-to-device of as-deposited as well as annealed samples.

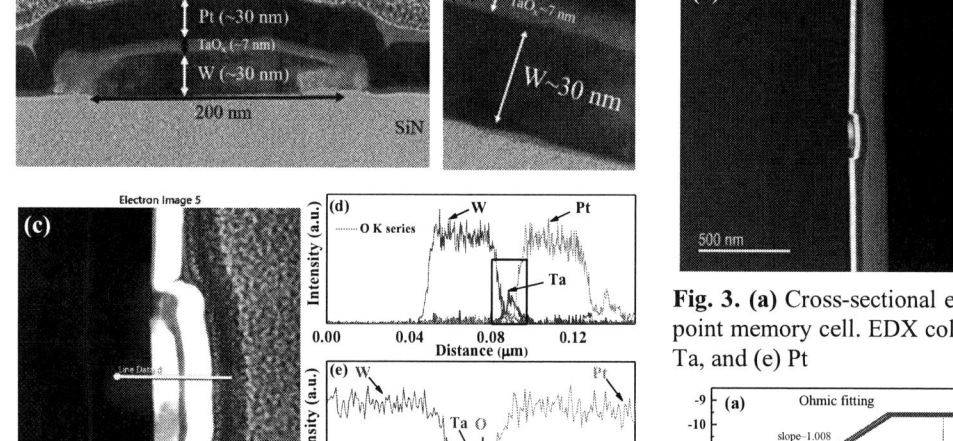

Fig. 2. (a) Cross-sectional TEM image of $W/TaO_x/Pt$ cross-point memory with 200 nm cell size. **(b)** HRTEM image revealing 7 nm-thick TaO_x switching layer. **(c)** TEM image indicating EDX scan line. **(d)** EDX line scan indicating $W/TaO_x/Pt$ structure. **(e)** EDX line scan in switching layer position indicating presence of TaO_x.

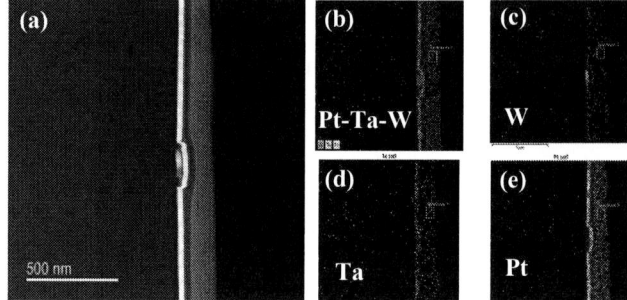

Fig. 3. (a) Cross-sectional electron image of $W/TaO_x/Pt$ cross-point memory cell. EDX colour map of **(b)** Pt-Ta-W **(c)** W, **(d)** Ta, and **(e)** Pt

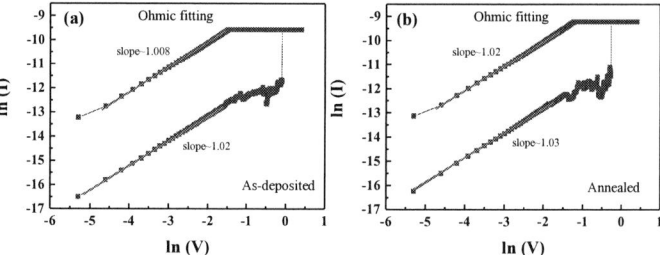

Fig. 5. Both **(a)** as-deposited and **(b)** annealed samples show Ohmic conduction in both LRS and HRS.

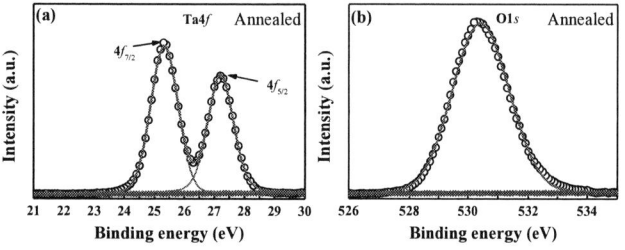

Fig. 4. XPS characteristics of **(a)** Ta$4f$ and **(b)** O1s

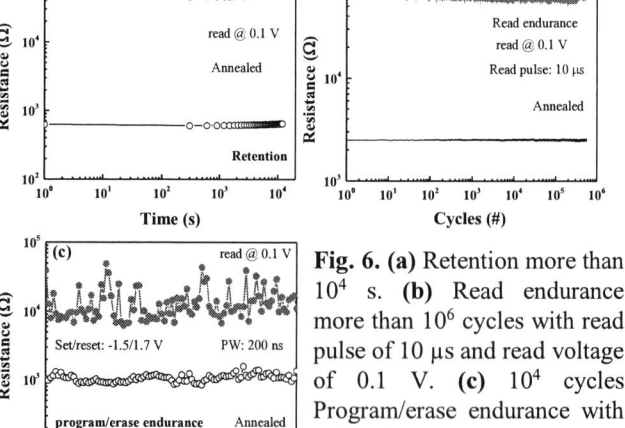

Fig. 6. (a) Retention more than 10^4 s. **(b)** Read endurance more than 10^6 cycles with read pulse of 10 μs and read voltage of 0.1 V. **(c)** 10^4 cycles Program/erase endurance with pulse width of 200 ns.

2020 IEEE Silicon Nanoelectronics Workshop

978-1-7281-9736-4/20 $31.00 © 2020 IEEE

Highly scalable 4F^2 cell transistor for future DRAM technology

Kyung Kyu Min[1,2], Sungmin Hwang[1], Jong-Ho Lee[1], and <u>Byung-Gook Park</u>[1]

[1]Inter-University Semiconductor Research Center, Department of Electrical and Computer Engineering, Seoul National University, Seoul 151-744, South Korea [2]SK Hynix Inc., Icheon 17336, South Korea
Email: bgpark@snu.ac.kr

Abstract — **A novel 4F^2 dynamic random access memory (DRAM) cell transistor structure was proposed that can solve various process problems and special failure modes that caused by floating body. The suitability of the transistor scheme for future DRAM technology nodes was also verified. Through this new structure, it can expect to realize 4F^2 DRAM and continuously expand DRAM technology node.**

Keywords: 4F^2 DRAM, Cell transistor, Vertical transistor

I. INTRODUCTION

For decades, the technology of memory devices has been developed by integrating more cells through size reduction. However, the existing scaling law has reached the limitation, and inventive attempts has been made to increase the density. Flash memory has solved the problem using 3D vertical structure [1], and dynamic random access memory (DRAM) has attempted to switch a design scheme from 6F^2 to 4F^2 to reduce the unit cell area. In the 4F^2 DRAM scheme, storage node (SN) and bit line (BL) are located at the top and bottom of the same space and connected by a vertical cell transistor [2], [3]. Due to its special configuration, floating body effects not appearing in 6F^2 DRAM occurs. Therefore, dynamic retention failure is caused by accumulated holes in the body region [3]. Recently, a novel 4F^2 DRAM cell transistor structure with the form of an inner gate was proposed [4]. In this work, the characteristics of the proposed DRAM cell and its applicability to future DRAM nodes are confirmed.

II. CHARACTERISTICS OF VERTICAL INNER GATE TRANSISTOR

Conventional 4F^2 DRAM (vertical outer gate; VOG) uses the gate-all-around cell transistor [2], [3]. This structure inevitably causes a voltage transfer problem because of the resistance of WL and a capacitor formation problem due to a small Si pillar area. The proposed vertical inner gate (VIG) cell can solve these problems because WL passes through Si (Fig. 1) [4]. To confirm the electrical characteristics of VIG cell, 3D technology computer-aided design simulation was performed on a 2x nm DRAM technology node.

VIG cell has better gate controllability and has good on-current characteristics compared with VOG cell (Fig. 2). These improved properties are due to the fringing field formed on the outside of the tri-gate. However, the junction depletion region at the off state is extended by the fringing field which degrades static retention (Fig. 3). It was confirmed that the overall DRAM margin can be improved in VIG cell by adjusting trade-off between retention time and on current through junction engineering as shown in Fig. 4.

The advantage of VIG cell is the improvement of not only on-off margin characteristics but also dynamic retention thanks to the transient bipolar effect (TBE) in 4F^2 DRAM (Fig. 5). Since the electric field in the off state reduces hole accumulation in a body, the parasitic bipolar junction transistor operation during BL toggling is weakened in the VIG cell.

III. DRAM TECHNOLOGY NODE EXTENSION WITH VIG CELL

As the technology node shrinks, the on current decreases, and the static retention deteriorates due to the increase of gate-induced drain leakage. However, in the downscaled 4F^2 DRAM, the dynamic retention characteristics are improved by reduced body dimension which suppresses hole accumulation.

Based on the simulation results considering the shrinkage of 4F^2 DRAM to 1x nm nodes (Fig. 6), it was proved that the properties of the VIG structure were maintained. Furthermore, with the TBE mitigating effect of the VIG structure, it is confirmed that the dynamic retention time is longer than the static retention time in the VIG of the 1x nm node (Fig. 7). Because the overall improvements are also effective in the shrink dimension, VIG cell is a suitable structure in terms of DRAM technology node expansion.

IV. CONCLUSION

In this work, the cell transistor structure suitable for 4F^2 DRAM was proposed. It improves the overall device characteristics and particularly has a great advantage in dynamic retention issue of 4F^2 DRAM. Through this, we can expect to realize 4F^2 DRAM and continuously expand DRAM technology node.

ACKNOWLEDGMENTS

This work was supported in part by the Brain Korea 21 Plus Project in 2020.

REFERENCES

[1] J. Jang et al., "Vertical cell array using TCAT (Terabit Cell Array Transistor) technology for ultra high density NAND flash memory," *2009 Symposium on VLSI Technology*, Honolulu, HI, 2009, pp. 192-193.

[2] S. Maeda et al., "Impact of a vertical /spl Phi/-shape transistor (V/spl Phi/T) cell for 1 Gbit DRAM and beyond," *IEEE Transactions on Electron Devices*, vol. 42, no. 12, pp. 2117-2123, Dec. 1995.

[3] H. Chung et al., "Novel 4F^2 DRAM cell with Vertical Pillar Transistor (VPT)," *2011 Proceedings of the European Solid-State Device Research Conference (ESSDERC),* Helsinki, 2011, pp. 211-214.

[4] K. K. Min et al., "Vertical Inner Gate Transistors for 4F^2 DRAM Cell," *IEEE Transactions on Electron Devices*, vol. 67, no. 3, pp. 944-948, Mar. 2020.

P1-1

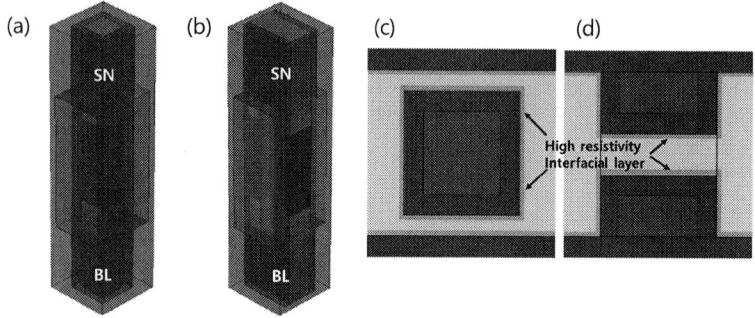

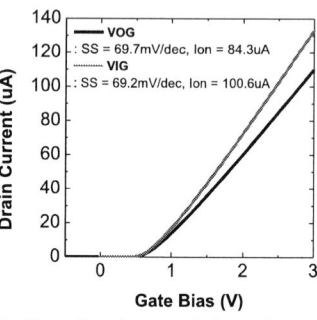

Fig. 1. Structure of 4F^2 DRAM (a) VOG cell transistor. (b) Proposed VIG cell transistor. Plan view of (c) VOG and (d) VIG with high resistivity interfacial layer. The WL resistivity of VIG is lower than that of VOG from reduced IL at the critical region.

Fig. 2. Transfer characteristics of VIG and VOG cell of V_d = 1.2 V. Lower subthreshold swing and higher on-current (at V_{th} + 2 V) can be obtain in VIG cell.

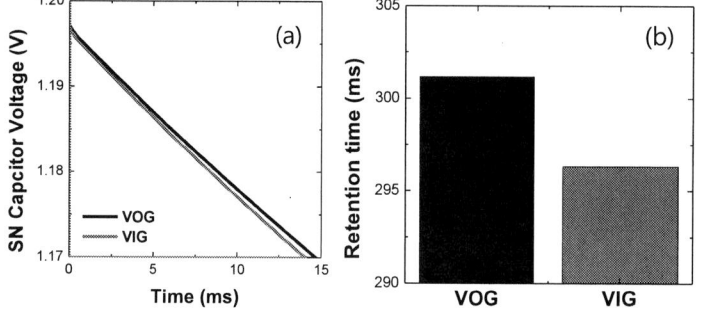

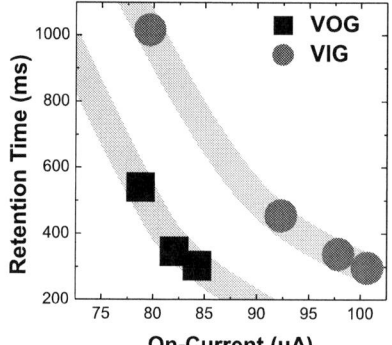

Fig. 3. Static retention properties of 4F^2 DRAM cells. (a) SN capacitor voltage through retention situation (b) Calculated retention time from (a). VIG shows degraded retention characteristics.

Fig. 4. Trade-off characteristics of on-current and retention time by junction overlap.

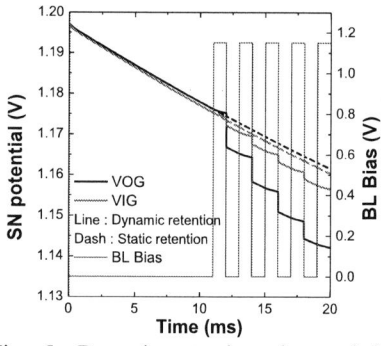

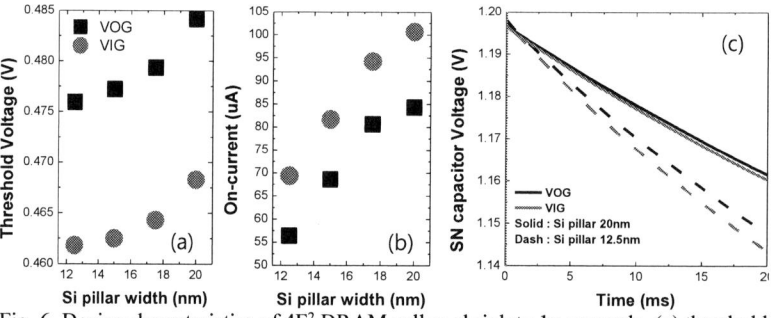

Fig. 5. Dynamic retention characteristics through BL toggling and WL off. VIG shows longer dynamic retention time than VOG.

Fig. 6. Device characteristics of 4F^2 DRAM cell as shrink to 1x nm node. (a) threshold voltage (b) on-current (at V_{th} + 2 V) (c) SN capacitor voltage vs time at static retention situation. VIG also shows improved gate controllability and on-current properties, but degraded static retention in scaled devices like 2x nm technology.

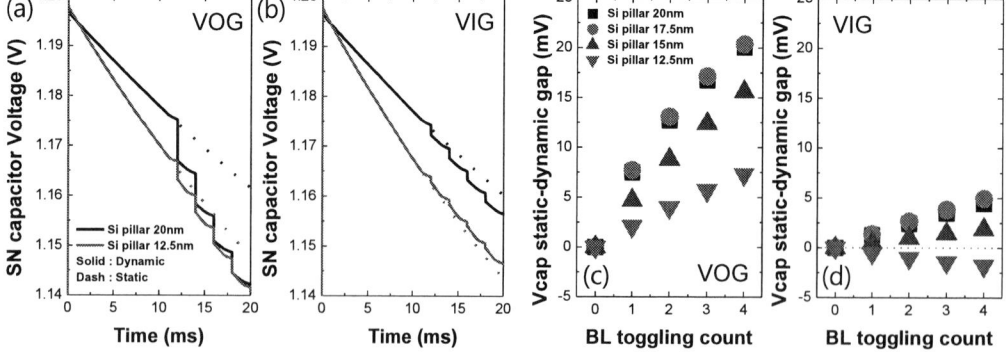

Fig. 7. Dynamic retention characteristic of scale 4F^2 DRAM cell. Static and dynamic retention characteristics of (a) VOG and (b) VIG. (c), (d) Capacitor voltage gap in static-dynamic retention situation by BL toggling count.

2020 IEEE Silicon Nanoelectronics Workshop

978-1-7281-9736-4/20 $31.00 © 2020 IEEE

Weighted synaptic behaviors of HfON based RRAM device by a novel waveform modulation method

Yuechi Ma[1], Ruiyi Li[2], Ao Yu[1], Zehao Wang[1], Xiangxiang Ding[2], Yulin Feng[2], Lifeng Liu[1,2*]

1 School of Software & Microelectronics, Peking University, Beijing 102600, China
2 Institute of Microelectronics, Peking University, Beijing 100871, China
*E-mail: lfliu@pku.edu.cn

Abstract —In this work, a novel waveform approach based on two synaptic plasticity learning rules was proposed to investigate the characteristic of RRAM for electronic synapse application. The connectivity among excitatory neurons is adapted by Spike-Timing-Dependent-Plasticity (STDP) learning, while the change of the weight is regulated by another learning rule: Spike-Rate-Dependent-Plasticity (SRDP). The effect of the pulse frequency applied on resistance was investigated. It indicates that higher frequency is beneficial to the change of the resistance of device. The novel waveform method also integrates the frequency and timing variables that affect the modulation of the weight.

Index Terms — RRAM, Spike-Timing-Dependent-Plasticity, Spike-Rate-Dependent-Plasticity, plasticity

I. INTRODUCTION

Resistive random-access memory (RRAM) is considered as one of next-generation memories due to its fast speed, CMOS compatibility and ultralow switching energy. Besides its possibility as a storage class memory, RRAM device also receives great attention as a synaptic device. Various bio-inspired learning rules including spike timing-dependent plasticity (STDP) and spike rate-dependent plasticity (SRDP) are under the extensive research to realize long-time potentiation (LTP) and depression (LTD)[1]. Another form of memory device , conductive bridge RAM (CBRAM) is employed to realize paired-pulse facilitation/depression (PPF/PPD) by modulating the pulse stimulation[2]. The common simulation methods only consider timing or frequency dependence, so a simulation waveform that integrates the timing and frequency variables is designed. In this paper, HfON based RRAM devices are fabricated with PVD HfON resistive switching layer. The relationship between the resistance and frequency was investigated for the electronic synapse applications. The experiment results indicate that the change of the resistance increases with the increase of pulse frequency, the potentiation or the depression of the change of resistance is determined by the polarity of the pulse. The principle of the new waveform and the simulation results on the model are described in this paper.

II. EXPERIMENTS

The RRAM device with TiN/HfON/Pt structure is fabricated. First, 100-nm Pt was deposited on SiO_2/Si substrate by DC sputtering as bottom electrode followed by 10-nm HfON. Then, 80-nm TiN was grown in Ar/N_2 atmosphere. Finally, UV lithography and dry etching processes are performed to form 50um×50um top electrodes. The electrical characteristics were measured with Agilent B1500A semiconductor parameter analyzer and Agilent 81160A pulse generator on the Cascade probe station.

III. RESULTS AND DISCUSSION

Pulse operation was performed and analyzed for the purpose of synaptic behavior measurement. From previous researches, an analog property of synaptic devices is essential in order to split the conductance levels in multiple states. A different frequency was applied to modulate the resistance change HfON based RRAM device. After adjusting and stabilizing the resistance to about 2.5kΩ, the same number continuous pulse (1.5V,1us) with different frequency (1Hz,5Hz,30Hz) (shown in Fig.1) were applied on the RRAM devices and the observed resistance changes were shown in Fig. 2.

As depicted in Fig. 2 (a)(c), the initial resistance of 2.5kΩ under positive bias pulse (1Hz,5Hz,30Hz) change to 2kΩ, 1.6kΩ, 1kΩ in SET process. The initial 2500Ω under negative bias pulse (1Hz,5Hz,30Hz) increase to 3.5kΩ, 4.6kΩ, 7.5kΩ in RESET process. Conductance is analogous to synapse weight. Fig. 2 (b)(d) shows the weight under pulse (1Hz,5Hz,30Hz) change to 27%, 67.5%, 147% and -27.5%, -45.7%, -66.7%. The mechanism of resistance change with frequency should be related to the forming and rupture of oxygen vacancy conducting filaments. As the time interval between the pulses becoming smaller, the combination and separation between oxygen ion and oxygen vacancy may be more active. The higher frequency pulse makes the resistance of device change on a larger scale.

According to the literature [3], inhibition and excitation of the central neural system depends on spike rate difference applied to pre and post-synapse. A waveform method that integrates the timing and frequency variables is designed. As depicted in Fig. 3. The waveform of Fig. 3.can integrate the spike rate and timing to affect the weight. The waveform is composed of several positive biased sine waves and one complete sine wave. The sine wave can make the amplitude of the superimposed pulse continuous. When the pre-spike rate is greater than post-spike rate, the superimposed signal is positive biased, the rate varies with the difference between pre and post-spike. Δt>0, the superimposed waveform produces a larger positive amplitude. The positive biased voltage makes device in SET progress, reduce the resistance and increase the weight. When the pre-spike rate is lower than post-spike rate, the superimposed signal is negative biased, Δt<0, the superimposed waveform produces a larger negative amplitude. The negative biased voltage makes device in RESET process, increase the resistance and reduce the weight. Other cases produce different superimposed waveforms that vary the weight by degree. Simulation results are shown in Fig.4.

Fig. 4 (a) shows the resistance state when waveform is applied on RRAM model hspice platform. The resistance change from 150Ω to 10^5Ω with frequency (±1Hz, ±5Hz, ±30Hz) and timing (-20ms~20ms) variation. Fig. 4 (b) shows the weight curve drawn according to the change of model resistance.

2020 IEEE Silicon Nanoelectronics Workshop

978-1-7281-9736-4/20 $31.00 © 2020 IEEE

IV. CONCLUSION

In this work, we analyzed the effect of pulse frequency on resistance, a waveform containing frequency and timing variables was designed and the simulated data was tested on RRAM model. By doing so, different timing and frequency interval of pre-post pulses were applied to RRAM to demonstrate the synaptic plasticity.

ACKNOWLEDGEMENT

This work was supported in part by the National Natural Science Foundation of China (61874006, 61421005, 61334007, and 61834001) and the 111 Project (B18001).

REFERENCES

[1] Milo V, et al. Demonstration of hybrid CMOS/RRAM neural networks with spike time/rate-dependent plasticity. *IEEE IEDM Tech Dig* 2016. P. 16.8.1-4.

[2] Wang Z, et al. Memristors with diffusive dynamics as synaptic emulators for neuromorphic computing. *Nat Mater* 2016;16(1):101–10.

[3] Galarreta M, Hestrin S. Frequency-dependent synaptic depression and the balance of excitation and inhibition in the neocortex. *Nat Neurosci* 1998;1(7):587–94.

[4] Peking University Resistive-Switching Random Access Memory Verilog-A Model-Version 2.1.1.

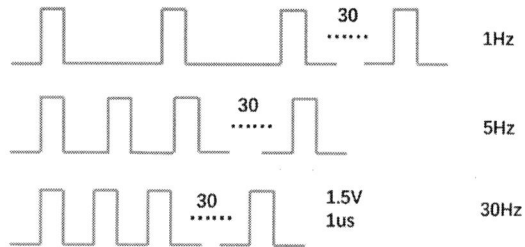

Fig. 1.   Pulse waveform diagram (1.5V,1us,1Hz,5Hz,30Hz)

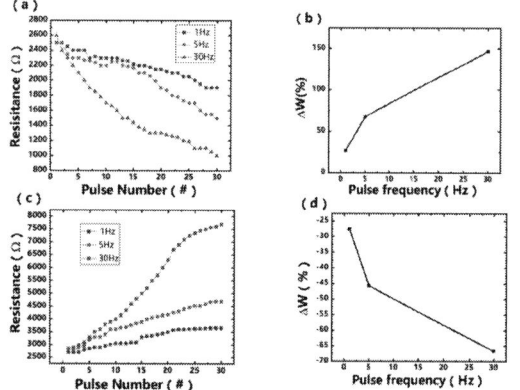

Fig. 2.   (a) Variation of resistance with the positive bias pulse at different frequency. (b)(d) The ratio of weight changes at different frequency. (c) Variation of resistance with the negative bias pulse at different frequency.

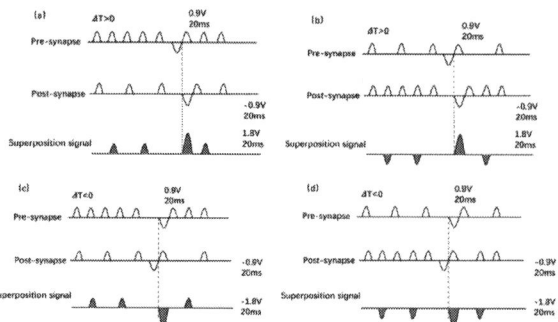

Fig. 3.   (a) pre-spike rate greater than post-spike rate and Δt>0. (b) pre-spike rate less than post-spike rate and Δt>0. (c) pre-spike rate greater than post-spike rate and Δt<0. (d) pre-spike rate less than post-spike rate and Δt<0.waveform diagram.

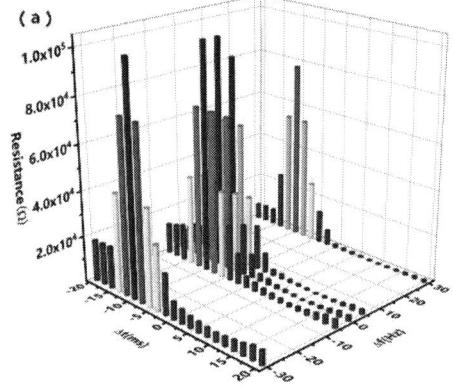

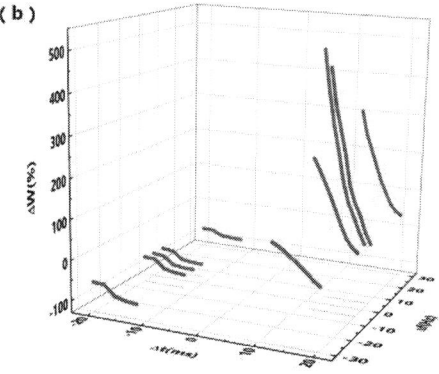

Fig. 4.   (a) The simulated data distribution on the RRAM model[4]. (b) The weight curve drawn according to the change of model resistance.

Ferroelectric Few Layer Black Phosphorus Field-Effect Transistors for SRAM Application

Cheng-Hsien Yang, Yun-Fang Chung, Kuan-Ting Chen and Shu-Tong Chang*

Department of Electrical Engineering, National Chung Hsing University, Taichung, Taiwan

*E-mail:stchang@dragon.nchu.edu.tw

Abstract—Two-dimensional materials such as few layer black phosphorus owing to the unique electronic properties of the atomically thin two-dimensional layered structure, can be made into the future metal–oxide–semiconductor field-effect device technology. In this paper, Ferroelectric HZO (HfZrOx) applied on few layer black phosphorus FETs are modeled and demonstrated with improvement on subthreshold swing (SS) to apply on SRAM design.

Keywords—Subthreshold Swing (SS), Ferroelectric HZO, SRAM

I. INTRODUCTION

Black phosphorus (BP) is candidate channel materials for an ultra-thin-body (UTB) transistor device with its two-dimensional super-thin atomic structure and has been broadly studied in past years. In this work, the modified effective mass approximation (MEMA) band model can fit first principle band structure of BP and be used for further shortening the calculation time without losing the accuracy. Negative Capacitance (NC) effect has been intensively and extensively investigated by ferroelectric Zr doped in HfO_2 as gate stack to integrate with FETs [1]. The requirement for scaling down supply voltage V_{DD} and power consumption for low power device in IoT (Internet of Things) era may lead the progress speed-up for following current CMOS architectures and feasible ALD (atomic layer deposition) super cycle approach [2]. Recently, the NC technology integrated with 2D material as FET has attracted lots of attention [3]. In this work, we investigate ferroelectric few layer BP FETs with better subthreshold swing and good performance of SARM.

II. CALCULATION METHOD

As you know that current EMA band model could not accurately fit the band structure of multilayer two-dimensional materials such as BP in larger energy and k-space regions. For this reason, it is necessary to develop the more correct compact band model of multilayer two-dimensional materials such as BP for mobility calculation, and current SPICE model [4] applied to transistor devices with 2D semiconducting materials is restricted to multilayer BP. It is therefore necessary to develop SPICE model with correct compact band model applicable to transistor devices with multilayer BP. In this work, we follow higher order correction for EMA mode as discussed in the text book [5] edited by Dr. Ridley to propose EMA compact band model for few layer BP considering up to the second order nonparabolic factors including α and β correction (MEMA band model). Assume that a nonparabolic dispersion relation for in-plane k space is presented as below.

$$E(1 + \alpha E + \beta E^2) = \frac{\hbar^2}{2}\left(\frac{k_x^2}{m_x} + \frac{k_y^2}{m_y}\right) \qquad (1)$$

In this work we fellow Prof. Tony Low's idea [6] to adopt an average of experimental [7] and theoretically [7] predicted out-of-plane masses. We also follow Prof. Gamiz's research work about carrier mobility calculation in semiconductor materials to use Kubo–Greenwood Mobility formula [8] for carrirt mobility calculation of multilayer BP. The structure of the ferroelectric few layer BP FET is designed to have a channel length Lg of 15 nm as shown in Fig. 1(a). TCAD simulation for the ferroelectric few layer BP FET will consider various physical models as mentioned in ref. [3]. The FE material is 7 nm HZO $(HfZrO_2)$ which could be simulated by the Landau–Khalatnikov (LK) model [9].

$$V_{FE}(P) = T_{FE} \cdot (\alpha P + \beta P^3 + \gamma P^5) \qquad (2)$$

where V_{FE} is the voltage across the ferroelectric material, T_{FE} is the thickness of the ferroelectric material, and P is the polarization. The parameters such as α, β, and γ are tuned in this work to fit the experimentally measured P-V data (polarization -voltage) of metal / FE-HZO / metal as shown in Fig. 1(b).

This modeling concept is based on the continuity of the electric field between the FE and dielectric layers and the relationship between the voltage across the ferroelectric material (V_{FE}) and the polarization given by the LK model. The relationship between the voltage across the MOS (V_{MOS}) and V_{FE} is given by

$$V_G = V_{MOS} + V_{FE} \qquad (3)$$

The electrical characteristics without the FE layer are obtained by the TCAD simulation, and the transfer characteristics of the I_D–V_G curve with the FE layer are modulated with the one-dimensional (1D) LK model.

III. RESULT AND DISCUSSION

Fig. 2 shows the band structure result for monolayer and bilayer BP using first principle method. Results from EMA model considering the second order nonparabolic correction (MEMA model) is also included for comparison.The calculated carrier mobility from MEMA will be utilized in TCAD device simulation for ferroelectric few layer BP FET. Fig. 3(a) show the I_D-V_G transfer characteristics of 1-layer, 3-layer BP nFET. As NC effect is considered, the I_D-V_G transfer characteristic curve is pulled to the small voltage region overall (refer to eq. (4)), so it can reduce both threshold voltage and subthreshold swing. It brings out ferroelectric few layer BP FET to achieve low voltage steep slope transistors in the development of sub-3 5 nm technology. The equation of voltage amplification is shown below.

$$Av = \frac{\partial V_{MOS}}{\partial V_G} = \frac{|C_{FE}|}{|C_{FE}| - C_{MOS}} \qquad (4)$$

where V_{MOS} is the voltage across the MOS, and C_{FE} and C_{MOS}

are the capacitances of the ferroelectric layer and MOS, respectively. Note that CMOS is calculated in the accumulation region. C_{FE} could be matched to C_{MOS} ($|C_{FE}|$ ~C_{MOS}), and will occur hysteretic behavior while $|C_{FE}|$ < C_{MOS} with a negative voltage gain. Fig. 3(b) shows the charge density v.s. capacitance characteristics for different layers NC BP nFET. We can see that 3-layer BP demonstrates the best capacitance matching. According to eq. (4), the better capacitance matching also causes large voltage amplification of 3-layer BP. In order to achieve more voltage amplification, C_{FE} can be tuned by changing the FE thickness (T_{FE}). On the other hand, gate oxide thickness and buried oxide thickness could be optimized. Fig. 4 showed that the impacts of layer number of BP on the normalized performance of few layer SRAM (against monolayer SRAM). With 4 L of BP, the performance of few layer SRAM cell can be ~30% better than that of monolayer SRAM cell.

IV. CONCLUSION

In this study, MEMA band model is used for band structure and carrier mobility calculations of few layer BP. Impact of the layer number issue and the NC effect on performance of ferroelectric BP FET and SRAM application are also discussed. This work will help the design of future nanoscale transistor device technology and design of SRAM using few layer BP.

ACKNOWLEDGMENT

The authors are grateful for the funding support from the Ministry of Science and Technology (MOST 108-2622-8-002-016 & MOST 106-2221-E-005 -097 -MY3), and computing support was received by the National Center for High-Performance Computing (NCHC), Taiwan.

REFERENCES

[1] M. Si, C. Jiang, C.-J. Su, Y.-T. Tang, L. Yang, W. Chung, M. A. Alam and P. D. Ye, "Sub-60 mV/dec Ferroelectric HZO MoS2 Negative Capacitance Field-effect Transistor with Internal Metal Gate: the Role of Parasitic Capacitance," IEDM Tech. Dig., pp. 573-576, 2017. .

[2] M. H. Lee et al., "Ferroelectric Al:HfO2 Negative Capacitance FETs, " IEDM Tech. Dig., pp. 565-568, 2017.

[3] Wei-Xiang You and Pin Su, "Design Space Exploration Considering Back-Gate Biasing Effects for 2D Negative-Capacitance Field-Effect Transistors," IEEE Transactions on Electron Devices, vol. 64, pp. 3476-3481, 2017.

[4] Eric Pop et al., Stanford 2D Semiconductor model, 1nd ed. Stanford University, 2016.

[5] Brian K. Ridley, Quantum Processes in Semiconductors, 5nd ed. Oxford University Press, 2013.

[6] Yongjin Jiang, Rafael Rold´an, Francisco Guinea, and Tony Low, Physcial Review B, 92, 085408 (2015).

[7] S.-I. Narita, S.-I. Terada, S. Mori, Y. Muro, Y. Akahama, and S. Endo, J. Phys. Soc. Jpn. 52, 3544 (1983).

[8] J. M. Gonzalez-Medina, F.G. Ruiz, E.G. Marin, A. Godoy, F. Gamiz, "Simulation study of the electron mobility in few-layer MoS2 metal–insulator-semiconductor field-effect transistors,m" Solid-State Electronics, vol. 114, pp. 30-34, 2015.

[9] S. Salahuddin and S. Datta, "Use of Negative Capacitance to Provide Voltage Amplification for Low Power Nanoscale Devices, " Nano Lett., vol. 8, No. 2, pp. 405-410, 2008.

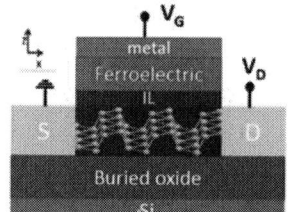
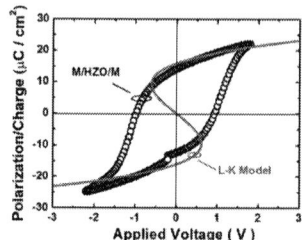

Fig. 1 (a) the structure of the ferroelectric few layer BP FET. (b) P-V hysteresis loop for a M/FE-HZO/M structure, and the red curve models the NC using the Landau–Khalatnikov (LK) model.

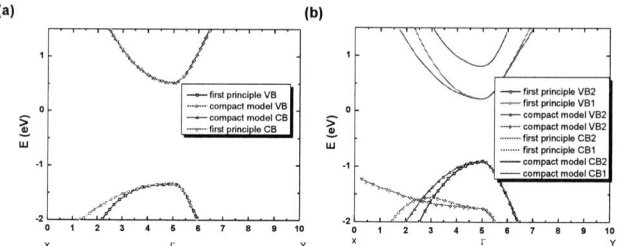

Fig. 2 Band structure for monolayer and bilayer BP from first principle method and compact model (MEMA), respectively. Note that tri-layer and bulk not show here. Therefore, regarding this compact analysis we can predict four low energy bands for bilayer BP.

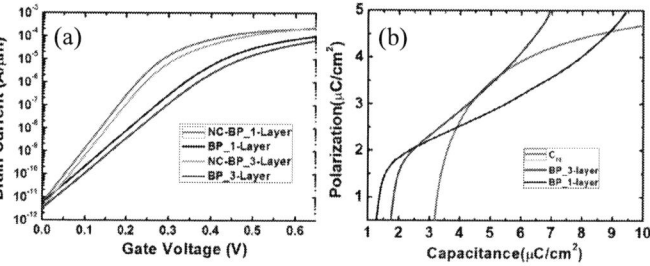

Fig. 3 (a) I_D-V_G transfer characteristic for multilayer BP. (b) Polarization versus capacitance for multilayer BP NC-FET. Note that only n-type BP FETs are shown.

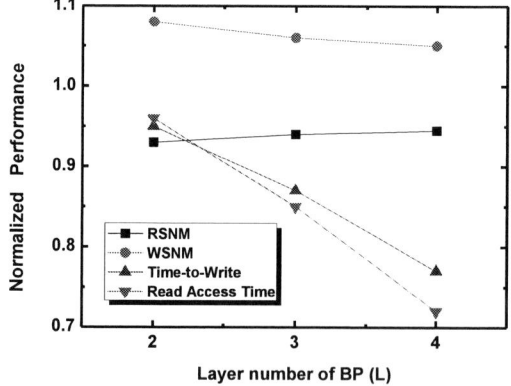

Fig. 4 Impacts of layer number of BP on the normalized performance such as RSNM, WSNM, Time-to-Write, and Read Access Time of few layer SRAM with respect to the monolayer one.

2020 IEEE Silicon Nanoelectronics Workshop

Bulk-limited Effect in Gradual Conductance Switching Behaviour of HfO_x-based Memristive Devices for Analog Synaptic Device Applications

Putu Andhita Dananjaya[1], Desmond Loy Jia Jun[1,2], Mun Yin Chee[1], Samuel Chow Chen Wai[1,2], Jia Min Ang[1,2], Kunqi Hou[1], Jia Rui Thong[1,2], Somsubhra Chakrabarti[1], Yong Chiang Ee[1] and Wen Siang Lew[1]

[1] School of Physical and Mathematical Sciences, Nanyang,Technological University, Singapore
[2] Globalfoundries, Singapore
Email: andhita1@e.ntu.edu.sg (PAD), wensiang@ntu.edu.sg (WSL)

Abstract — **In this paper, conductance switching behavior of Pt/HfO_x/Ti redox-based memristive devices has been thoroughly investigated. The conduction mechanisms involved during the device operation can be associated with the trap-controlled SCL conduction mechanisms. The extracted parameters suggest the gradual switching behavior achieved by the devices was due to the transition between different trap level within trap-fill SCL regime, which is beneficial for the device implementation as analog synapse in the neuromorphic computing platform.**

Keywords: memristor, synapse, neuromorphic

I. INTRODUCTION

Redox-based memristive devices have emerged as one of the most promising candidates for analog synaptic element in an artificial neural network (ANN). However, the non-linear and asymmetric behaviour of the device conductance update hinder its hardware implementation [1]. While these challenges can be mitigated by write-verify scheme under ex-situ training, the same treatment is less practical for in-situ training approach. Most of the reported redox-based memristive devices exhibit abrupt switching (SET) from low (LCS) to high conductance state (HCS), while having gradual RESET. Material and programming engineering have been implemented to mitigate this issue [2].

In this study, gradual switching behaviour is demonstrated during SET and RESET operation of the Pt/HfO_x/Ti devices and the physical origin of this excellent device property is also extensively discussed.

II. EXPERIMENTAL METHOD

Cross-point structure with $10 \times 10 \ \mu m^2$ effective device area was fabricated using conventional UV-lithography. All metal layers, i.e., Ti and Pt, were deposited by DC sputtering of pure metallic targets with 50 W power and 2 mTorr operating pressure. On the other hand, the HfO_2 switching layer was deposited by RF-sputtering of HfO_2 ceramic target under 50 W power and 1.5 mTorr sputtering pressure. The electrical characterization of the devices was performed by using Keithley 4200A SCS Analyzer.

III. RESULTS AND DISCUSSION

The as fabricated devices exhibit stable bipolar switching behaviour under DC IV sweep, as depicted on the Fig. 1. The forming and subsequent SET operations were performed in the positive bias polarity under 200 μA compliance current (CC), while the RESET operations were carried out in the negative bias sweep up to -1.5 V. The device was also operated under optimized varying SET and RESET pulse amplitudes with constant 200 ns duration. Gradual conductance update, shown in Fig. 2., was successfully achieved during SET/RESET process. Excellent conductance state distribution was obtained, as shown in the Fig. 3. This type of behaviour has not been reported in any simple HfO_x/Ti memristor. Thus, the underlying physical mechanisms responsible for it was further investigated.

From the IV characteristic of the device, the conduction mechanism during the device operation can be thoroughly examined. The first key observation is the symmetrical IV behaviour observed under opposite bias polarity despite the large difference in work function of Pt and Ti electrode [3]. Theoretically, this configuration should result in significantly different current level under opposite bias polarity due to the decrease in barrier height at HfO_x/Ti interface. Thus, instead of the electrode-limited type of conduction, the device is driven by bulk-limited mechanism.

Under the double logarithmic plot (log I – log V), the device exhibits linear correlation with different slopes within different voltage regimes, as depicted on the Fig. 4. In the low current regime, ohmic conduction dominates the current flow via thermally activated electrons, followed by trap-unfill SCL governed by electrode-injected electrons that hop in between the oxide traps, and trap-fill SCL driven by traps filling below the quasi-Fermi energy level (follows eq. (1), (2), and (3) respectively). These conduction mechanisms have been reported on the same structure previously with ALD-grown HfO_2 [3]. However, the proposed switching mechanism is solely based on formation and rupture of filamentary defects, which cannot be used to explain the gradual switching behaviour of the device observed during SET process. Similar switching behaviour demonstrating multilevel conductance states with bipolar gradual switching characteristics has also been reported in Ag/LCMO/Pt heterostructure [4]. However, its underlying switching mechanism is ascribed to bulk-like limited effect due to traps-dominated conduction at the Ag/LCMO interface instead of electric field-induced defects modulation during the forming process.

The switching mechanism of HfO_x/Ti devices have been widely accepted based on the presence of electric field-induced oxygen vacancy defects within the dielectric. The trap-controlled SCL conduction and gradual conductance switching behaviour strongly suggest the switching dynamics based on the transition between different carrier

2020 IEEE Silicon Nanoelectronics Workshop

978-1-7281-9736-4/20 $31.00 © 2020 IEEE

trapping levels in the trap-fill SCL regime with oxygen vacancy defects serving as charge carrier traps after the initial forming process. The conductance change from LCS to HCS during SET process can be associated with the decrease in voltage exponent τ in the trap-fill space charge limited regime right before and after the switching process (from ~7.09 to ~3.20), while the RESET process corresponds to an increase in voltage exponent (from ~2.67 to ~7.08).

IV. CONCLUSION

Gradual conductance switching behaviour under varying pulse amplitudes during SET and RESET operation has been demonstrated in simple Pt/HfO$_x$/Ti structure. The underlying mechanisms behind this desired device property can be attributed to the transition between different carrier trapping levels in trap-fill SCL conduction regime. Different conductance switching operation of the device can be associated with an increase or decrease of the voltage exponent near the switching voltages. The understanding on the origin of this gradual switching characteristic of the device is highly beneficial for its implementation as an analog synaptic element to enable in-situ learning of ANN.

ACKNOWLEDGMENTS

The work was supported by the RIE2020 ASTAR AME IAF-ICP Grant (No. I1801E0030).

REFERENCES

[1] S. Yu, "Neuro-inspired computing with emerging nonvolatile memorys," *Proceedings of the IEEE,* vol. 106, no. 2, pp. 260-285, 2018.

[2] R. Islam *et al.*, "Device and materials requirements for neuromorphic computing," *Journal of Physics D: Applied Physics,* vol. 52, no. 11, p. 113001, 2019/01/18 2019.

[3] A. S. Sokolov *et al.*, "Influence of oxygen vacancies in ALD HfO$_{2-x}$ thin films on non-volatile resistive switching phenomena with a Ti/HfO$_{2-x}$/Pt structure," *Applied Surface Science,* vol. 434, pp. 822-830, 2018/03/15/ 2018.

[4] D. S. Shang, Q. Wang, L. D. Chen, R. Dong, X. M. Li, and W. Q. Zhang, "Effect of carrier trapping on the hysteretic current-voltage characteristics in Ag/La$_{0.7}$Ca$_{0.3}$MnO$_3$/Pt heterostructures," *Physical Review B,* vol. 73, no. 24, p. 245427, 06/22/ 2006.

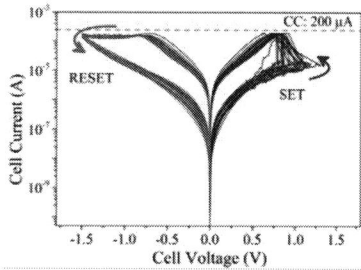

Fig. 1. The IV characteristics of first 25 cycles of the device under CC of 200 uA. The initial forming process was achieved at ~2.1 V

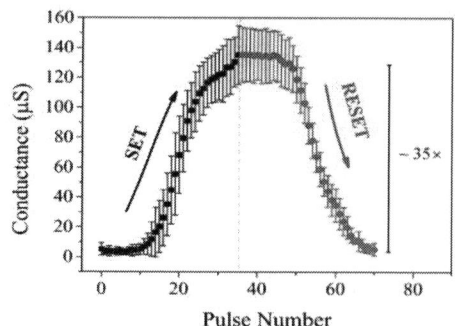

Fig. 2. Gradual switching characteristic of the device during SET and RESET operation under varying pulse amplitude.

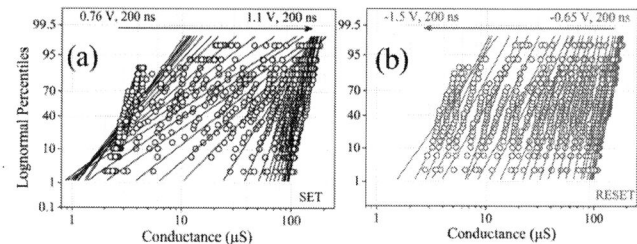

Fig. 3. Excellent conductance states distribution can be obtained under non-identical pulse programming scheme for both SET (a) and RESET (b) operation.

$$J_{Ohm} = \frac{q\mu_N N_0}{L}V, \qquad (1)$$

$$J_{unfill} = \frac{9\varepsilon_0\varepsilon_R\mu_N\theta}{8L^3}V^2, \qquad (2)$$

$$J_{fill} = \frac{q^{1-\tau}\mu_N n_c}{L^{2\tau+1}}\left(\frac{2\tau+1}{\tau+1}\right)^{\tau+1}\left(\frac{\tau}{\tau+1}\frac{\varepsilon_0\varepsilon_R}{n_t}\right)^{\tau}V^{\tau+1}, \qquad (3)$$

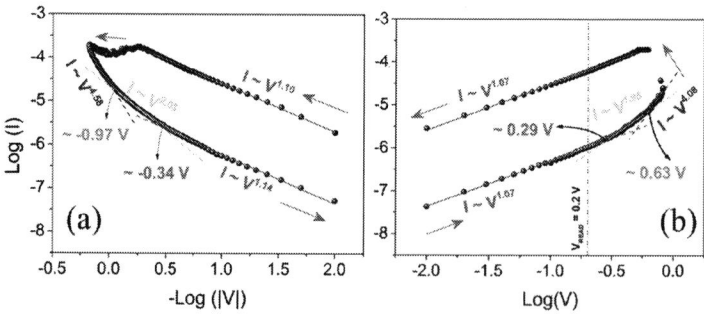

Fig. 4. Trap-controlled SCL conduction mechanism fits well with the experimental data. Different voltage exponent of the IV characteristic can be extracted under different voltage regime

Floating-gate transistor at cryogenic temperature:

Characterization and modelling of tunnelling and hot electrons injection

Michele Castriotta[1], Enrico Prati[2], Giorgio Ferrari[1],

[1]Dipartimento di Elettronica, Informazione e Bioingegneria, Politecnico di Milano, Italy
[2]Istituto di Fotonica e Nanotecnologie, Consiglio Nazionale delle Ricerche, Italy
Email: giorgio.ferrari@polimi.it

Abstract — **We demonstrate the cryogenic operation of a floating-gate device fabricated using an inexpensive commercial CMOS process. Device architecture and basic characteristics at 15K and at room temperature are presented. A fine tuning of the stored charge is obtained by combining channel hot electron injection and Fowler-Nordheim tunneling. This programmability is compatible with an operation of the proposed floating-gate as long-term accurate analog memory, as required by cryogenic neuromorphic systems or precise analog circuits operating at few Kelvin.**

Keywords: floating-gate transistor, analog memory, CMOS, cryogenics, quantum computer

I. INTRODUCTION

Solid state-based quantum information processing involving both silicon [1] and superconductive qubits calls for silicon integrated circuits for controlling gate operations [2][3]. In particular, quantum machine learning involves neuromorphic architecture logics [4] programmed on quantum computers at a software level, but it could take benefit of temporary memory storage for setting synapse weight at a hardware level in quantum neural networks. Silicon proves to fulfill all the requirement to assess neuromorphic architecture circuit [5], as floating gate memories can act of hardware permanent synapses, at room temperature [6]. In order to address silicon neuromorphic at cryogenic temperature to support hardware-based quantum machine learning, we explore commercial floating-gate devices at cryogenic temperature, so to extend the operation range to hybrid quantum-classical solid-state systems.

Floating-gate transistor is a type of MOSFET where the gate is electrically isolated from the surrounding, creating a floating gate node. A second gate, known as control gate, is capacitively coupled to the previous one in order to control and bias the transistor. The floating gate, surrounded by insulating material, enables the storage of electron charge to program the threshold voltage V_{Th} of the transistor. The control of the amount of the charge can be achieved by Fowler-Nordheim tunnelling [7] and hot electron injection [8]. In the former, a large electric field across the fine oxide of the transistor increases the quantum transparency of the potential barrier allowing electrons to tunnel through the oxide. In the latter, a large accelerated electric field in the transistor channel can create hot electrons which can gain enough energy to surmount the oxide barrier and increase the number of electrons in the floating gate.

Here we report the electrical characterization both at 300K and 15K of a floating gate fabricated in standard CMOS technology. The Fowler-Nordheim tunnelling and hot electron injection mechanisms allow to increase or decrease the amount of charge in the floating gate.

II. RESULTS AND MODEL

Figure 1 shows the schematic of a p-type floating-gate cell, realized in standard 350-nm CMOS technology by AMS. It includes a p-type MOS capacitor, M_C, for electron tunnelling and a p-type MOS transistor, M_T, for hot electron injection and threshold voltage, V_{Th}, assessment. The voltage of the floating gate, FG, node can be capacitively controlled by an input capacitance C_{CG}.

In order to show how the operations are carried by the device, Figure 2 displays the variation of V_{Th} of transistor M_T after an electron tunnelling event of duration equal to 100ms (dots) through the oxide of M_C both at room temperature and 15K. The tunnelling mechanism is activated by applying a pulse at the MOS capacitor M_C with an amplitude greater than 6.5V. Every 100ms the threshold voltage of M_T is measured by applying a voltage sweep to the control gate CG and then a new pulse is applied to M_C. The threshold voltage of M_T increases due to the reduction of the amount of electron charge in the floating gate node. We find that the model ordinarily used at room temperature holds at such cryogenic temperature as well. Indeed, the measurements are also compared with a model prediction based on Fowler-Nordheim current model (continuous lines) of equation [9]:

$$I_{Tun} = WLA \left(\frac{V_{Tun} - V_{FG}}{t_{ox}} \right)^2 e^{-B \frac{t_{ox}}{V_{Tun} - V_{FG}}}$$

(1)

where t_{ox} is the oxide thickness of M_C, WL is its area, V_{tun} the voltage applied to the terminal Tun and V_{FG} the voltage of the floating gate node. In the Eq. (1) A and B respectively act as two fitting parameters which depend on temperature. From these measurements we demonstrate the effectiveness of tunnelling also at low temperature, although at RT the current I_{Tun} is large due to a higher thermal energy of the electrons.

Next, we move our interest to hot electron injection. Figure 3 shows the decrease in threshold voltage, V_{Th}, (dots) of M_T after each impact-ionized hot electron injection pulse of duration equal to 100ms. Holes are accelerated in the channel of transistor M_T by the horizontal electric field. At the drain side they can generate electron-hole pairs. Therefore, the impact-ionized electrons can gain enough energy to surmount the oxide energy barrier and reach the floating gate node. Furthermore, in the same Figure is shown the comparison with a semi empirical model based on the equation [8][10][11]:

$$I_{Hot} = C(V_{SD} - \delta V_{ov})^3 I_{Drain} e^{-\frac{D}{(V_{SD} - \delta V_{ov})}}$$

where V_{ov}, V_{SD} and I_{Drain} are the overdrive voltage, source-drain voltage, and the drain current of transistor M_T, respectively. C, D and δ are fitting parameters which depend on temperature. The hot electron injection is a mechanism more effective at high temperature due to the higher electron thermal energy. Note that the threshold voltage of the transistor can be tuned with a resolution better than 1 mV. The excellent agreement between the model and the experimental data opens the possibility of precise simulations of circuits based on the proposed cryogenic floating gate.

To conclude, we demonstrate that commercial silicon CMOS technology can supply memory hardware for neuromorphic architectures at cryogenic temperature below 20 K and it is therefore compatible with liquid helium cryogenic equipment. Floating gate devices can be operated in cryostats where quantum hardware is routinely exploited, toward quantum artificial intelligence applications.

ACKNOWLEDGMENTS

This work was supported by QUASIX Grant from Italian Space Agency.

REFERENCES

[1] L. Petit *et al, Nature* **580**, 355, 2020.
[2] B. Patra *et al, IEEE Inter. Solid-State Circuits Conf,* 304, 2020.
[3] B. Patra *et al, IEEE J. Solid-State Circuits,* pp. 1–13, 2017.
[4] E. Prati, *Journal of Physics: Conference Series* **880**, 2017.
[5] E. Prati, *Int. J. of Nanotech.* **13**, 509, 2016.
[6] L. Danial et al., *Nat. Electron.* **2**, 596, 2019.
[7] M. Lenzlinger, *Journal of Applied Physics* **40**, 278, 1969.
[8] S. Tam *et al, IEEE Trans. on Electron Devices* **31**, 1116, 1984.
[9] A. Bhattacharyya *et al, Solid-state electronics* **27**, 899, 1984.
[10] T-C Ong *et al, IEEE Trans. on Electron Devices* **37**, 1658, 1990.
[11] Y. El-Mansy *et al, Inter. Electron Devices Meeting,* 11, 1975.

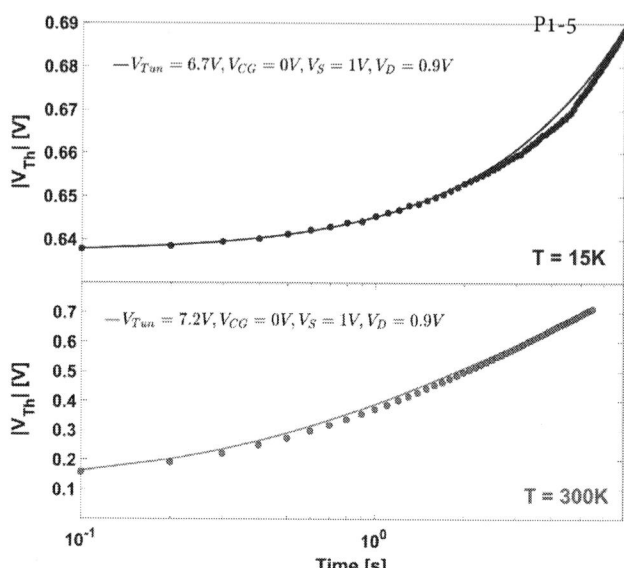

Fig.2: Threshold voltage V_{Th} evolution after each tunneling event measured (dots) and modeled (continuous line) with Eq.1 both at room temperature (bottom) and at 15K (top). The tunnelling events are obtained by applying the reported voltages for a duration of 100ms.

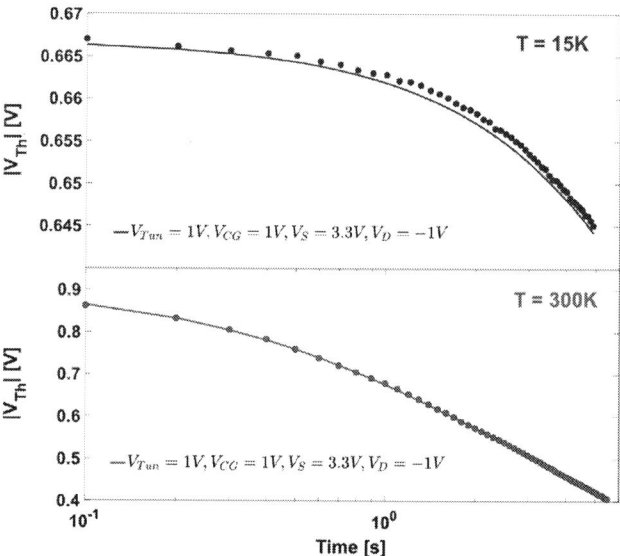

Fig.3: Threshold voltage V_{Th} evolution after each impact-ionized hot electron injection event measured (dots) and modeled (continuous line) with Eq.2 both at room temperature (bottom) and at 15K (top). The hot electron injection events are obtained by applying the reported voltages for a duration of 100ms.

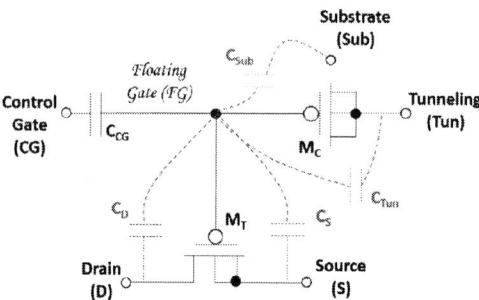

Fig.1: Schematics of a p-type FG cell realized in standard CMOS technology. Both p-MOS devices have a width of 400 nm and a length of 350 nm. The value of the control gate capacitor is 50fF.

Impact of Stopping Voltage and Hopping Conduction on the Oxygen Vacancy Concentration of Multi-Level HfO₂-Based Resistive Switching Devices

Desmond Jia Jun Loy[1,2], Putu Andhita Dananjaya[1], Somsubhra Chakrabarti[1], Kuan Hong Tan[1], Samuel Chen Wai Chow[1,2], Mun Yin Chee[1], Jia Rui Thong[1,2], Kunqi Hou[1], Jia Min Ang[1,2], Gerard Joseph Lim[1], Yong Chiang Ee[1], Eng Huat Toh[2], Wen Siang Lew[1]

[1]School of Physical and Mathematical Sciences, Nanyang Technological University 21 Nanyang Link, Singapore 637371, Singapore, [2]Globalfoundries Singapore Pte Ltd, 60 Woodlands Industrial Park D Street 2, Singapore 738406, Singapore
Email: DLOY002@E.NTU.EDU.SG (Desmond Jia Jun Loy), WENSIANG@NTU.EDU.SG (Wen Siang Lew)

Abstract — A multi-level state HfO₂-based resistive switching model is reported, where the increase in stopping voltage (V_{stop}) and thus activation energy (E_{AC}) is attributed to the depletion of oxygen vacancy (V_o) concentration (n_c) during reset. Hopping conduction fittings also indicated a depletion of n_c due to an increase of V_{stop} as shown by an increase of trap-to-trap distance (a) and trap energy (ϕ_T).

Keywords: RRAM, HfO₂, multi-level states, filament evolution, hopping conduction

I. INTRODUCTION

Resistive random access memory (RRAM) is one of the most promising non-volatile memories, owing to its simple metal-insulator-metal (MIM) structure, low power, high scalability and CMOS compatibility [1]. As the world becomes more data-centric, the search for faster and more compact memories becomes increasingly demanding. In this work, we investigated the multi-level states of Pt/HfO₂/Ti resistive switching devices by varying the reset stopping voltages (V_{stop}). We observed that the activation energy (E_{AC}) at various temperatures increases with increasing V_{stop} and decreasing oxygen vacancy (V_o) concentration (n_c). Therefore, a switching model was proposed while hopping conduction further validated this relationship with increasing trap-to-trap distance (a) and trap energy (ϕ_T).

II. EXPERIMENTAL METHOD

Cross-point 10 x 10 um² Pt/HfO₂/Ti resistive switching devices were fabricated using UV lithography. Pt and Ti were deposited using DC sputtering at 50W and 2mTorr while HfO₂ was deposited using RF at 50W and 1.5mTorr. The results were obtained using both 4200A Keithley and Keysight B1500A Semiconductor Parameter Analyzers.

III. RESULTS AND DISCUSSIONS

A consistent switching process was observed in **Fig. 1(a)** from the current-voltage (I-V) characteristics of Pt/HfO₂/Ti resistive switching devices and its box plots in **Fig. 1(b)**. The TEM analysis in **Fig. 2(a)** exhibited uniform deposited layers of the RRAM and the inset of **Fig. 2(a)** shows an amorphous HfO₂ structure. The Hf4f peaks indicated well separated spin-orbit components while the O1s peaks indicated broad and multiple overlapping components as shown in **Fig. 2(b) and 2(c)**.

Temperature studies were performed at -40°C, 25°C, 75°C and 125°C where the high resistance states (R_{HRS}) were extracted at 0.4V and fitted using an Arrhenius equation (eq.(1)), as shown **Fig. 3**. ln (R_{HRS}) is plotted against $1/T$ and R_0 was found to be 3036.44 and E_{AC} was found to be 0.047eV. The E_{AC} of HfO₂ was further investigated in its relationship with n_c, which is related to the V_{stop}. Multi-level state investigations were performed by varying the V_{stop} from -0.9V to -1.3V at intervals of -0.1V and the cumulative probability plot of the multi-level states were shown in **Fig. 4**. It was observed that the HRS increased at an increasing rate as V_{stop} increases, and hence an exponential relationship model was proposed in eq.(2). ln (R_{HRS}) was plotted against V_{stop} in **Fig. 5** where α was found to be 86.34 and β was found to be 5.42, which also relates E_{AC} and V_{stop} directly as shown in eq.(3) and eq.(4). As E_{AC} was reported to linearly decrease with n_c,[2] eq.(4) in turn shows that V_{stop} was also inversely related to n_c. This relationship was also shown by the proposed switching model in **Fig. 6**, explaining that a reduction in the n_c resulted in a higher R_{HRS}.

The electron migration at different V_{stop} of **Fig. 6** was further investigated using conduction mechanism studies. In the conduction mechanism fitting analysis, it was found that the hopping conduction (eq.(5)) was the dominant conduction mechanism for V_{stop} = -0.9V to -1.3V at intervals of -0.1V as shown in **Fig. 7** [3,4]. Parameters such as a and ϕ_T were extracted and exhibited in Table 1 from eq.(6) and eq.(7), as indicated by the gradient and vertical intercept respectively. These findings show that as V_{stop} increases, a and ϕ_T both increase as well. This suggests that as V_{stop} increases and n_c decreases, V_o are further apart, while a is larger and hence require a higher trap energy, ϕ_T to overcome the trap-to-trap barrier.

IV. CONCLUSION

In this work, a switching model has been proposed to describe the multi-level switching dependent on stopping voltage. The E_{AC} obtained from the Arrhenius plot was found to exhibit a linearly increasing relationship with V_{stop} and linearly decreasing relationship with n_c. As V_{stop} increases, the n_c decreases, which in turn resulted in a higher E_{AC}. In the hopping conduction fitting, the decrease in n_c also suggested that the V_o are further apart and resulted in a larger a and larger ϕ_T to be overcome as electrons hop from trap to trap, thus validating the relation between E_{AC} and V_{stop}.

ACKNOWLEDGMENTS

The work was supported by the RIE2020 ASTAR AME IAF-ICP Grant (No. I1801E0030).

2020 IEEE Silicon Nanoelectronics Workshop

978-1-7281-9736-4/20 $31.00 © 2020 IEEE

P1-6

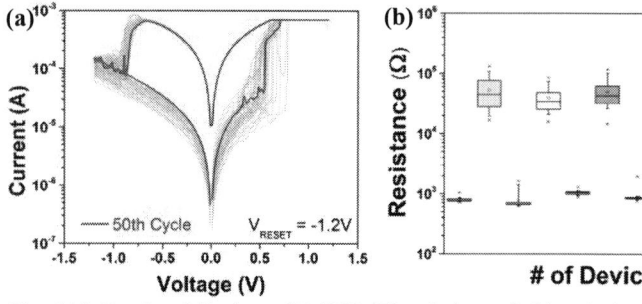

$$R_{HRS} = R_0 \exp\left(\frac{E_{AC}}{kT}\right) \quad (1)$$

$$R_{HRS} = \alpha \exp\left(\beta V_{stop}\right) \quad (2)$$

$$E_{AC} = \left[\ln(\alpha) - \ln(R_0)\right]kT + \beta V_{stop}kT \quad (3)$$

$$E_{AC} = \gamma + \phi V_{stop} \quad (4)$$

$$J = qanv \exp\left(\frac{qaE - \phi_T}{kT}\right) \quad (5)$$

$$M = \frac{qa}{kTd} \quad (6)$$

$$C = \ln(qanvA) - \frac{\phi_T}{kT} \quad (7)$$

Fig. 1(a). Semilog I-V plots of Pt/HfO₂/Ti resistive switching devices exhibiting the forming and first 50 I-V cycles with an ON/OFF ratio of about 50x, (b) resistance box-plot of Pt/HfO₂/Ti resistive switching devices exhibiting device-to-device variability.

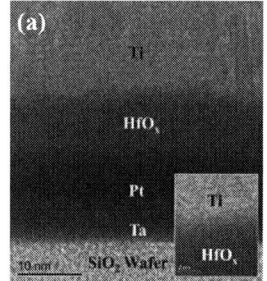

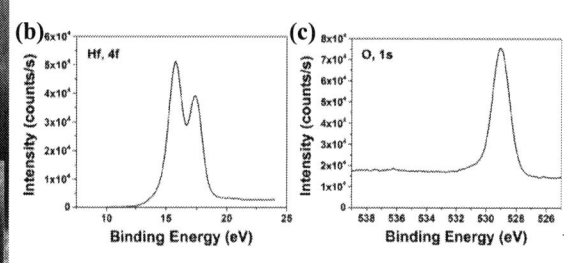

Fig. 2. (a) TEM of Pt/HfO₂/Ti resistive switching devices, (b) XPS of Hf4f and (c) XPS of O1s of HfOₓ films.

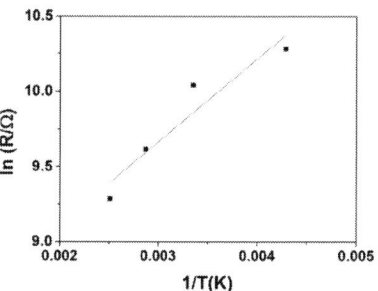

Fig. 3. An Arrhenius relationship, ln (R) vs. 1/T with a linear fit of the average I-V results of each temperature at 0.4V read.

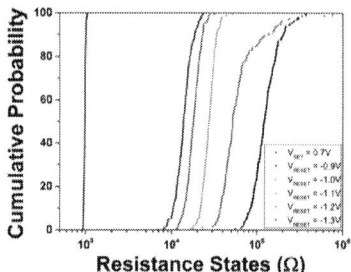

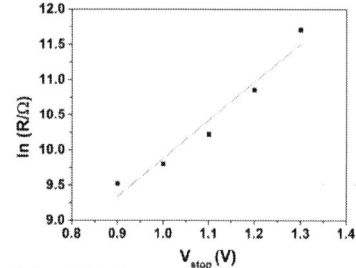

Fig. 4. Cumulative probability plots at V_read = -0.1V of 200 I-V cycles for V_stop = -0.9V to -1.3V at intervals of -0.1V.

Fig. 5. ln (R) vs. V_stop with a linear fit indicating a positive exponential relationship between R and V_stop.

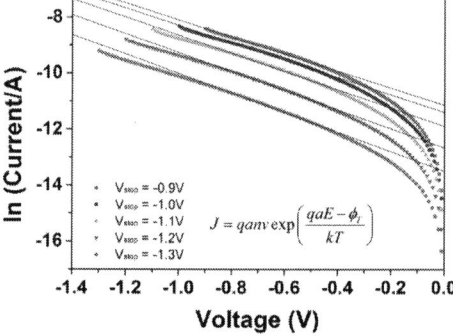

Fig. 7. Hopping conduction analysis of V_stop from -0.9V to -1.3V at intervals of -0.1V.

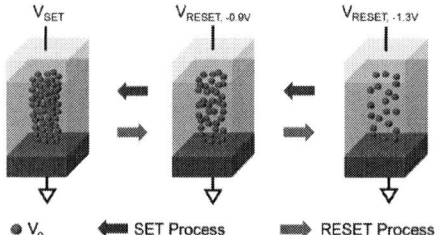

Fig. 6. A filament evolution model exhibiting nc decreases with higher V_stop as described by the SET and RESET processes.

TABLE I

PARAMETERS (a & ϕ_T) EXTRACTED FROM FIG. 10 SHOWN TO INCREASE AT INCREASING V_stop.

V_stop [V]	-0.9	-1.0	-1.1	-1.2	-1.3
Trap-to-trap Distance, a [nm]	0.782	0.797	0.834	0.871	0.893
Trap Energy, ϕ_T [eV]	0.459	0.465	0.476	0.495	0.515

REFERENCES

[1] E. Ambrosi, A. Bricalli, M. Laudato, and D. Ielmini, "Impact of oxide and electrode materials on the switching characteristics of oxide ReRAM devices," Faraday Discussions, vol. 213, no. 0, pp. 87-98, 2019.

[2] S. Larentis, F. Nardi, S. Balatti, D. C. Gilmer, and D. Ielmini, "Resistive Switching by Voltage-Driven Ion Migration in Bipolar RRAM—Part II: Modeling," IEEE Transactions on Electron Devices, vol. 59, no. 9, pp. 2468-2475, 2012.

[3] T. V. Perevalov. Vladimir A. Gritsenko, Damir R. Islamov, "Electronic properties of hafnium oxide: A contribution from defects and traps," Physics Reports, vol. 613, pp. 1-20, 2016.

[4] O. Dominguez, T. L. McGinnity, R. K. Roeder, and A. J. Hoffman, "Optical characterization of polar HfO₂ nanoparticles in the mid- and far-infrared," Applied Physics Letters, vol. 111, no. 1, p. 011101, 2017.

2020 IEEE Silicon Nanoelectronics Workshop

Ru Conducting Filament Based Cross-Point Resistive Switching Memory for Future Low Power Operation

Siddheswar Maikap[1,2] and Asim Senapati[1,2]

[1]Thin Film Nano Tech. Lab., Department of Electronics Engineering, Chang Gung University (CGU), Taoyuan, 33302,Taiwan, [2]Division of Gynecology-Oncology, Department of Obstetrics-Gynecology, Chang Gung Memorial Hospital (CGMH), Taoyuan, Taiwan 33302.
Email: sidhu@mail.cgu.edu.tw

Abstract — Here we have reported Ru conducting filament-based novel $Ru/Ta_2O_5/W$ cross-point resistive switching memory structure for future efficient data-storage application. The device can sustain $>10^3$ repeated DC endurance cycles at very low self current compliance (CC) of 100 μA. At low field region HRS current follows the Schottky conduction whereas at high field region HRS current follows the hopping conduction. The device exhibits long program / erase (P/E) cycle endurance of $>5 \times 10^8$ at very low programming current of 100 μA with very fast pulse width of 100 ns. The initial memory characteristics are very much promising, and the device could be potentially suitable for low power resistive-switching operation.

Keywords: Resistive switching, Ru, Cross-points, Conducting filament.

I. INTRODUCTION

In the era of artificial intelligence (AI) and internet of things (IoT), a huge number of data are being transmitted at a time [1]. Non-volatile memory technologies have been almost reached to its technological saturation in terms of energy efficiency and scalability [1]. Conducting-filament (CF) based memristor or resistive random-access memory (RRAM) is one of the most suitable candidates for the next generation non-volatile memory technologies as well as brain-inspired parallel computing architecture to fulfil the bottleneck [2]. The switching mechanism of the CF based memristors is basically rely on the reduction-oxidation (redox reaction) of top-electrode (TE) metal [3]. By applying a sufficient bias on the TE the filament formation and ruptured can be controlled. Although Cu, Ag, Co are the most well known TE, serve as the active electrodes, however recently Yoon et al have reported that Ru could be the potential candidate as the active electrode in terms of low power switching operation as under low current compliance conducting filament based RRAM usually suffers from the poor retention [4]. In this work, Ru CF based novel $Ru/Ta2O5/W$ cross-point memory device performances in 9 × 9 array have been evaluated.

II. EXPERIMENT

The cross-point RRAM device has been fabricated according to the following process. At first, Radio Corporation of America (RCA) cleaning was performed to clean the 4" p-type Si wafer. Then, using e-beam evaporation 200 nm SiO_2 layer was grown. Next, the bottom electrode (BE) bars were patterned by conventional photolithography. Using RF sputter, 10 nm titanium (Ti) as adhesive metal and 100 nm tungsten (W) as BE was deposited consecutively. Another photolithography step was performed to pattern TE which oriented at right angles to the BE. After that, 15 nm $Ta2O5$ as SM was deposited. 100 nm Ru as TE was deposited by RF sputtering. Finally lift-off was done to obtain the 9 × 9 cross-point devices with 10 μm × 10 μm device area. Fig. 1 shows the optical microscopic image of successfully developed cross-point devices in our university laboratory.

III. RESULTS AND DISCUSSION

The cross-point device shows forming-free bipolar resistive switching characteristics. Fig. 2 shows the typical I-V characteristics of the device under low self CC of 100 μA, following the voltage sweeping path of 1→4. The device can sustain more than 1000 repeated DC switching cycles without any degradation. To evaluate the conduction mechanism current conduction mechanisms are being analysed. Fig.3(a) confirms that at low field region Schottky emission dominates over the HRS current whereas at high field region Hopping conduction is responsible (Fig. 3(b)). The obtained Schottky barrier height is around 0.55 eV and hopping distance is about 0.80 nm. Ru active electrode based cross-point device exhibit very long and stable $>5 \times 10^8$ P/E endurance cycles at very low 100 μA programming current and very high-speed pulse width of 100 ns (Fig.4).

IV. EXPERIMENT

Ru CF base cross-point memory exhibit promising memory characteristics in terms of cycle endurance and data retention utilizing very low energy. This could be potentially capable to fulfil the energy efficient desired memory applications in near future.

ACKNOWLEDGMENTS

This work was supported by Ministry of Science and Technology (MOST), Taiwan under contract numbers: MOST-107-2221-E-182-041 and MOST-108-2221-E-182-026.

REFERENCES

[1] M. King, B. Zhu, and S. Tang, "Optimal path planning," *Mobile Robots*, vol. 8, no. 2, pp. 520-531, March 2001.
[2] H. Simpson, *Dumb Robots*, 3rd ed., Springfield: UOS Press, 2004, pp.6-9.

[3] J.-G. Lu, "Title of paper with only the first word capitalized," *J. Name Stand. Abbrev.*, in press.

[4] Y. Yorozu, M. Hirano, K. Oka, and Y. Tagawa, "Electron spectroscopy studies," *IEEE Translated J. Magn. Japan*, vol. 2, pp. 740-741, August 1987 [*Digest 9th Annual Conf. Magnetics Japan*, p. 301, 1982].

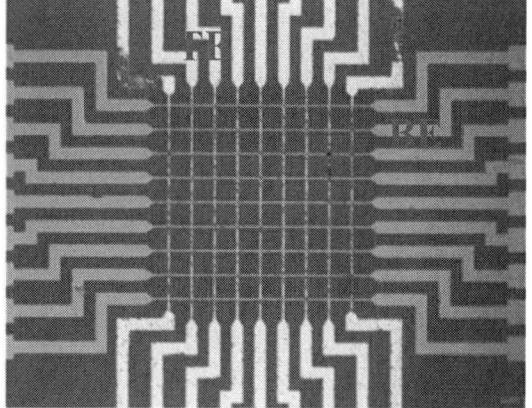

Fig. 1. OM image of beautifully developed 9 × 9 cross-point devices with the device size of 10 μm × 10 μm.

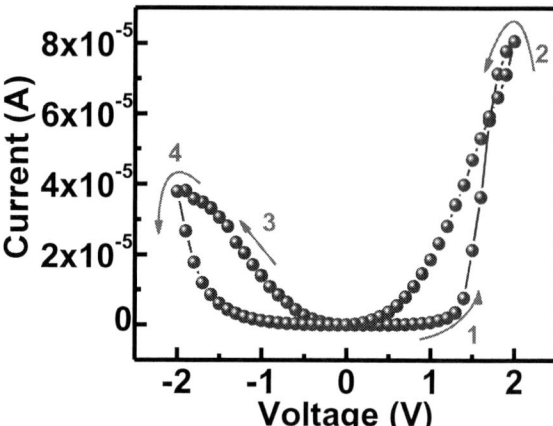

Fig. 2. Forming free I-V characteristics of the cross-point memory at low self CC of 100 μA .

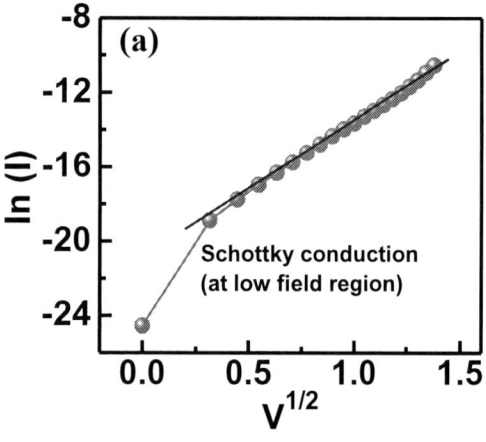

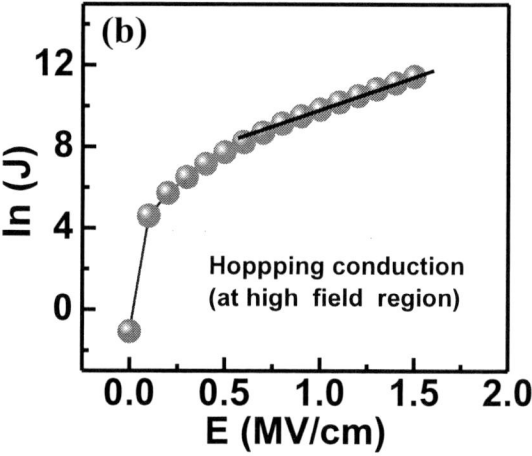

Fig. 3. Current conduction mechanism of HRS current. (a)Schottky conduction at low field region. (b) Hopping conduction at high field region.

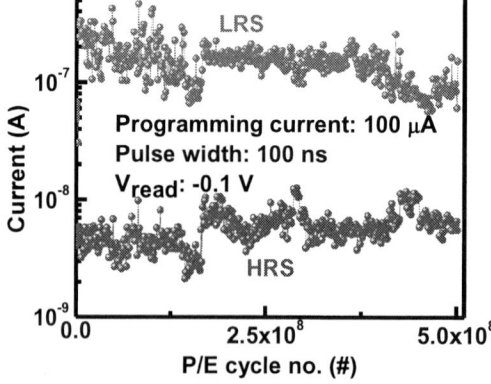

Fig. 4. More than 5 × 10^8 P/E endurance cycle endurance characteristic after 1000 DC cycle measurements at very low operation current of 100 μA and very high speed pulse width of 100 ns.

2020 IEEE Silicon Nanoelectronics Workshop

Influence of Gate to Drain Underlap on Negative Differencial Resistance in Ferroelectric FET

Kitae Lee, Sihyun Kim, and Byung-Gook Park

Inter-University Semiconductor Research Center (ISRC) and
Department of Electrical and Computer Engineering (ECE), Seoul National University, Korea
E-mail: bgpark@snu.ac.kr

Abstract — **The negative differential resistance in ferroelectric FET was investigated through TCAD device simulation. The gate to drain underlap forms depletion effectively during drain sweep, negative differential resistance is observed remarkably in the large gate to drain underlap.**

I. INTRODUCTION

Ferroelectric field effect transister (FeFET) has been researched for low power non-volatile memory devices [1]. FeFETs have distinct characteristic which is negative differential resistance (NDR) in output characteristics. In this work, we investigate influence of gate to drain underlap on NDR in FeFET using technology computer-aided design (TCAD) simulation. Through this study, the effective condition for observing NDR is confirmed.

II. SIMULATION METHODS

Synopsis Sentaurus^{TM} TCAD device simulator was used for 2-D n-type FeFET. The schematic diagram of FeFET and specific simulation parameters are presented in Fig. 1. The ferroelectric material parameters are selected based on experimental value, such as saturation polarization (P_s), remanent polarization (P_r), and coercive field (E_c) [2]. Especially, in this simulation, default gate to drain overlap length (L_{ov}) is 10 nm whereas L_{ov} is negative value in case of underlap condition.

III. SIMULATION RESULTS

The P-E curve of ferroelectric material used in this simulation is shown in Fig. 2. This P-E curve represents saturation loop which is extracted by metal-ferroelectric-metal (MFM) capacitor structure. In order to verify influence of gate to drain overlap/underlap condition, L_{ov}s are split from -10 to 10 nm. Fig. 3 shows the output characteristics of FeFETs with respect to L_{ov}. The NDR is confirmed in negative L_{ov} (gate to drain underlap condition) and disappears in positive L_{ov}. The ΔI_D ($I_{D,(VD = 1.5 V)} - I_{D,(VD = 0.5 V)}$) is presented in Fig. 4, which exhibits that gate to drain underlap range includes NDR range. In order to explicate the results, the P-E curve is extracted in drain-side ferroelectric layer during gate sweep for initializing polarization and drain sweep

(Fig. 5). In the FeFET, P-E curve is represented as minor loop because electric field across the ferroelectric layer is not sufficient for saturation of polarization. Because NDR resulting from transient negative capacitance (TNC) is presented in forward sweep with +P_s (low V_{th}) initialization, initializing gate voltage is set to 4 V [3]. Fig. 6 shows a magnified part of minor loop for drain sweep. In case of, TNC is observed in gate to drain underlap condition (L_{ov} = -10 nm), whereas not observed in overlap condition (L_{ov} = 10 nm). Because TNC considering MOS structure appears in depletion after accumulation or inversion, the underlap condition is more suitable for observing the TNC than the overlap condition [2].

IV. CONCLUSION

In this paper, the influence of gate to drain underlap on NDR characteristic in FeFET is explicated by using TCAD simulation. Since NDR is affected by depletion in substrate, the large underlap length is effective variable to observe NDR phenomenon in output characteristics.

ACKNOWLEDGEMENT

This work was supported in part by The Brain Korea 21 Plus Project in 2020, in part by the Future Semiconductor Device Technology Development Program (10067739) funded by Ministry of Trade, Industry and Energy (MOTIE) and Korea Semiconductor Research Consortium (KSRC), in part by Synopsys Inc.

REFERENCES

[1] K. Ni, P. Sharma, J. Zhang, M. Jerry, J. A. Smith, K. Tapily, R. Clark, S. Mahapatra, and S. Datta, "Critical Role of Interlayer in Hf0.5Zr0.5O2 Ferroelectric FET Nonvolatile Memory Performance," *IEEE Trans. Electron Devices*, vol. 65, no. 6, pp. 2461-2469, Apr. 2018.

[2] K. Lee, J. Lee, S. Kim, R. Lee, S. Kim, M. Kim, J. -H. Lee, S. Kim, and B. -G. Park, "Negative capacitance effect on MOS structure: Influence of electric field variation," *IEEE Trans. Nanotechnology*, vol. 19, pp. 168-171, Feb. 2020.

[3] C. Jin, T. Saraya, T. Hiramoto, and M. Kobayashi, "Transient Negative Capacitance as Cause of Reverse Drain-induced Barrier Lowering and Negative Differential Resistance in Ferroelectric FETs," in *2019 Symposium on VLSI Technology*, pp. T220-T221, Jun. 2019.

P2-1

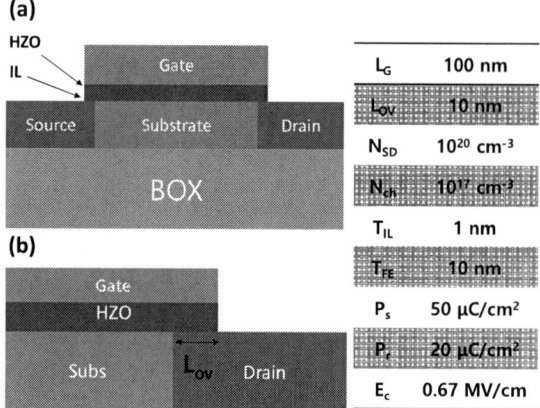

Fig. 1. (a) The overall and (b) the drain-side schematic diagram of simulated FeFET and specific parameters.

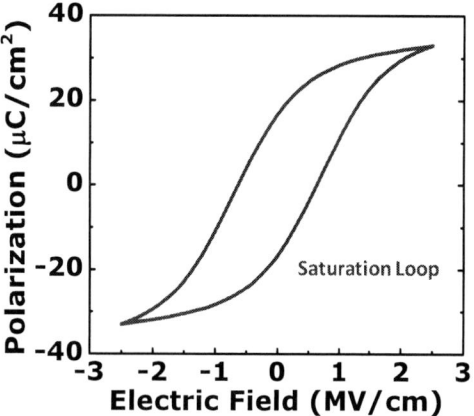

Fig. 2. *P-E* curve of ferroelectric material used in simulation. Saturation loop is extracted by MFM capacitor.

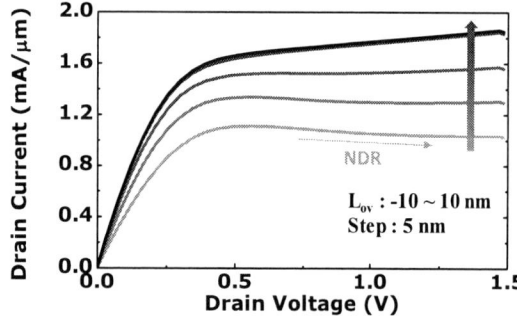

Fig. 3. The output characteristics of FeFETs with respect to gate to drain overlap length. NDR occurs in underlap condition.

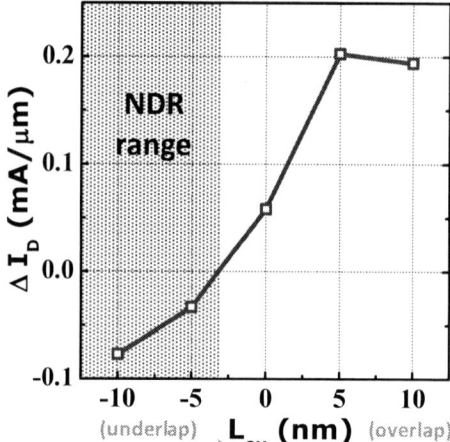

Fig. 4. ΔI_D ($I_{D, (VD = 1.5 V)} - I_{D, (VD = 0.5 V)}$) with respect to gate to drain overlap length.

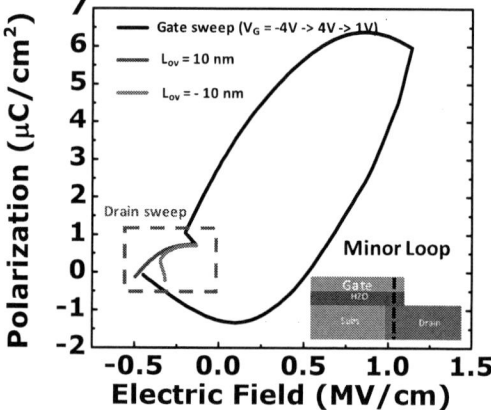

Fig. 5. Minor loop of *P-E* curve in drain-side ferroelectric layer for gate sweep (black line) and drain sweep (red and blue line).

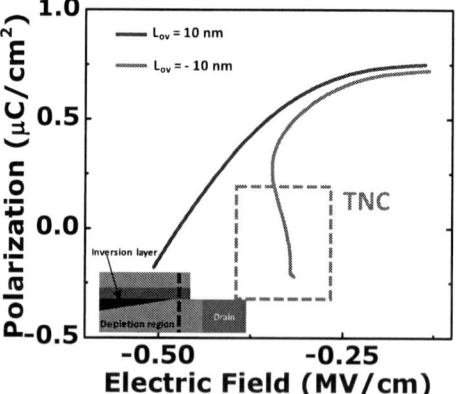

Fig. 5. A magnified part of minor loop for drain sweep. In case of underlap, depletion under drain-side ferroelectric layer results in TNC.

2020 IEEE Silicon Nanoelectronics Workshop

Gap in pagination due to formatting issues.

Pages 97-98

Study on Etch Slope in Fin and Source/Drain Etch Process of Vertically-Stacked Nanosheet Gate-All-Around MOSFET

Sihyun Kim, Kitae Lee, and Byung-Gook Park

Inter-University Semiconductor Research Center (ISRC) and
Department of Electrical and Computer Engineering (ECE), Seoul National University, Korea
E-mail: bgpark@snu.ac.kr

Abstract — **The electrical behavior of vertically-stacked nanosheet (NS) gate-all-around MOSFET (GAAFET) having slanted NS channel and source/drain resulted from the etch profile in reactive ion etching processes was investigated through TCAD device simulation. It was observed that the off current (I_{OFF}) and threshold voltage (V_{TH}) were variable depending on the etch slope (*E/S*) and the number of stack (n_{stack}).**

I. INTRODUCTION

Vertically-stacked nanosheet (NS) gate-all-around MOSFET (GAAFET) has been regarded as a promising candidate for FinFET in sub-5 nm technology node thanks to its superior electrostatic gate controllability and compatibility to the conventional FinFET fabrication [1-3]. One of the key issues in stacked GAAFET is the layer-to-layer variation resulting from the etch profiles. Due to the etch slope (*E/S*) of reactive ion etching (RIE) processes including Fin and source/drain (SD) etching, the nanosheet width (W_{NS}) and the effective channel length (L_{eff}) can be variable depending on the channel location. In this paper, the effect of etch profile in RIE processes on the electrical characteristics of stacked NS GAAFET was analysed through TCAD device simulation.

II. METHODS

Fig. 1 summarizes the process flow of the stacked NS GAAFET, which represents *E/S* in the Fin and SD RIE etching. To confirm the etch profile in Si RIE, photolithography and RIE were conducted on the bulk Si wafer [scanning electron microscopy (SEM) images in Fig. 2]. To reflect the etch profile in RIE processes, 3D structure of stacked NS GAAFET having slanted shaped NS channels and SD was designed as described in Fig. 3(a). The physical parameters are summarized in Table. I, which has been referred to IRDS 2018 [4]. The cross-sectional images across the channel and the gate are shown in Fig. 2(b), exhibiting the layer-to-layer variation of W_{NS} and the L_{eff}. The W_{NS} and the L_{eff} of lower stacks are larger than top channel whose W_{NS} and L_{eff} are equal to the designed size. To investigate the effect of etch profile in Fin and SD etching on the electrical characteristics of stacked NS GAAFET,

Synopsis Sentaurus™ TCAD device simulator was used.

III. RESULTS

The transfer characteristics of GAAFETs depending on the *E/S* (82-90°) and the number of stack (n_{stack} = 1-5) are presented in Fig. 4 (a) and (b), respectively. Off current (I_{OFF}) decreases as *E/S* decreases since the L_{eff}s of lower stacks are increased, resulting in the reduction of junction leakage current. Nevertheless, the on current (I_{ON}) loss was not noticeable, since the reduction of I_{ON} was almost compensated by W_{NS} increase. In addition, I_{OFF} (per n_{stack}) reduction in GAAFETs of larger n_{stacks} were observed in Fig. 4(b), which can be explained as same. For detailed analyses, I_{OFF} and threshold voltage (V_{TH}) of GAAFETs according to *E/S* and n_{stack} were extracted [Figs. 5(a)-(d)]. Here, the V_{TH} was extracted by constant current method [$V_{TH} = V_{GS}$ @ I_{DS} = (100 nA) $\times$ (L_{eff}/W_{eff}), where W_{eff} represents an effective channel width]. The I_{OFF} decreased by about 46% at *E/S* = 82° [Fig. 4(a)] and 40% n_{stack} = 5 [Fig. 4(b)]. On the other hand, the V_{TH} increased by about 3.7% due to the I_{OFF} reduction as *E/S* decreases and n_{stack} increases as shown in Figs. 4 (c) and (d).

IV. CONCLUSION

The electrical characteristics of vertically stacked NS GAAFET having slanted NS channels and SD profile resulted from the *E/S* in Fin and SD etching were investigated. It was confirmed that the I_{OFF} and the V_{TH} were variable depending on the *E/S* and n_{stack}, which should be considered in compact modeling the stacked GAAFETs.

ACKNOWLEDGEMENT

This work was supported by the Brain Korea 21 Project in 2019, in part by the Future Semiconductor Device Technology Development Program (10067739) funded by Ministry of Trade, Industry and Energy (MOTIE) and in part by Synopsys Inc.

REFERENCES

[1] D. Yakimets, G. Eneman, P. Schuddinck, B. Trong Huynh, M. G. Bardon, P. Raghavan, A. Veloso, N. Collaert, A. Mercha, D. Verkest, A. Voon-Yew Thean, and K. De Meyer, "Vertical GAAFETs for the Ultimate CMOS Scaling," *IEEE Trans. Electron Devices*, vol. 62, no. 5, pp. 1433-1439, 2015.

2020 IEEE Silicon Nanoelectronics Workshop

[2] N. Loubet, T. Hook, P. Montanini, C.-W. Yeung, S. Kanakasabapathy, M. Guillom, T. Yamashita, J. Zhang, X. Miao, and J. Wang, "Stacked nanosheet gate-all-around transistor to enable scaling beyond FinFET," in *Proc. Symp. VLSI Technol.*, 2017.

[3] H. Mertens, R. Ritzenthaler, A. Hikavyy, M.-S. Kim, Z. Tao, K. Wostyn, S. A. Chew, A. De Keersgieter, G. Mannaert, and E. Rosseel, "Gate-all-around MOSFETs based on vertically stacked horizontal Si nanowires in a replacement metal gate process on bulk Si substrates," in *Proc. Symp. VLSI Technol.*, 2016.

[4] (2018). *International Roadmap for Devices and Systems (IRDS)*. Accessed: Feb. 2020. [Online]. Available: https://irds.ieee.org/editions/2018

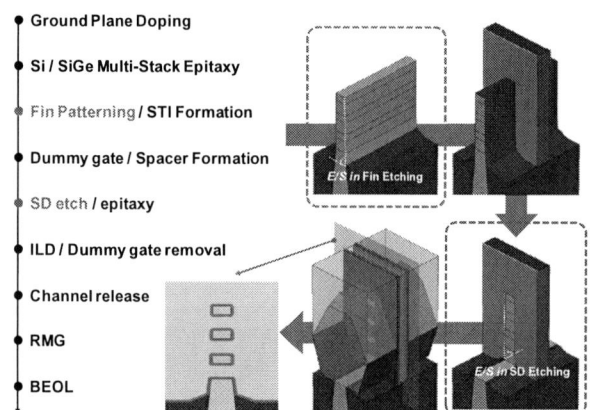

Fig. 1. Process flow of vertically stacked NS GAAFET.

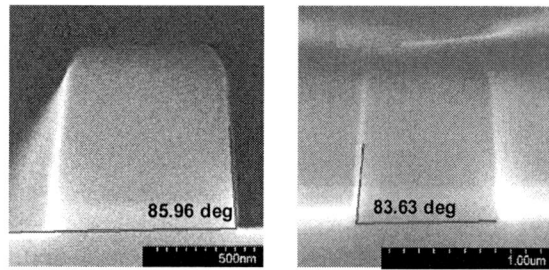

Fig. 2. (a) SEM image presenting *E/S* in silicon RIE.

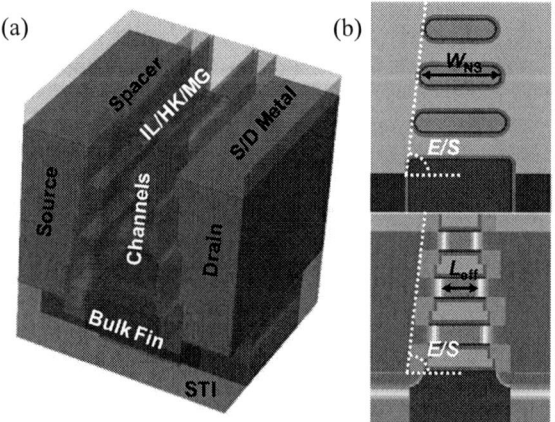

Fig. 3. (a) 3D and (b) cross-sectional image of GAA FET having slanted Fin and SD etch profile.

TABLE I. Physical Parameters for NS GAAFET

Parameter	Value	Parameter	Value
L_G (top)	14 nm	N_{SD}	1×10^{20} cm^{-3}
W_{NS} (top)	25 nm	N_{CH}	5×10^{16} cm^{-3}
t_{NS}	7 nm	SD overlap	20 %
L_{Spacer}	6 nm	*E/S*	82 - 90°
EOT	1 nm	n_{stack}	1 - 5

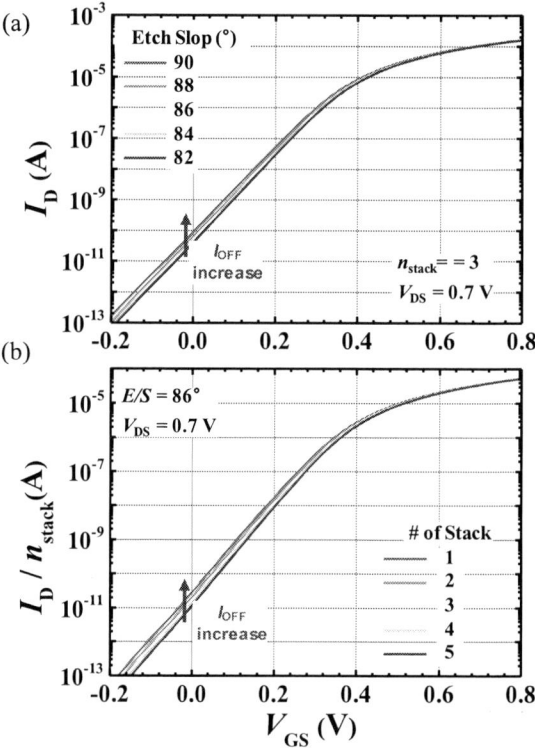

Fig. 4. Transfer curves of GAAFETs according to (a) *E/S* (@n_{stack} = 3) and (b) n_{stack} (@*E/S* = 86°).

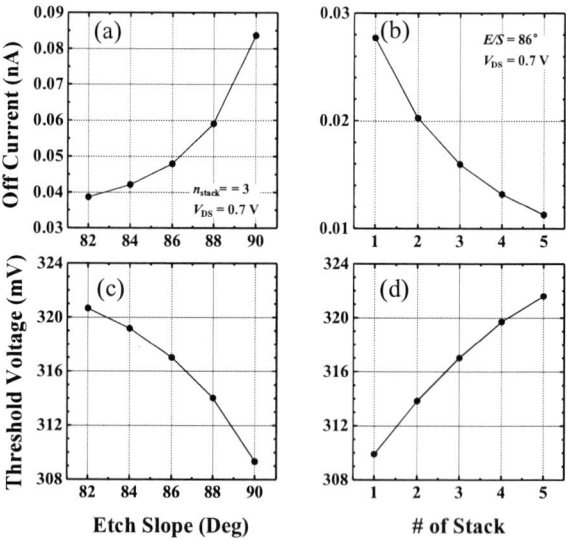

Fig. 5. I_{OFF} as a function of (a) *E/S* and (b) n_{stack}. V_{TH} as a function of (c) *E/S* and (d) n_{stack}.

2020 IEEE Silicon Nanoelectronics Workshop

Reliability analysis of P-type SOI FinFETs with multiple SiGe channels on the degradation of NBTI

Tzuting Cho, Renrong Liang, Guofang Yu, and Jun Xu

Department of Microelectronics and Nanoelectronics, Tsinghua University, China

Email: liangrr@mail.tsinghua.edu.cn and junxu@tsinghua.edu.cn

Abstract —SOI FinFETs are a multi-gate structure that can not only reduce the leakage current and improve the control of the device, but also inhibit the outflow of heat. Heat dissipation is an important issue for small chips, because high temperatures can let them unstable and even cause them to fail. In this paper, P-type multi-gate SOI FinFETs are used to analyze the reliability of negative bias temperature instability (NBTI) under room temperature. The results indicate that the estimated voltage for 10 years of lifetime is 0.0895V, and the threshold voltage (V_{TH}) has a high dependence on the stress time. Some researchers have found that one cause of device degradation is the presence of interface defects (N_{it}) and oxide defects (N_{ot}). The appearance of N_{it} and N_{ot} caused by NBTI can be explained by the reaction-diffusion model (R-D model) [1]. In this experiment, the defects are successfully extracted by linear extrapolated I_D to the midgap current (I_{mg}). The results show that both N_{it} and N_{ot} exist in the oxide layer, and N_{ot} is the major factor affecting degradation at room temperature.

Keywords: NBTI, SOI FinFETs, N_{ot}, N_{it}

I. INTRODUCTION

The characteristic's stability of SOI FinFETs has been the issue of many scientific researches [2]. The stability of thin oxide layer SOI FinFETs devices will be less stable as the device temperature increases. Scientists use NBTI to estimate the reliability of the device. Bias temperature instability (BTI) degradation of MOSFETs parameters is an important reliability issue [3]. Although scientists have carried out extensive research, the exact degradation mechanism of the device is still not conclusively confirmed. In addition, most of the past research on NBTI has focused on planar P-MOSFETs devices. This paper will investigate the characteristic degradation of P-type multi-gate SOI FinFETs under NBTI effect.

II. PROCESS AND EXPERIMENT

The classical structure of P-type SiGe–channels junctionless SOI FinFETs is shown in **Fig. 1a**, which has multiple parallel channels to enhance its operation capability. **Fig. 1b** shows the cross sectional transmission electron microscope (TEM) image of the measured device whose cross sectional area of single Fin is about 37 nm x 35nm. And the designed length of the channel is about 30 nm. Due to the tri-gate structure of FinFETs, the ratio of width and length (W/L) need to be calculated by eq.(1). The effective channel width is 440 nm, so the W/L is about 440/30. In this work, DC characteristics are measured by Agilent B1500A Semiconductor Parameter Analyzer. The

test condition is under the gate stress voltage of -1V, -1.5V and -2V, the drain voltage is -0.5V and the stress time is 0s, 5s, 10s, 50s, 100s, 500s and 1000s.

III. RESULTS AND DISCUSSION

Fig. 2a-2c show the time dependence of different voltage parameters of P-SOI FinFETs. It is observed that V_{TH} is most sensitive to stress time. In addition, **Fig. 3a-3c** show the trend of device degradation. In this paper, using 10% as the degradation standard and extracting the V_{TH} to gain the lifetime trend chart for the 10 years voltage prediction. The E_{ox} model (electric field in the oxide model) [4] and the power law model are used for lifetime analysis, as shown in **Fig. 3d**. The predicted voltage of the E_{ox} model is -2.51V and that of the power law model is 0.0895V. The result shows that the power law model can predict the lifetime of device under low voltage, so the result of device lifetime is close to the reality [5].

Effective defect density (ΔN_{th}) can be calculated by the eq.(2) C_{ox} is the oxide capacitance and q is the amount of charge. In addition, the sub-threshold swing (SS) method is used to calculate defect variation. The effective interface defects density (ΔN_{it}) can be calculated by the eq.(3). Where $E_{it}=KT/q\ln(N_B/n_i)$ is energy rage in which the active interface states distributed. NBTI introduces both the interface defects and oxide defects, which causes the shift of the threshold voltage. By extracting midgap voltage (V_{mg}) into the midgap current (I_{mg}) [6] by linear extrapolation I_D, the subthreshold current can be calculated using the surface band bending eq.(5). In addition, the gate voltage displace at the midgap point (V_{mg} at $\varphi_s = \varphi_F$), where the contribution of N_{it} is negligible [7]. φ_s is silicon surface potential, φ_F is Fermi potential and L_D is the Debye length, as shown in **Fig. 4a**. Therefore, the value of N_{ot} can be calculated according to eq.(4). **Fig. 4b** is the relation diagram of defects concentration and stress time according to the calculated results. It is found that N_{it} and N_{ot} exist, and N_{ot} is the major factor affecting degradation at room temperature.

IV. CONCLUSION

In summary, the experimental results show that the P-type junctionless SOI FinFETs with multi-gate SiGe-channels will degrade the device characteristics under the influence of NBTI, especially the effect on threshold voltage. The N_{ot} is the main effect of degradation at room temperature. In addition, the power law model is more realistic in predicting the electrical properties of devices over a 10-year lifetime at low voltage.

2020 IEEE Silicon Nanoelectronics Workshop

ACKNOWLEDGMENTS

This work was supported by the National Key Research and Development Program of China under Grant 2016YFA0200400 and Grant 2016YFA0302300, the National Science and Technology Major Project of China under Grant 2016ZX02301001, the National Natural Science Foundation of China under Grant 61306105.

REFERENCES

[1] S.Chakravarthi, A. T.Krishnan, V.Reddy, and S.Krishnan, "Probing negative bias temperature instability using a continuum numerical framework: Physics to real world operation,"Microelectron. Reliab, vol. 47, no. 6, pp. 863–872, 2007.

[2] S. H.Shin, S. H.Kim, S.Kim, H.Wu, P. D.Ye, and M. A.Alam, "Substrate and layout engineering to suppress self-heating in floating body transistors,"Tech. Dig. - Int. Electron Devices Meet. IEDM, no. 765, pp. 15.7.1-15.7.4, 2017.

[3] S.Mahapatra and M. A.Alam, "A predictive reliability model for PMOS bias temperature degradation," Tech. Dig. - Int. Electron Devices Meet, pp. 505–508, 2002.

[4] J. W.McPherson andH. C.Mogul, "Underlying physics of the thermochemical e model in describing low-field time-dependent dielectric breakdown in SiO2 thin films," J. Appl. Phys., vol. 84, no. 3, pp. 1513–1523, 1998.

[5] H.Aono et al., "Modeling of NBTI saturation effect and its impact on electric field dependence of the lifetime," Microelectron. Reliab, vol. 45, no. 7–8, pp. 1109–1114, 2005.

[6] P.J.McWhorter and P. S.Winokur, "Simple technique for separating the effects of interface traps and trapped-oxide charge in metal-oxide-semiconductor transistors,"Appl. Phys. Lett., vol. 48, no. 2, pp. 133–135, 1986.

[7] N.Silicon-oxynitride and Y.Wang, "Effects of Interface States and Positive Charges on," vol. 8, no. 1, pp. 14–21, 2008.

$$\frac{W}{L} = n \times \frac{(2H_{fin} + W_{fin})}{L} \qquad (1)$$

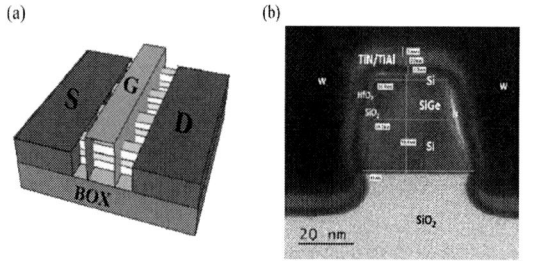

Figure 1. (a) The classical structure of the SOI FinFETs. (b) The TEM of measured device.

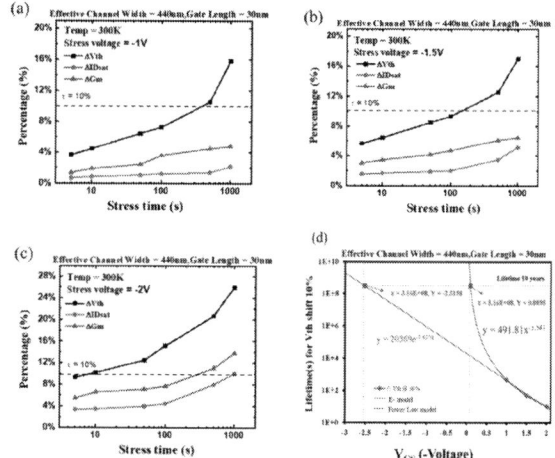

Figure 3. (a), (b) and (c) The degradation percentage of device parameters at stress voltages of -1V, -1.5V and -2V. (d) The trend chart of V_TH prediction for 10 years of device lifetime.

$$\Delta Nth \approx \frac{\Delta V_{TH} C_{ox}}{q} \qquad (2)$$

$$\Delta Nit \approx \frac{\Delta SS C_{ox} Eit}{2.3KT} \qquad (3)$$

$$\Delta Not \approx \frac{\Delta Img C_{ox}}{q} \qquad (4)$$

$$I_{mg} = \sqrt{2}\mu \frac{W}{2L} \frac{qN_A}{\beta} L_D \left(\frac{ni}{N_A}\right)^2 \exp(\beta\varphi_s)(\beta\varphi_s)^{-\frac{1}{2}} \qquad (5)$$

$$\varphi_s = \varphi_F \,, \quad \varphi_F = \frac{KT}{q}\ln(\frac{ni}{N_D})$$

$$\beta = \frac{q}{KT} \,, \quad L_D = (\frac{\varepsilon_r \varepsilon_0}{\beta q N_A})^{\frac{1}{2}}$$

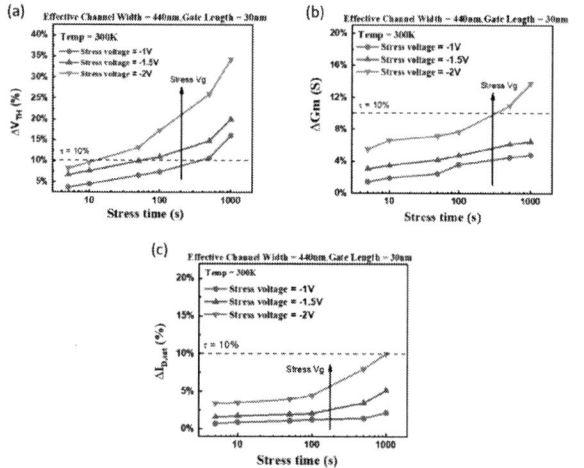

Figure 2. Variation Percentage versus time for (a) threshold voltages, ΔV_{TH}, (b) transconductance, ΔGm, and (c) saturation current, $\Delta I_{D,sat}$ of the device.

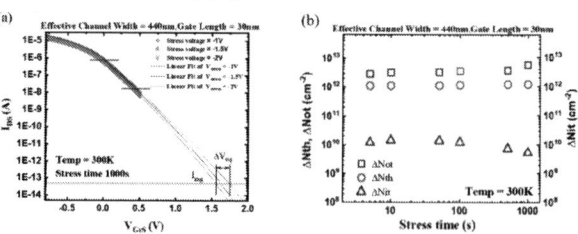

Figure 4. In figure (a), midgap voltages (ΔV_{mg}) are calculated by extrapolating the midgap current (ΔI_{mg}) equation. In figure (b) is a comparison of defects concentration at room temperature.

2020 IEEE Silicon Nanoelectronics Workshop

Impacts of Biaxial Tensile Strain in Double-gate Tunneling Field-effect-transistor (DG-TFET) with a Monolayer WSe₂ Channel

Qianwen Wang, Pengpeng Sang, Yuan Li, Jiezhi Chen*

School of Information Science and Engineering, Shandong University, Qingdao, P. R. China

*Email: chen.jiezhi@sdu.edu.cn

1. Abstract

In this work, biaxial tensile strain effects are systematically investigated in double-gate tunneling field-effect-transistor (DG-TFET) with a monolayer WSe₂ channel. Results of the first-principles calculations show that decreased bandgap and reduced effective masses can be obtained when the biaxial tensile strain is applied to WSe₂ channel. However, in terms of the strain-induced channel-current modulations, it is found that the strain should be well optimized for the largest performance enhancement. Based on our calculations, it is shown that, under 2% biaxial tensile strain, the ON current and transconductance can be improved by a factor of 8.05 and 8.9, respectively.

2. Introduction

In the last few years, with the rapid development of process and device technology, ultra-scaled low power field-effect transistors (FETs) have drawn increasing attention. In particular, tunneling field-effect transistors (TFETs) have been widely studied [1,2] due to the large On-Off current ratio and steep subthreshold swing (SS). Owing to the smooth surfaces and diverse electrical properties, two-dimensional (2D) materials, especially monolayer transition-metal dichalcogenides (TMDs), such as WSe₂, have garnered great interest in device applications [3-5]. WSe₂ is used as the channel materials because of its good electrical conductivity and perfect interface without dangling bonds and trap state. Nevertheless, it is difficult to achieve high ON current and On-Off current ratio for WSe₂-based TFETs at the same time, which still remains a critical issue to be solved. It was previously demonstrated that atomic-defect can assist carrier tunneling [7] and enhance the channel tunneling current.

Here, *ab initio* simulations are carried out to investigated the impacts of biaxial tensile strain on monolayer WSe₂ double-gate TFET (DG-TFET). Compared with unstrained DG-TFET, there are smaller bandgap and effective mass under biaxial tensile strain. SS is reduced to 63.7mV/dec with 5% of strain, but the ON current (I_{on}) and the transconductance (g_{max}) reach their highest values with 2% strain. In simple words, the strain should be well optimized for better device performance.

3. Device Structure and Calculation Approach

The device simulations are performed with the Atomistic Toolkit (ATK), by using the density functional theory (DFT) in combination with the nonequilibrium Green's function (NEGF). The local density approximation (LDA) is employed as the exchange correlation potential. As shown in Fig.1(b), the WSe₂ monolayer DG-TFET simulation is set up with 6.1nm channel length (L_{ch}), 0.5nm equivalent oxide thickness (EOT) and 0.72V supply voltage (V_{ds} = 0.72 V). The real space grid techniques are used with an energy cutoff of 150Ry in the numerical integrations. On the one hand, the intrinsic monolayer WSe₂ is employed as the channel materials and the p-doped/n-doped as the source/drain. On the other hand, to keep the carrier concentration and reduce the influence of tensile strain, strains with different concentrations by maintaining the energy gap between the channel CBM and Fermi level. The geometry of monolayer WSe₂ DG-TFET is optimized until the residual force was less than 0.05eV/ Å on each atom. This optimized WSe₂ monolayer structure is in good agreement with that of previous publications [6]. The flowchart of our calculations is shown in Fig.2.

4. Results and Discussions

Fig.3(a) shows the atomic structure of WSe₂ monolayer, in which the smallest rectangular structure used for biaxial tensile strain calculation, is within the solid black line. The band gaps of WSe₂ monolayer with unstrained, 3% and 5% biaxial tensile strained shown in Fig.3 is 1.64eV, 1.17eV and 0.79eV, respectively. The WSe₂ DG-TFET presents a reducing trend in both bandgap and electron effective mass of CBM (Fig.4(a)) with increasing biaxial tensile strain, as shown in Fig.4(b). Therefore, it is speculated that the currents of WSe₂ channel can be modulated by strain engineering. To verify this, the transfer characteristics and transconductance of the monolayer WSe₂ DG-TFET for various biaxial tensile strain were calculated and the results have been summarized in Fig.5. As shown in Fig.5(a), there exists an obvious shift in threshold voltage. The I-V curves were calculated from 0% to 5% of strains. With increasing strains, the threshold voltage shifts negatively from 1.45eV to 0.72eV, respectively (Fig.6(a)). On the other hand, SS describes the switching behavior of FETs in the subthreshold region, which is another key indicator of device performance. The smaller SS, the better gate controllability. Fig.5(a) presents that when the strain is up to 5%, SS is reduced to 63.7mV/dec.

To understand the effects of biaxial tensile strain on I_{on} and transconductance, we analyzed the g_{max} and tunneling barrier width with projected local density of state (PLDOS). High g_{max} guarantees a sharp switching of the transistor from off-state to on-state, which is defined as $g_m = \partial I_d / \partial V_G$ (Fig.5(b)). In terms of the channel-current modulations of WSe₂ DG-TFET, it is found that both the I_{on} and g_{max} can be improved effectively. As shown in Fig.6, I_{on} and g_{max} with 2% biaxial tensile strain are respectively 8.05 and 8.9 times larger than those of unstrained DG-TFET. Moreover, the carrier mobility is also enhanced at the same time. As shown in Fig.7, g_{max} increases with larger biaxial tensile strain due to the ultra-low tunneling barrier width from 0% to 2% of strain. Owing to the increased width of the tunneling barrier, g_{max} decreases with increasing strain from 2% to 5%. Besides, the I_{on} is defined as the current associated with the overdrive voltage $V_{over} = V_{GS} - V_{th}$ at 0.5V of the DG-TFET device. As increasing the gate voltage, the minimum of the conduction band of the strained WSe₂ is lower in energy than the maximum of the valence band of the intrinsic WSe₂ in the channel. The ON currents achieve the maximum with 2% biaxial tensile strain for the same reason.

5. Conclusions

In summary, the impacts of biaxial tensile strain on the transport characteristics of monolayer WSe₂ DG-TFETs are studied theoretically. With biaxial strain applied in the WSe₂ channel, lower SS can be obtained with higher strain while the largest enhancement of I_{on} and g_{max} occur at a certain condition (2% strain), higher strain will cause I_{on} and g_{max} degradation on the contrary. Our results underline the importance of strain engineering in enhancing the properties of WSe₂ DG-TFET.

Acknowledgment: This work is supported by China Key Research and Development Program (2016YFA0201802) and the National Natural Science Foundation of China (91964105, 61874068).

2020 IEEE Silicon Nanoelectronics Workshop

References:
[1] A. M. Ionescu and H. Riel, Nature 479, 329 (2011).
[2] U. Avci, R. Rios, K. Kuhn, and I. A. Young, VLSI Symp., 2011, p. 124.
[3] Y. Guo, et al., ACS Appl. Mater. Interfaces, p.33316, 2018;
[4] F. K. Perkins, et al., Nano Lett., p.668, 2013;
[5] T. Cheng, et al., Energy Storage Materials, p.282, 2018;
[6] Jixuan Wu and Jiezhi Chen, Applied Physics Express 11, 054001 (2018)
[7] X.-W. Jiang, J.-W. Luo, S.-S. Li, and L.-W. Wang, IEDM Tech. Dig.,2015, p. 309.

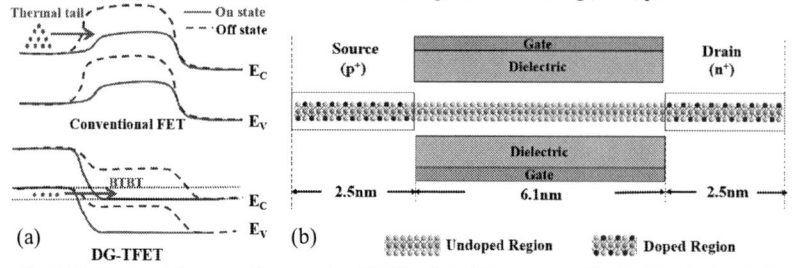

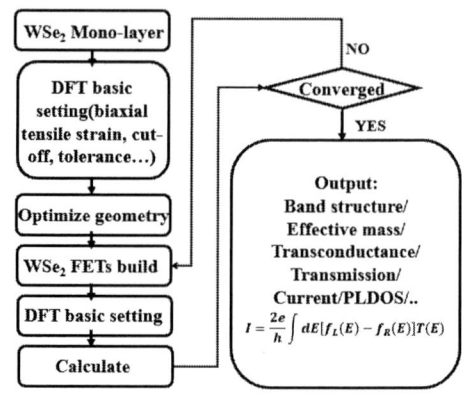

Fig.1 (a) The band diagram of conventional FET and double-gate tunneling FET (DG-TFET) for off-state and on-states. (b) Monolayer WSe₂ DG-TFET device structure, where p-type source and n-type drain are designed with a doping level of 0.007 e/atom (equivalently 1.15×10^{20}).

Fig. 2 Flow chart of simulating WSe₂ DG-TFET with biaxial tensile strain.

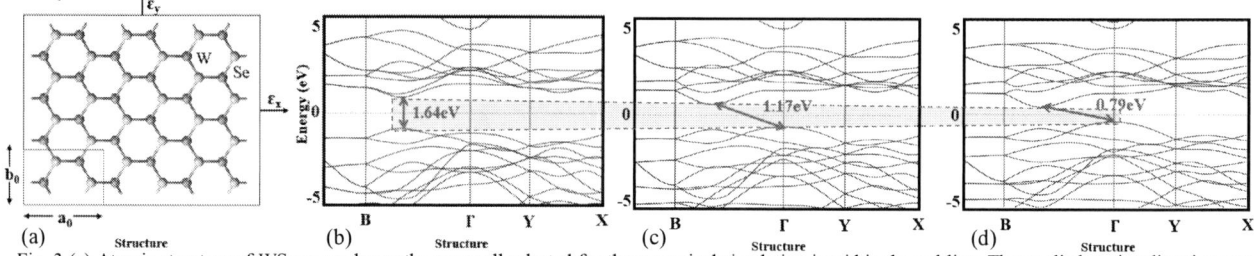

Fig. 3 (a) Atomic structure of WSe₂ monolayer: the supercell selected for the numerical simulation is within the red line. The applied strains directions are indicated by black arrows. $\varepsilon_x \equiv \Delta a_0/a_0$ and $\varepsilon_y \equiv \Delta b_0/b_0$ are defined as the components of strain, where Δa_0 and Δb_0 denote the increase of the lattice constants a_0 and b_0. Band structures of monolayer WSe₂ corresponding to (b) 0%, (c) 3%, and (d) 5% of symmetrical biaxial tensile strains, respectively.

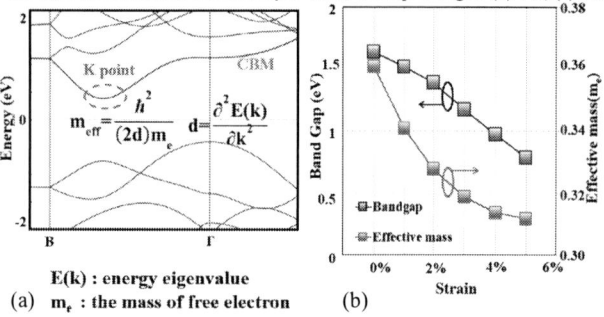

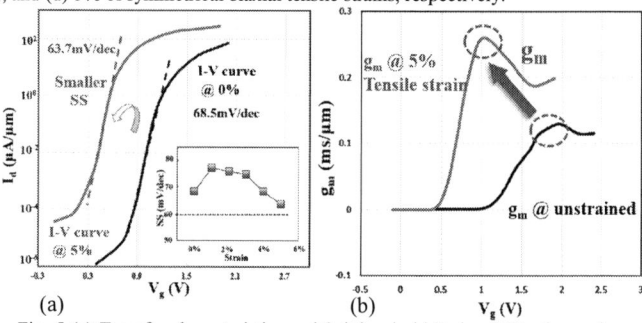

Fig. 4 (a) Method of calculating the effective mass. (b) The relationship bandgap and effective mass of monolayer WSe₂ DG-TFET as a function of the strain.

Fig. 5 (a) Transfer characteristics and Subthreshold Swing (SS) of monolayer WSe₂ DG-TFET with biaxial tensile strain from unstrained to 5% strain. (b) Calculated transconductance (g_{max}) from unstrained to 5% biaxial tensile strain in WSe₂ DG-TFET.

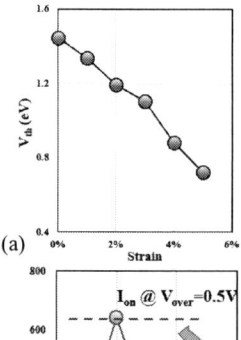

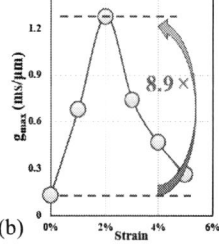

Fig. 6 Calculated (a) V_{th} shift, (b) g_{max}, (c) I_{on} in monolayer WSe₂ DG-TFET with diverse biaxial tensile strain. With increasing strains, the threshold voltage shifts negatively. I_{on} and g_{max} with 2% biaxial tensile strain are respectively 8.05 and 8.9 times larger than those of unstrained DG-TFET.

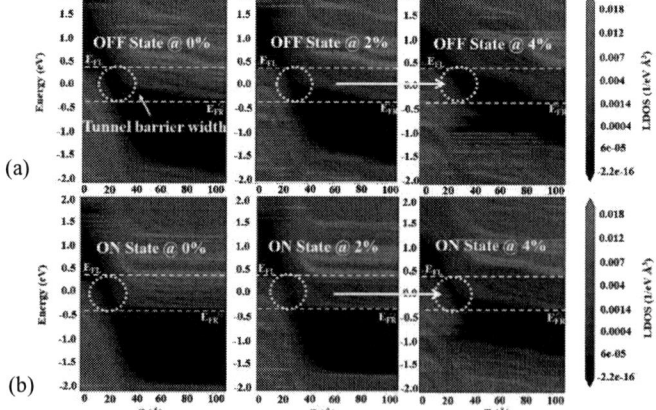

Fig. 7 PLDOS of (a) WSe₂ DG-TFET in off-state and (b) on-state with 0%, 2% and 4% biaxial tensile strain, respectively. From 0% to 2% of strain, ultra-low tunneling barrier width with larger biaxial tensile strain. The increased width of the tunneling barrier for increasing strain from 3% to 5%.

2020 IEEE Silicon Nanoelectronics Workshop

Towards Novel Channel Doping Profiles in Short Channel Bulk MOSFETs for OFF-State Current Reduction and Superior Channel Electrostatics

Harshit Kansal[1*] and Aditya Sankar Medury[1,]

[1]Department of Electrical Engineering and Computer Science, Indian Institute of Science Education and Research Bhopal,
Bhopal By-pass Road, Bhauri, Bhopal, 462066, Madhya Pradesh, India,
*Email: harshit16@iiserb.ac.in, Fax No: +91-755-2692391, Tel No: +91-755-2692645

Abstract — **For short channel MOS Transistors, leakage current flowing underneath the Si/SiO$_2$ interface (sub-surface leakage) both within the source-drain junction depth, as well as the current flowing through the bulk, are found to significantly contribute to the OFF-State current (I_{OFF}). Through TCAD Simulations, we firstly identify the region most susceptible to sub-surface leakage and then compare channel doping profiles such as the Uniformly doped (UD) and Super-Steep Retrograde (SSR), with a view to reducing I_{OFF}. This comparison enables us to propose a novel doping profile in the channel, demonstrating a higher ON-State current (I_{ON}) compared to the SSR channel doping, while showing a comparable OFF-State current, both for $V_{ds} = 0.05$ V and $V_{ds} = 1$ V.**

I. INTRODUCTION

For short channel MOSFETs, due to increasing influence of the drain, the OFF-State current (I_{OFF}) is significant and impacts the static power leakage [1]. One of the important ways to enhance the I_{ON}/I_{OFF} ratios, is to reduce the OFF-State current. In this paper, through detailed TCAD simulations, we examine the efficacy of different channel doping profiles in reducing the sub-surface leakage current and hence the OFF-State current [2], while maintaining high ON-State current, in both the linear ($V_{ds} = 0.05$ V) and saturation ($V_{ds} = 1$ V) regions. We initially compare the results for the channel current obtained, between the UD and a suitably designed SSR channel doping profile [2], where the SSR channel doping profile shows significantly lower I_{OFF}, while also showing a lower I_{ON} compared to the UD channel. Based on this comparison, we then propose, a novel channel doping profile, where the I_{OFF} obtained is comparable to the SSR doping profile, with a relatively higher I_{ON}.

II. RESULTS AND DISCUSSION

In this work, we have used Sentaurus 3-D TCAD to simulate the short-channel Bulk MOSFETs, shown in Figure 1. As a first step, under strong accumulation bias, using TCAD simulations, we identify a region of high electron concentration, away from the Si/SiO$_2$ interface, when $V_{ds} = 0$ V, shown in Figure 2. This region of high electron concentration is identified as the sub-surface leakage region. We then sweep the gate voltage bias from -2 V (strong accumulation bias) to 2 V (strong inversion bias), where the effect of the gate voltage on the short channel Bulk MOSFET (shown in Figure 1) electrostatics are investigated under two cases: (i) A Uniformly Doped (UD) channel ($N_A = 10^{17}$/cm^3) whose doping profile along the depth, is shown in Figure 3(a).

(ii) A Super-steep retrograded (SSR) channel, as shown in Figure 3(a), where the doping near the Si/SiO$_2$ interface and up to the precise location of the onset of sub-surface leakage (as identified from Figure 2) is constant and equal to 10^{17}/cm^3 (similar to the Uniformly doped channel), beyond which the channel doping abruptly increases to a high value of 5×10^{19}/cm^3. This comparison between the UD and SSR cases are shown in Figure 3(b). From this figure, it is clearly seen that the SSR doping profile shows significant (11 order) reduction of OFF-state current, while achieving a slight reduction in ON-state current (4 to 5 order), thus significantly increasing the I_{ON}/I_{OFF} ratio. In order to further improve the ratio, we propose a novel channel doping profile, as shown in Figure 4(a), termed as the gate doping (GD) profile, where a high doping region (similar to the SSR doping profile) coincides with the width of the sub-surface leakage region (within the source-drain junction depth and slightly beyond), thus reducing the OFF-state current (I_{OFF}). On the other-hand, Figure 4(a), also shows a region of very low doping in the region between the Si/SiO$_2$ interface and the onset of the sub-surface leakage region and also another region of constant but moderate doping in the substrate beyond the sub-surface leakage region. The choice of doping in the high doping region enables reduction of the sub-surface leakage current and hence the OFF-State current, while the doping in the low-doped region controls the ON-State current. On the other-hand the moderately doped region (in the substrate) enables control of the threshold voltage and the sub-threshold swing. Figure 4(b) shows the channel current comparison (in log-scale) for the gated doping (GD) profile and the SSR doping profile, for the linear and saturation region, clearly demonstrating nearly identical OFF-State currents, but enhanced ON-State currents for the proposed GD profile. **In summary**, this work has thus enabled us to propose a channel doping methodology (GD channel doping profile), meant to mitigate the effects of OFF-State current leakage while ensuring relatively high ON-State currents, thus resulting in superior channel electrostatics for short channel Bulk MOSFETs over both the linear and saturation region.

ACKNOWLEDGEMENTS

We acknowledge financial support from DST, (SERB), Government of India (grant No: ECR/2017/000011).

REFERENCES

[1] K. Roy, S. Mukhopadhyay and H. Mahmoodi-Meimand, Proceedings of IEEE, Vol.91, NO. 2, February 2003.
[2] I. De and C. M. Osburn, IEEE Transactions on Electron Devices, Vol. 46, No. 8, AUGUST 1999, Pp. 1711-1717.

P2-6

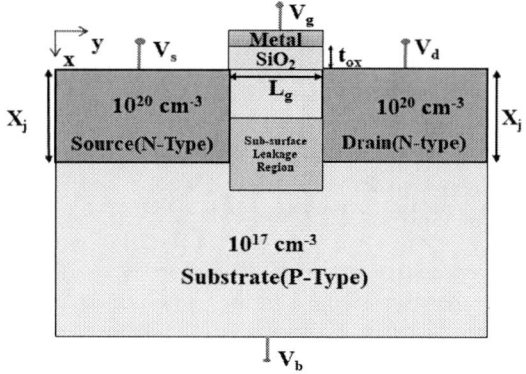

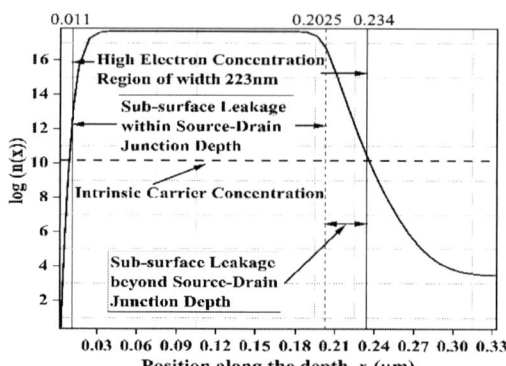

Fig 1: Schematic of a n-channel Bulk MOSFET showing a sub-surface leakage region away from the Si/SiO$_2$ interface.

Fig 2: Electron concentration (in log-scale) along the depth of the substrate showing a region of high electron concentration away from the Si/SiO$_2$ interface which is defined as the sub-surface leakage region.

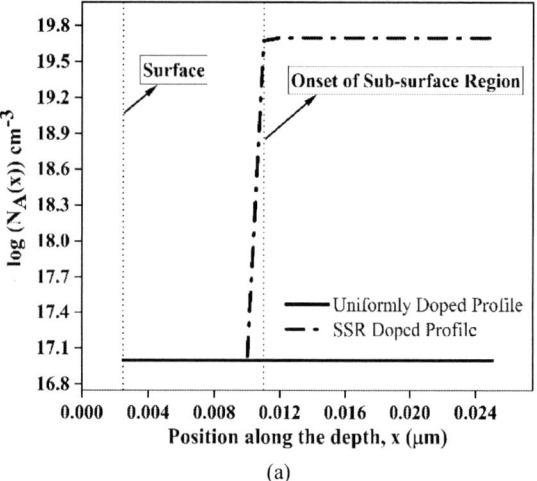

(a)

(b)

Fig 3: (a) Doping profile (in log-scale) along the depth of the substrate showing a comparison between the UD and SSR channel doping profile where the retrograding in the SSR doping profile coincides with the position along the depth of the substrate, shown in Figure 2, where the electron concentration becomes significant. (b) Comparison of the channel currents (in log-scale) between the UD and SSR channel doping profiles for V$_{ds}$ = 0.05 V and 1 V, respectively. I$_{ON}$ is defined as the I$_{ds}$ where $\frac{d^2(log(I_{ds}))}{dV_g^2}$ is minimum, while I$_{OFF}$ is defined as the I$_{ds}$ where $\frac{d^2(log(I_{ds}))}{dV_g^2}$ is maximum. This definition of I$_{ON}$ is a numerical approach to determine the channel current (I$_{ds}$) corresponding to the transition from weak to strong inversion, similarly I$_{OFF}$ corresponds to the transition from accumulation/depletion to weak inversion.

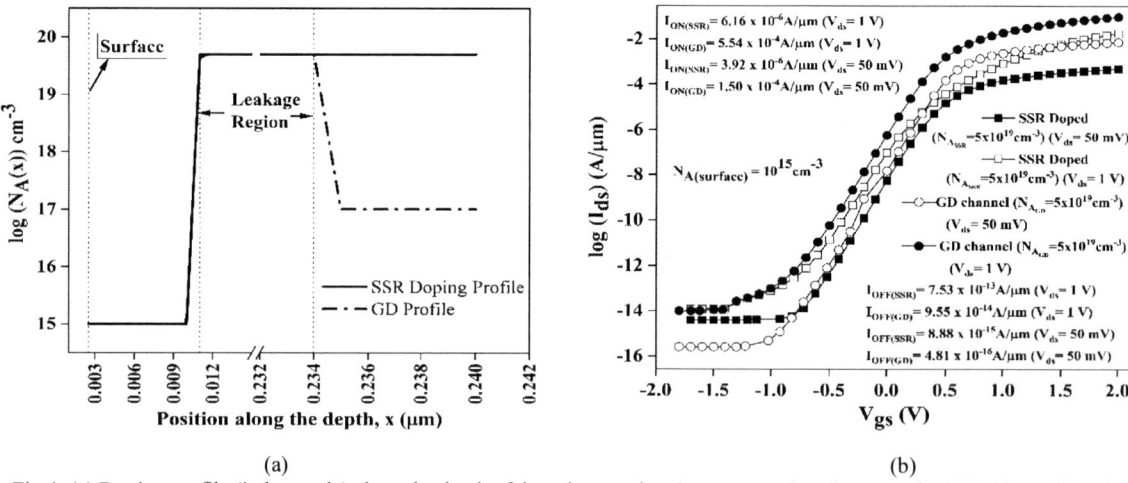

(a)

(b)

Fig 4: (a) Doping profile (in log-scale) along the depth of the substrate showing a comparison between the SSR Channel Doping profile and the proposed gated doping (GD) profile (b) Comparison of the channel currents (in log-scale) between the SSR and GD channel doping profiles for V$_{ds}$ = 0.05 V and 1 V, respectively, where for different drain voltages, the I$_{OFF}$ in the GD profile is nearly identical to the SSR profile, while I$_{ON}$ is higher in the GD profile compared to the SSR doping profile.

2020 IEEE Silicon Nanoelectronics Workshop

978-1-7281-9736-4/20 $31.00 © 2020 IEEE

A Mobility Stress Response Model of FinFET: Silicon vs Germanium

Cheng-Hsien Yang, Yun-Fang Chung, Kuan-Ting Chen, and Shu-Tong Chang*

Department of Electrical Engineering, National Chung Hsing University, Taichung, Taiwan

*E-mail:stchang@dragon.nchu.edu.tw

Abstract—**FinFET, which could be applied to sub 22 nm technology nodes, is the novel 3D transistor device structure to effectively inhibit short channel effect for the continuous microminiaturization of transistor components. This study focuses on the comparison of silicon and germanium, as the channel material, to FinFET for calculating carrier distribution and carrier mobility on inversion layer as well as the stress response. A mobility stress response model (piezomobility model) which could rapidly and accurately predict carrier mobility changes under different stress conditions is also investigated.**

Keywords—FinFET, piezomobility model, stress

I. INTRODUCTION

Since Intel, in 2003, proposed to add the technology using SiGe alloy embedded source/drain on Si channel for generating the strain of compressive stress into the 90 nm technology node to enhance the carrier mobility, 65, 45, 32, 22, ,14 (16), 10, 7 and 5 nm technology was successively promoted for introducing to strain engineering. Mechanical bending used for generating mechanical strain was a comparatively direct method to directly apply mechanical stress to semiconductor devices so that device bending generated strain inside the substrate. To rapidly predict the effect of applied stress on carrier mobility of semiconductor device, the piezomobility model published by Professors Smith [1] and Bufler [2] is referred in this study. By simulating the changes in band structure and carrier mobility of components under different stress conditions, the more accurate and compact nine-parameter second-order piezomobility model is used to extract the piezomobility coefficients of P-type FinFET as well as to analyze the importance of strain engineering to future nano transistor device design.

II. CALCULATION METHOD

A piezomobility model could be used for rapidly predicting the relationship between mobility change and stress that it could be applied to the compact model for mobility calculation. The first piezomobility model was proposed by C. S. Smith in 1954 [1]. Although first-order piezomobility model is still broadly used nowadays, it would appear inaccurate prediction because it simply considers the effect of uniaxial stress at each direction on carrier mobility change, but ignores the mutual effect between each stress. For this reason, it is necessary to consider the coupling in the stress in the first-order piezomobility model to promote the accuracy. The second-order piezomobility model is further derived, with the following equation [4].

$$\frac{\Delta\mu}{\mu_0} = -\begin{pmatrix} \pi_x\sigma_{xx} + \pi_y\sigma_{yy} + \pi_z\sigma_{zz} + \\ \pi_{xy}\sigma_{xx}\sigma_{yy} + \pi_{xz}\sigma_{xx}\sigma_{zz} + \pi_{yz}\sigma_{yy}\sigma_{zz} \end{pmatrix} \quad (1)$$

Dr. F. M. Bufler proposed a higher order piezomobility model with stress dependent pirzocoefficients. [2], in 2009. Bufler's

model could not accurately predict the carrier mobility change when large biaxial stress simultaneously applies to x and y. The original six-parameter piezomobility model [4] revised in [5] is therefore applied, and three uniaxial stress square term piezomobility coefficients are used for the correction, with the following equation.

$$\frac{\Delta\mu}{\mu_0} = -\begin{pmatrix} \pi_x\sigma_{xx} + \pi_y\sigma_{yy} + \pi_z\sigma_{zz} + \pi'_x\sigma_{xx}\sigma_{xx} + \pi'_y\sigma_{yy}\sigma_{yy} + \pi'_z\sigma_{zz}\sigma_{zz} \\ + \pi_{xy}\sigma_{xx}\sigma_{yy} + \pi_{xz}\sigma_{xx}\sigma_{zz} + \pi_{yz}\sigma_{yy}\sigma_{zz} \end{pmatrix} \quad (2)$$

Different from Bufler's model, the nine piezomobility coefficients in [5] are constant, and the xy second-order piezomobility coefficient, which is lack in Bufler's model, is taken into account. The hole mobilities of silicon and germanium are calculated with Kubo-greenwood equation as reported in Ref. [6].

III. RESULT AND DISCUSSION

Hole mobility is calculated and extracted in this study, P-type Si or Ge FinFET structure used for extracting piezomobility coefficients (PC) are shown in Figure 1. The width of FinFET is 6nm, the substrate doping concentration is $10^{17}cm^{-3}$, and the applied uniaxial and biaxial stress is ranged from -2Gpa to 2GPa. The carrier mobility calculated with the inversion carrier concentration $5\times10^{12}cm^{-2}$ is used for extracting the piezomobility coefficient. The relationship between the mobility gain of Si P-type FinFET and stress applied is shown in Fig. 2. The solid line in figure is TCAD simulation results, and the hollow is the result of nine-parameter piezomobility model. The six lines contain 3 for uniaxial stress and 3 for biaxial stress conditions, respectively. Table I shows the extracted piezomobility coefficients of Fig.2. The relationship between the mobility gain of Ge P-type FinFET and stress applied is shown in Fig. 3 and Table II shows the extracted piezomobility coefficients of Fig.3. From Figs 2 and 3, we discovered that the result of nine-parameter piezomobility model fits pretty well with result calculated from TCAD. Fig. 4 shows PC as a function of width for Si FinFET with the same structure condition as mentioned in Fig. 1. Linear PC, π_x, is major contribution factor to mobility stress response than other two PR coefficients, π_y, and π_z. Linear PR coefficient, π_y, is relatively less sensitive to mobility stress response. Width dependent of linear PR coefficients is not obvious.

IV. CONCLUSION

The nine-parameter piezomobility model could merely accurately predict the mobility gain under six stress applied conditions for P-type FinFET devices. Hole mobility stress response of Ge FinFET is better than Si one. Such a work would assist in the future design of Si- and Ge-based CMOS devices and prove the importance of strain engineering in future transistor device.

ACKNOWLEDGMENTS

This work was supported by the National Science Council, Taiwan, R.O.C., under contract Nos. of MOST 108-2622-8-002-016 and MOST 106-2221-E-005-096 -MY3

REFERENCES

[1] C. S. Smith "Piezomobility effect in Ge and silicon, " *PHYSICAL REVIEW*, vol. 94, no. 1, pp. 42-49, Apr. 1954.

[2] F. M. Bufler, A. Erlebach, and M. Oulmane "Hole Mobility Model With Silicon Inversion Layer Symmetry and Stress-Dependent Piezoconductance Coefficients, " *IEEE Electron Device Lett.*, vol. 30, no. 9, pp. 996-998, Sep. 2009.

[3] G. Sun, Y. Sun, T. Nishida, and S. E. Thompson, "Hole mobility in silicon inversion layers: Stress and surface orientation," *Journal of Applied Physics,* vol. 102, no. 8, p. 084 501, Oct. 2007.

[4] C.-F. Lee, R.-Y. He, K.-T. Chen, S.-Y. Cheng, and S.-T. Chang, "Strain engineering for electron mobility enhancement of strained Ge NMOSFET with SiGe alloy source/drain stressors," *Microelectronic Engineering*, vol. 138, no. 4, pp. 12-16, Jan. 2015.

[5] K.-T. Chen, C.-F. Lee, R.-Y. He, and S.-T. Chang, "A Stress Response Model for Hole Mobility in the Inversion Layer of Ge MOSFETs," *Journal of Nanoscience and Nanotechnology*, vol.17, no. 11, pp. 8511-8515, Nov. 2017.

[6] M.V. Fischetti, Z. Ren, P.M. Solomon, M. Yang, K. Rim, J. Appl. Phys, vol. 94, no. 2, pp. 1079–1095, 2003.

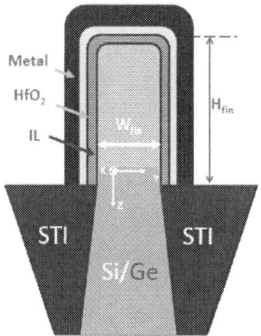

Fig. 1 2D cross-section schematic diagram of P-type FinFET with Si or Ge as channel material. X direction is [110], Y direction is [-110], and Z direction is [001].

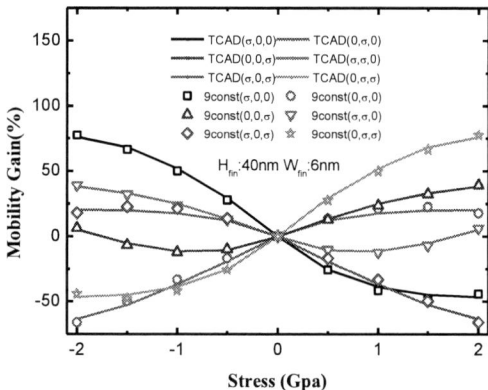

Fig. 2 Relation diagram of mobility gain in Si P-type FinFET to applied stress. Fin width is 6nm. Solid lines are the TCAD simulation results, and hollow points are the results of nine-parameter piezomobility model.

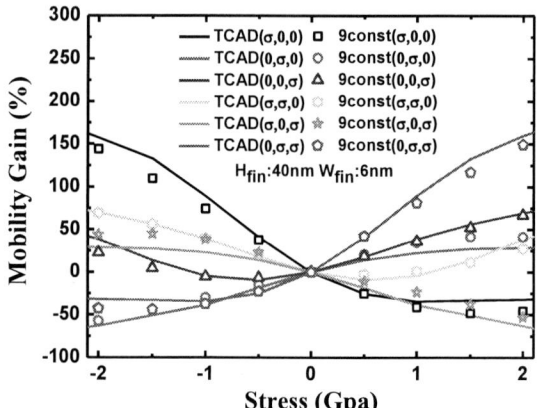

Fig. 3 Relation diagram of mobility gain in Ge P-type FinFET to applied stress. Fin width is 6nm. Solid lines are the TCAD simulation results, and hollow points are the results of nine-parameter piezomobility model.

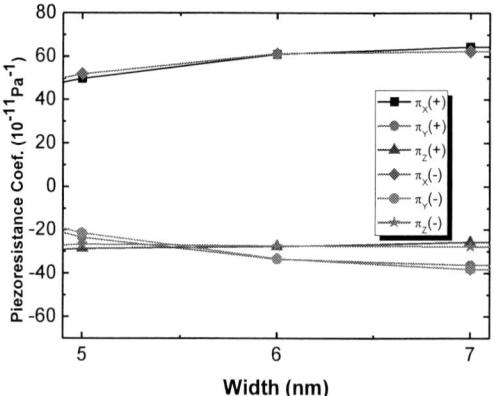

Fig. 4 Piezoresistance coefficient as a function of width for Ge FinFET. Note that only linear PR coefficients such as π_x, π_y, and π_z are included and others are not shown.

	π_x	$(\pi_x)^t$	π_y	$(\pi_y)'$	π_z	$(\pi_z)^t$	π_{xy}	π_{xz}	π_{yz}
Stress-	61.06539	11.1449	-33.559	-0.23206	-27.6604	-15.3997	-6.898	16.51855	-3.90672
Stress+	61.32202	-19.599	-33.2866	12.1892	-27.4117	3.937194	-8.1902	15.2237	-5.17256

Table I The piezomobility coefficients extracted with nine-parameter piezomobility model for Fig. 2, where – and + respectively stand for negative and positive applied stress being substituted into piezomobility equation.

	π_x	$(\pi_x)^t$	π_y	$(\pi_y)'$	π_z	$(\pi_z)'$	π_{xy}	π_{xz}	π_{yz}
Stress-	76.5	2.27	-31.1	-1.3	-22.1	-16.7	4.34	30.6	2.1
Stress+	59.6	-18.4	-48.5	13.9	-39	2.83	-8.01	18.5	-10.3

Table II The piezomobility coefficients extracted with nine-parameter piezomobility model for Fig. 3, where – and + respectively stand for negative and positive applied stress being substituted into piezomobility equation.

2020 IEEE Silicon Nanoelectronics Workshop

Coulomb-Blockade Charge-Transport Mechanism in Band-to-Band Tunneling in Heavily-Doped Low-Dimensional Silicon Esaki Diodes

G. Prabhudesai,[1*] K. Yamaguchi,[1,2] M. Tabe,[1] and D. Moraru[1,2**]

[1]Research Institute of Electronics, Shizuoka University, 3-5-1 Johoku, Hamamatsu 432-8011, Japan
[2]Graduate School of Integrated Science and Technology, Shizuoka University, 3-5-1 Johoku, Hamamatsu 432-8011, Japan
Email: * prabhudesai.gaurang.pramod@cii.shizuoka.ac.jp ** moraru.daniel@shizuoka.ac.jp

Abstract — **Band-to-band tunneling (BTBT), with its key role as a transport mechanism in Esaki (tunnel) diodes, has been studied extensively for about 60 years. In such devices, it is expected that energy states of ionized dopants in the depletion-layer can affect the BTBT mechanism. In this paper, we introduce the observation and analysis of a novel transport mechanism in Si Esaki diodes: single-charge BTBT transport mediated by donor-cluster quantum-dots, statistically expected in the depletion-layer of nanoscale Esaki diodes. This demonstration can open new pathways for band-to-band tunneling devices.**

I. INTRODUCTION

Quantum-mechanical tunneling was first demonstrated about 60 years ago in heavily-doped *pn* junctions (so-called Esaki diodes) [1,2]. Subsequently, several studies have helped clarify the main charge-transport mechanisms in the band-to-band tunneling (BTBT) in Si Esaki diodes [2-4], including phonon-assisted tunneling (at low forward and reverse biases) and the 'excess' current via gap states (at intermediate-high forward biases). Recently, it has been shown that resonant tunneling via localized energy states of facing Phosphorus (P) and Boron (B) atoms in the depletion-layer of low-dimensional Esaki diodes leads to sharp enhancement of the BTBT current, even at low forward biases [5].

In this paper, we present new transport mechanisms in Si Esaki diodes. First, we show the role of dopant-cluster quantum dots (QDs), formed in the depletion-layer of low-dimensional Esaki diodes, in the charge transport based on the Coulomb blockade mechanism [6]. Then, we analyze the possibility of BTBT transport mediated by pairs of donors and acceptors at the edges of the depletion-layer as a background BTBT charge-transport mechanism.

II. DESIGN OF LOW-DIMENSIONAL SI ESAKI DIODES

The devices under study are designed as low-dimensional Esaki diodes in a thin SOI layer (final thickness, $t_{SOI} \approx 10$ nm). As shown in **Fig. 1(a)-(b)**, an overlapped doping profile was used to ensure the formation of the *pn* junction within a nanoscale constriction. The estimated doping concentrations are $N_D \approx 2.7 \times 10^{20}$ cm^{-3} (P) and $N_A \approx 0.9 \times 10^{20}$ cm^{-3} (B). It is expected that the depletion-layer width is 10~20 nm, considering also the low-dimensionality of the structure. The formation of the *pn* junction could be confirmed by transport characteristics at low temperatures (T=5.5 K), as the observation of a negative differential conductance (NDC) region in the forward bias and a high reverse-bias current. The current-voltage characteristics (I_p-V_p) of a representative device are shown in **Fig. 2**. The arrows indicate the known manifestations of phonon-assisted tunneling in Si Esaki diodes [3]. Different from these well-known Esaki diode

properties, we observed step-like features in the I_p-V_p characteristics at low forward biases, as described next.

III. COULOMB BLOCKADE CHARGE TRANSPORT

In about half of the devices measured, step-like features as shown in **Fig. 3(a)** were seen at low forward bias, at V_p positions different from the expected phonon-assisted tunneling features described before. In addition, at lowest biases, a low-current region is typically observed.

These features are similar to the so-called Coulomb staircase observed in the output characteristics of single-electron tunneling transistors. These are typically ascribed to single-electron tunneling transport based on the Coulomb blockade effect through QDs with asymmetric tunnel barriers. To confirm the occurrence of the Coulomb blockade effect, we measured the effect of changing the substrate voltage (V_{SUB}) that acts as a back-gate. The I_p-V_{SUB} characteristics (analogous to the transfer characteristics of transistors) show indeed the oscillation of the current I_p as a function of V_{SUB} (**Fig. 3 (b)**, black scatter points).

The step-like features in the I_p-V_p characteristics and the oscillations in the I_p-V_{SUB} characteristics provide strong support for our model that single-charge (electron) band-to-band tunneling (SC-BTBT) mechanism is observed in these low-dimensional Si Esaki diodes. The oscillations in the I_p-V_{SUB} characteristics were also corroborated, with a reasonable match, with Monte Carlo simulations based on the orthodox Coulomb blockade theory (**Fig. 3(b)**, line plot). These are also evidenced by the observation of (half) Coulomb diamonds in the forward bias, in experiment and simulation (**Fig. 3(c)-(d)**).

Next, extensive statistical analyses of randomly-distributed donors and acceptors, placed at substitutional lattice sites in the depletion-layers, were carried out. A picture of the depletion-layer with atomistic distribution of dopants is shown in **Fig. 4(a)-(b)**, before and after compensation between P-donors and B-acceptors found in close proximity ($d_{PB} < r_B$, with r_B the Bohr radius for a dopant in Si, approximately 2.5 nm). It is found that the QDs mediating the SC-BTBT mechanism are formed by 7-10 ionized donor atoms (as shown schematically in **Fig. 4(c)**, top panel) [6]. These QDs are formed due to the clustering (coupling) of ionized donors, left non-compensated by the counterpart dopants, in the nanoscale depletion-layers.

Different from these dopants inside the depletion-layer, another key role can be expected from pairs of P-donors and B-acceptors at the opposite edges of the depletion-layer; this analysis is still under way. It is, however, expected that such donor-acceptor pairs can mediate electron-hole recombination in BTBT transport, leading to a fine-pitch oscillation in the differential conductance.

2020 IEEE Silicon Nanoelectronics Workshop

978-1-7281-9736-4/20 $31.00 © 2020 IEEE

IV. CONCLUSIONS

In this work, we mainly presented a novel charge transport mechanism in low-dimensional Si Esaki diodes: single-charge band-to-band tunneling (SC-BTBT) mediated by quantum-dots formed by clusters of ionized dopant-atoms located in the nanoscale depletion-layers. We found that typical donor-cluster quantum-dots are formed by 7-10 ionized donors, statistically formed inside such depletion-layers. These results open new pathways for BTBT transport mediated by dopants in depletion-layers, which can be extended to dopants at the edges of the depletion-layers.

ACKNOWLEDGEMENTS

The authors thank L. T. Anh, M. Manoharan, H. Mizuta for support in simulations, and M. Hori and Y. Ono for valuable discussions. This work was partially supported by Grant-in-Aid for Scientific Research (I19K045290) from MEXT, Japan, a Cooperative Research Project of Research Institute of Electronics, Shizuoka University.

REFERENCES

[1] L. Esaki, Physical Review **109**, 603–604 (1958).
[2] L. Esaki and Y. Miyahara, Solid State Electron. **1**, 13–21 (1960).
[3] A. G. Chynoweth, R. A. Logan, and D. E. Thomas, Phys. Rev. **125**, 877 (1962).
[4] A. G. Chynoweth, W. L. Feldmann, and R. A. Logan, Phys. Rev. **121**, 684–694 (1961).
[5] M. Tabe *et al.*, Appl. Phys. Lett. **108**, 093502 (2016).
[6] G. Prabhudesai *et al.*, Appl. Phys. Lett. **114**, 243502 (2019).

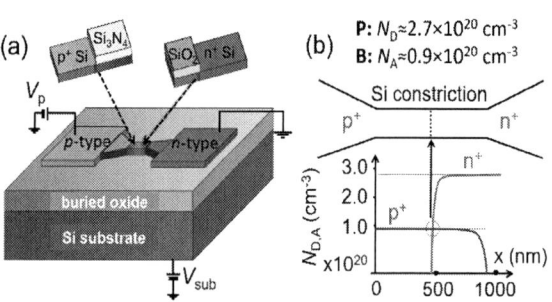

Fig. 1. (a) Low-dimensional Si Esaki diode in a thin SOI layer (doping masks are schematically shown in the insets). **(b)** Schematic representation of the doping profile in the Si constriction.

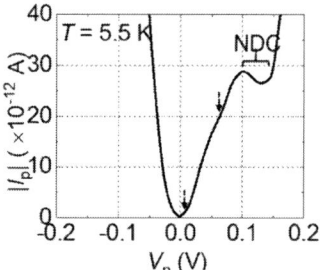

Fig. 2. Typical low-temperature ($T = 5.5$ K) I_p-V_p characteristics of a typical Esaki diode with high reverse-bias current, NDC region and signatures of phonon-assisted tunneling at forward bias.

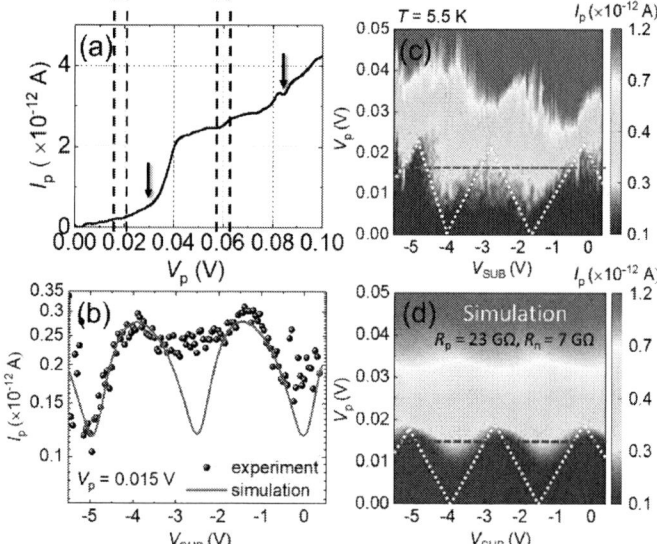

Fig. 3. (a) Low-temperature (T=5.5 K) I_p-V_p characteristics similar to a Coulomb staircase (OFF-current near zero bias and steps at values different from the phonon-assisted tunneling signatures). **(b)** Oscillation in the I_p-V_{SUB} characteristics from experiment (black scatter points) and simulation (red curve). **(c)-(d)** Half Coulomb diamonds in the forward bias (T=5.5 K) measured experimentally (**Fig. 3(c)**) and simulated by the orthodox Coulomb blockade theory (**Fig. 3(d)**).

Fig. 4. (a)-(b) Random dopant distribution of ionized donors and acceptors in the depletion-layer before compensation (**Fig. 4(a)**) and after compensation ($d_{comp} = d_{PB} \leq 2.5$ nm) (**Fig. 4(b)**). **(c)** The largest isolated quantum-dot formed by ionized donors in the depletion-layer after compensation. **(d)** Schematic representation of the Coulomb-blockade charge transport mechanism mediated by a donor-cluster quantum-dot in the depletion-layer, leading to SC-BTBT.

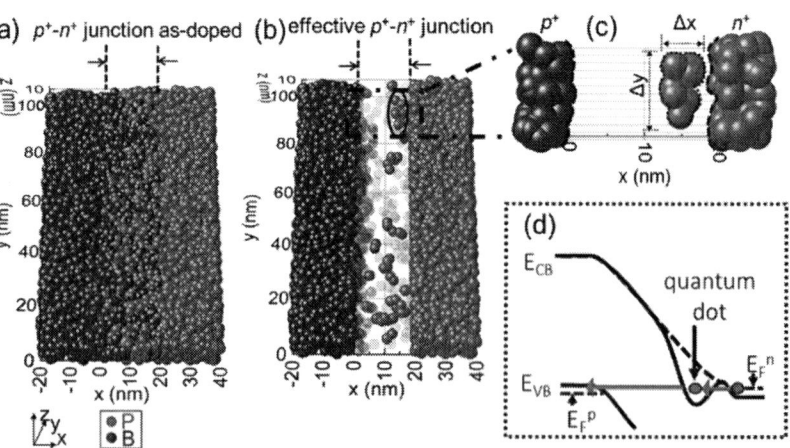

Effects of Co-doping on the Transport Characteristics of Nanoscale *n*-type Silicon-on-Insulator Transistors

C. Pandy,[1,2] A. Debnath,[2] K. Yamaguchi,[2] T. Teja Jupalli,[1,2] G. Prabhudesai,[2]
Ramakrishnan V N[3], Y. Neo,[2] H. Mimura,[2] and D. Moraru[1,2*]

[1]Graduate School of Science and Technology, Shizuoka University, 3-5-1 Johoku Hamamatsu 432-8011, Japan
[2]Research Institute of Electronics, Shizuoka University, 3-5-1 Johoku Hamamatsu 432-8011, Japan
[3]School of Electronics Engineering, VIT University, Vellore, India
[*]Email: moraru.daniel@shizuoka.ac.jp

Abstract — In conventional downscaled Si transistors, the role of discrete dopants in nano-channels becomes critical; uniform high doping is desirable. Even in such highly-doped transistors, however, quantum dots formed by clusters of donors may work for single-electron tunneling (SET), and can be further assisted by co-doping. Here, we first analyze transport characteristics of co-doped Si nanowires by simulations, extending then the analysis to experiments on co-doped SOI-FETs with nanoscale channels. SET behavior is reported, and B-acceptors may contribute to the formation of tunnel barriers in such channels, suggesting a simple technique for Si SET devices.

I. INTRODUCTION

As Si transistors are continuously miniaturized and become scaled down to nanometer-scale, discrete impurities (dopants) start to play a more and more critical role, even as quantum dots (QDs) [1,2], but mainly at low temperature. Using a selective-doping technique, with control over a few tens of nanometers, we have demonstrated that clusters of a few P-donors can form QDs that sustain single-electron tunneling (SET) operation even at elevated temperatures (~150 K) [3]. However, the abrupt doping required by this technique, as well as for fabricating conventional *npn* MOSFETs, is challenging in nanoscale, and a uniform doping is desirable. It can be expected that co-doping further enhances the possibility of formation of QDs.

Over the past decade, a new simple device concept of *junctionless* transistor (JLT) has been developed [4], but the SET functionality has not been sufficiently addressed. Here, we propose and present a basic analysis for *n*-type JLTs, doped with P-donors at concentrations $>10^{20}$ cm^{-3}, and also co-doped with B-acceptors. We aim to analyze the role of co-doping and compensation effect in the definition of tunnel barriers. We first show the impact of B-acceptors by fundamental transport simulations, and then provide a first experimental evidence of SET transport in nanoscale co-doped SOI-FETs at low temperature.

II. TRANSPORT SIMULATIONS OF DOPED AND CO-DOPED SI NANOSCALE TRANSISTORS

In order to clarify the fundamental role of B-acceptors in transport through P-donors in Si nanoscale transistors, device-transport simulations are carried out using the Quantum ATK tool [5]. **Figure 1** shows the atomistic structure of a Si nanowire transistor, built with a <100> orientation along transport direction, with the channel length of 5.43 nm; H atoms were used to passivate all the dangling bonds on the surface. The structure is converted into a device, with *n*-type source and drain regions with lengths of 1.086 nm each, electrostatically doped with $N_D=2\times10^{20}$ cm^{-3}, while P-donors and B-acceptors are discretely added only in the channel region. A metallic gate is placed around the channel. Here, all device characteristics are calculated using semi-empirical Extended-Huckel calculator at room temperature.

Figure 2 shows the simulated I_D-V_G transfer characteristics for transistors with channels undoped, single-P-doped and co-doped (single-P/single-B). Relative to the undoped-transistor, current is increased by introducing a P-donor, but it is significantly reduced by introducing also a B-acceptor (co-doping). This suggests the important role of co-doping. However, for comparison with our experimental devices (see Section III), cases with channels doped with 3 P-donors and 1 B-acceptor are considered next. In particular, the roles of the position of the B-acceptor relative to the 3-P-donor cluster and of the atomistic configuration are analyzed. As seen in **Fig. 3**, it is found that, although all cases exhibit the expected *n*-type transistor behavior, there is a significant current enhancement when the B-acceptor is near the source-side (or in the center) compared with the drain-side position. Furthermore, fine changes in the atomistic configuration lead to visible changes in the electrical characteristics (most visible in I_D-V_D curves, not shown). An analysis of DOS spectra infers that specific positions of satellite B-acceptors nearby a multiple-P-donor cluster can provide a significant current modulation, beyond simple compensation.

III. CO-DOPED NANOSCALE SOI-FETs: DEVICE STRUCTURE AND ELECTRICAL CHARACTERISTICS

For the experimental analysis, we fabricated SOI-FETs with thin channels ($t_{Si}\approx10$ nm) designed as point contacts, but which become narrow and short nanowires by final gate oxidation (≈9 nm); the top-Si layer is uniformly co-doped ($N_D\approx2.0\times10^{20}$ cm^{-3}, followed by $N_A\approx0.7\times10^{20}$ cm^{-3}, i.e., roughly a 3:1 ratio of P:B atoms). **Figure 4** shows the schematic device structure, circuit and channel doping. **Figure 5** shows I_D-V_G transfer characteristics measured at room temperature for SOI-FETs with smallest dimensions. It can be observed that, even for devices with designed width of 40 nm, the channel is highly conductive and current cannot be switched off even at highly-negative V_G. However, for devices with 30 nm width, a weak trend of switching off can be observed, while for even smaller devices (20 nm) no current can be detected.

In order to evaluate the impact of donor clusters, low-temperature measurements are taken (T=15 K). Data for one device (designed width of 30 nm) is shown here by the I_D-V_G characteristics [**Fig. 6(a)**] and by the stability diagram [**Fig. 6(b)**]. This device exhibits clearly the signatures of SET transport based

2020 IEEE Silicon Nanoelectronics Workshop

978-1-7281-9736-4/20 $31.00 © 2020 IEEE

on Coulomb blockade, as current peaks and, respectively, Coulomb diamonds. From the large period between peaks and assuming that a dominant QD is responsible for this behavior, under the model of a circular parallel-plate capacitor QD, its radius is estimated to be 2.4±0.6 nm. This is consistent with a cluster of a few P-donors strongly coupled even within the Bohr radius (r_B≈2.5 nm) of a single P-donor in Si. A rough analysis of the slopes of the Coulomb diamonds suggests that the QD is close to the center of the channel, between source and drain extensions.

The role of B-acceptors in this behavior is still under study. However, some B-acceptors surrounding the multiple-P-cluster may work to enhance the tunnel barriers. In fact, digitized current jumps (marked by arrows in the inset of **Fig. 6(a)**) are consistent with electron charging near the transport path, likely related to the B-acceptors remaining in the tunnel barriers.

IV. CONCLUSIONS

We analyzed the transport characteristics of co-doped Si nanoscale transistors, both by simulations and by experiments. We found from simulations that B-acceptors can significantly affect the transport via P-donor clusters and that the characteristics are sensitive to the position and configuration of dopants. From experiments, we revealed the possibility of SET operation in a simple design, with uniformly co-doped nano-channels, in which clusters of P-donors can work as QDs, and can be assisted by the presence of B-acceptors. This opens the way for further analysis to implement the concept of co-doping for obtaining SET functionalities in junctionless Si transistors.

ACKNOWLEDGMENTS We thank R. Tsujimura for contribution to device fabrication and M. Tabe for useful discussions. This work is supported by MEXT Kakenhi (I19K045290) and Coop. Res. of RIE (Shizuoka Univ.).

REFERENCES

[1] H. Sellier *et al.*, Phys. Rev. Lett. **97**, 206805 (2008).
[2] M. Tabe *et al.*, Phys. Rev. Lett. **105**, 016803 (2010).
[3] A. Samanta *et al.*, Appl. Phys. Lett. **110**, 093107 (2017).
[4] J. P. Colinge *et al.*, Nat. Nanotechnol. **5**, 225 (2010).
[5] K. Stokbro *et al.*, Phys. Rev. B **82**, 075420 (2010).

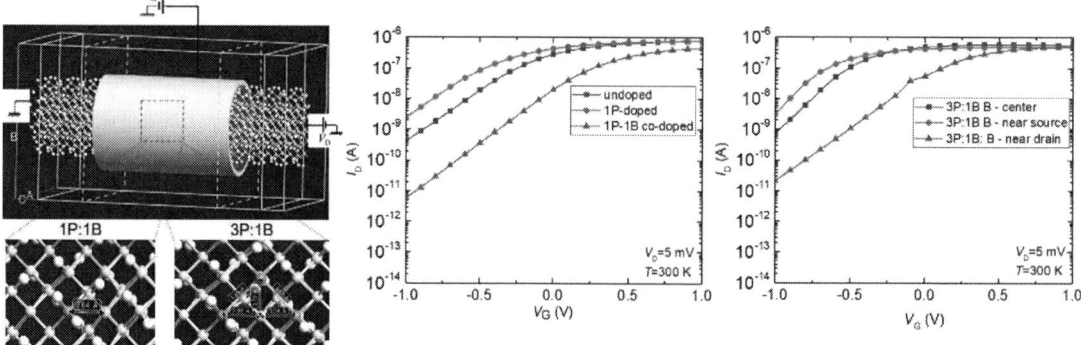

Fig. 1. Atomistic view of a Si nanowire gated device, with typical biasing circuit. Lower panels: zoom-in views of channel region co-doped with 1 P-donor and 1 B-acceptor (left); 3 P-donors and 1 B-acceptor (right).

Fig. 2. Simulated I_D-V_G characteristics at room temperature (T=300 K) and V_D=5 mV for undoped, single-P-doped and 1P:1B co-doped-channel cases.

Fig. 3. Simulated I_D-V_G characteristics at room temperature (T=300 K) and V_D=5 mV for 3P:1B co-doped-channels with different dopant configurations.

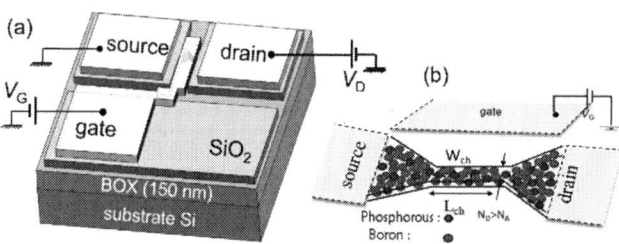

Fig. 4. (a) Device structure and biasing circuit for a nanoscale SOI-FET. (b) Schematic illustration of the channel region co-doped with P-donors and B-acceptors, roughly with a ratio of 3:1.

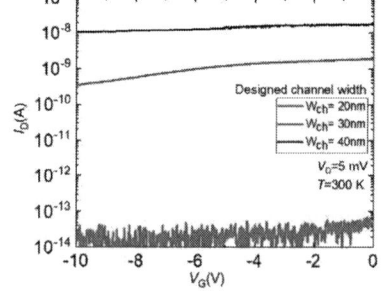

Fig. 5. I_D-V_G curves measured at room temperature (T=300 K) and V_D=5 mV for several SOI-FETs with small co-doped channels. For the SOI-FET with designed width W_{ch}=30 nm, weak gate control can be seen.

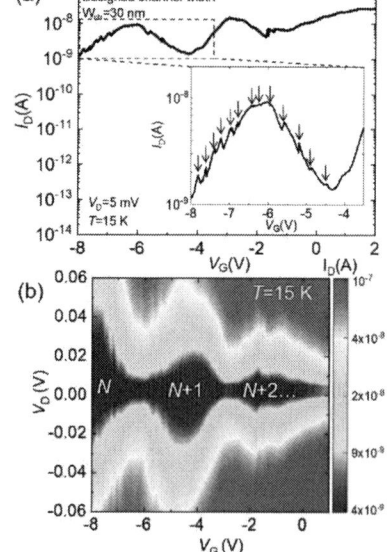

Fig. 6. (a) I_D-V_G characteristics (T=15 K) for co-doped nanoscale SOI-FET, with charging marked by arrows in inset. (b) Stability diagram, showing SET behavior with different numbers of electrons.

2020 IEEE Silicon Nanoelectronics Workshop

978-1-7281-9736-4/20 $31.00 © 2020 IEEE

High Speed Nanoantenna Thermopiles for Long-Wave Infrared Detection

Gergo P. Szakmany[1], Gary H. Bernstein[1], Edward C. Kinzel[2], Alexei O. Orlov[1], and Wolfgang Porod[1]

[1]Department of Electrical Engineering, University of Notre Dame, USA,
[2]Department of Aerospace and Mechanical Engineering, University of Notre Dame, USA,
Email: gszakman@nd.edu

Abstract — We study thermoelectric signal generation by suspended nanoantenna thermopiles in response to amplitude-modulated long-wave infrared radiation. We experimentally demonstrate response times on the order of a few µs with a signal bandwidth of about 200 kHz. Analytical calculation and simulation results show that the limiting factor of the response time is not the inverse RC time constant of the array but rather the thermal response of the nanostructure.

Keywords: nanoantenna, thermal detector, infrared

I. INTRODUCTION

Thermal long-wave infrared detectors (microbolometers and thermopiles) absorb IR energy and convert a temperature increase to an electrical signal. These devices are low-cost and uncooled; however, their response time and sensitivity are inversely proportional to each other and determined by the absorption volume of the detector. In this paper, we show that thermopiles can be made fast and sensitive when the IR radiation is resonantly absorbed by nanoscale antennas.

Antenna-coupled nanoantennas (ACNTCs) are constructed from a dipole antenna and a nanothermocouple (Fig. 1), and their operation is based on the resonant absorption of the IR radiation by the dipole antenna. The radiation-induced antenna currents are proportional to the IR energy and they resistively heat the hot junction of a nanothermocouple that produces a measurable open-circuit voltage, V_{OC}, due to the Seebeck effect [1]. The ACNTCs are suspended above a quasi-spherical cavity etched into the Si substrate (Fig. 2). The cavity thermally insulates the nanoantennas from the Si substrate and functions as an optical element that focuses the IR radiation to the antenna [2]. We have shown [3] that nanothermocouples can be constructed either from two different metals or from a single-metal layer with different cross sections exploiting the size-dependent Seebeck coefficient. Here we present the frequency-dependent response of nanoantenna thermopiles constructed from 52 suspended single-metal ACNTCs connected in series (Fig. 3).

The dipole antennas and the NTCs were patterned by electron beam lithography and formed from a 45-nm-thick Pd layer. The resonant length of the dipole antennas is 3.5 µm, as determined by COMSOL simulations. The narrow and wide segments of the single-metal NTCs are 50 nm and 300 nm wide, respectively, and the lead lines are 3.5 µm long. The cavities were etched into the Si substrate by using XeF_2 vapor etch at 3.5 torr for 25 s. As a result, cavities with 4.9 µm depth and 12 µm aperture were formed under each antenna in the thermopile, as shown in Figs. 3 and 4. In this case, the ACNTC response is maximized because the cavity focuses the IR radiation to the antenna, and the incident and reflected waves are in phase and constructive interference occurs [2].

II. RESULTS

The thermopiles were illuminated by a CO_2 laser beam operating at 10.6 µm and was square-wave-modulated by an acousto-optic modulator (AOM) between 1 Hz and 6 MHz. The V_{OC} response was first amplified by a broad-band differential pre-amplifier and measured by a lock-in amplifier synchronized to the chopping frequency.

Figure 5 shows the measured frequency-dependent response of a suspended nanoantenna thermopile. The measured signal is frequency-independent between 1 Hz and 70 kHz, and starts to roll-off at higher modulation frequencies reaching the -3 dB level at around 200 kHz. Importantly, the cutoff frequency of the thermopile is primarily determined not by its inverse RC time constant that was measured to be around 1.5 MHz. Instead, the cutoff frequency is due mainly to the thermal properties of the antenna and the NTC lead lines. The thermal mass of the antenna can be calculated as $m_t = c_p \rho V$, where c_p is the specific heat, ρ is the density, and V is the antenna volume. The thermal mass of the antenna in this paper is m_t=25 fJ/K, which is extremely low compared to microbolometers [4]. The thermal time constant of the system can be expressed as $\tau_t = m_t R_t$, where R_t is the thermal resistance and corresponds to a -3 dB cutoff frequency of $f_c = 1/2\pi\tau_t$. However, this model does not include the heat removed by the lead lines of the NTC, and unlike microbolometers this is significant for ACNTCs due to the very low thermal mass of the antenna, and therefore cannot be neglected. Finite element method simulations (ANSYS) were used to calculate the theoretical frequency-dependent response of nanoantenna thermopiles. Figure 6 shows the comparison of the analytical model to finite element simulations. At low frequencies the two models match. However, at higher frequencies the analytical model predicts a much higher cutoff frequency than observed in experiment (Fig. 5) and calculated in simulations since the model does not include the thermal mass of the lead lines, which is a significant portion of the system.

REFERENCES

[1] G. P. Szakmany, P. M. Krenz, A. O. Orlov, G. H. Bernstein, and W. Porod, "Antenna-Coupled Nanowire Thermocouples for Infrared Detection," *IEEE Trans. Nanotechnol.,* vol. 12, pp. 163-167, 2013.

[2] G. P. Szakmany, A. O. Orlov, G. H. Bernstein, and W. Porod, "Cavity-Backed Antenna-Coupled Nanothermocouples," *Sci. Rep.,* vol. 9, p. 9606, 2019.

[3] G. P. Szakmany, A. O. Orlov, G. H. Bernstein, and W. Porod, "Single-Metal Nanoscale Thermocouples," *IEEE Trans. Nanotechnol.,* vol. 13, pp. 1234-1239, 2014.

[4] G. D. Boreman and E. L. Dereniak, *Infrared detectors and systems*: New York : Wiley, 1996.

P3-1

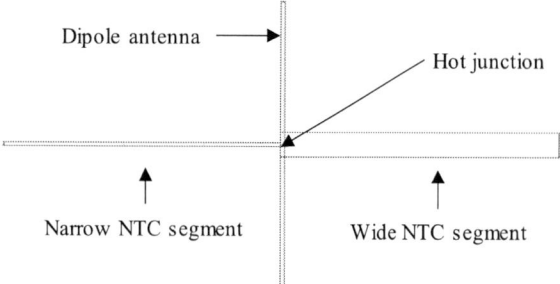

Fig. 1. Schematic of a single-metal antenna-coupled nanothermocouple.

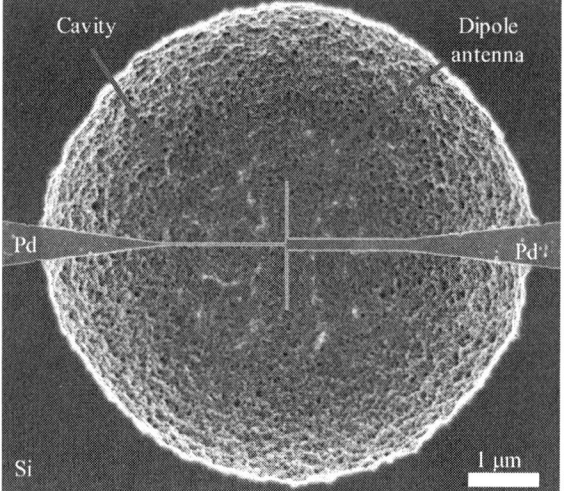

Fig. 2. SEM image of a suspended antenna-coupled nanothermocouple suspended over a cavity.

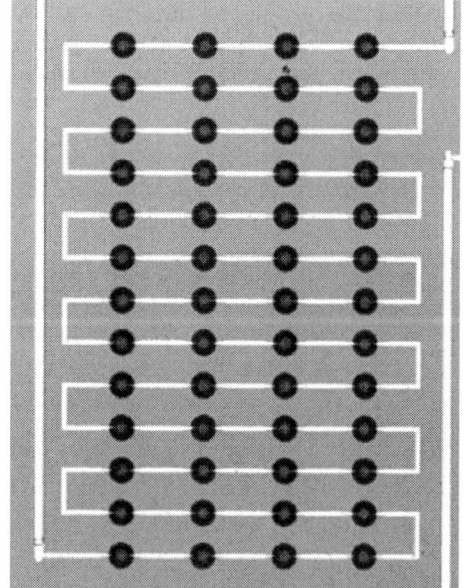

Fig. 3. Nanoantenna thermopile. Individual ACNTCs are connected in series by their cold junctions.

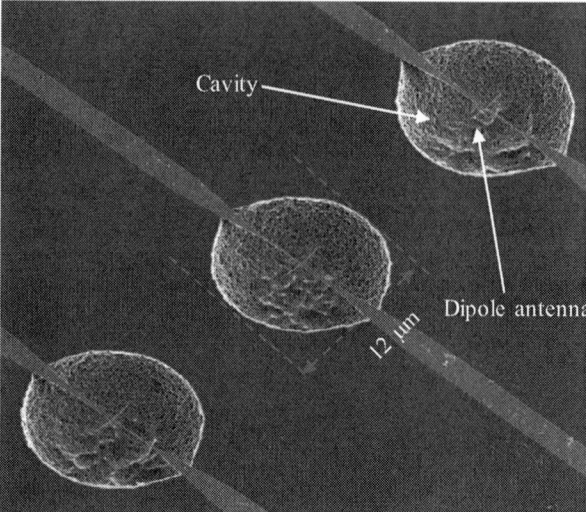

Fig. 4. Section of a nanoantenna thermopile. Each antenna possesses its own cavity.

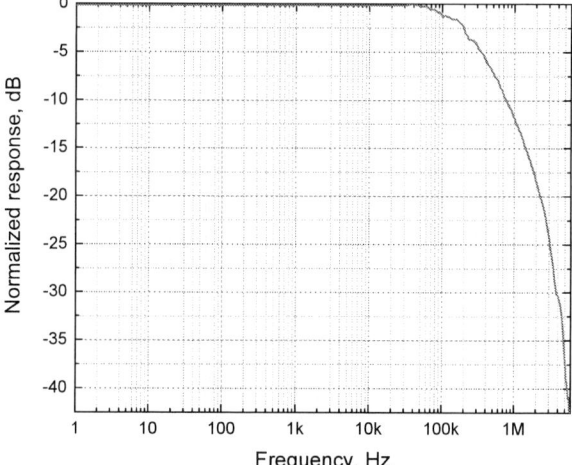

Fig. 5. Frequency-dependent response of the nanoantenna thermopile obtained from experiments.

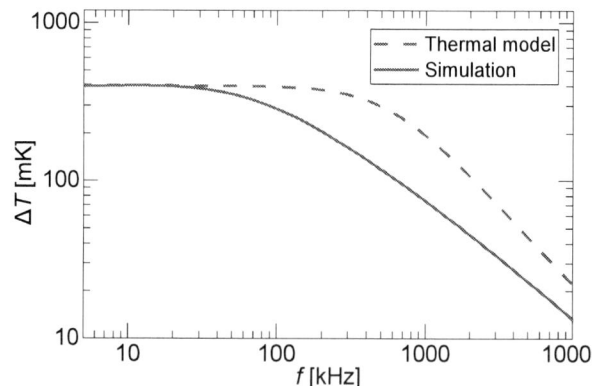

Fig. 6. Calculated and simulated frequency-dependent responses. The analytical calculation does not include the NTC lead lines, which significantly contribute to the thermal mass of the system.

2020 IEEE Silicon Nanoelectronics Workshop

Adiabatic Capacitive Logic using Voltage-controlled Variable Capacitors

Rene Celis-Cordova, Alexei O Orlov, and Gregory L Snider

Electrical Engineering Department, University of Notre Dame, USA

Email: rcelisco@nd.edu

Abstract — **Reversible computing is a promising approach to energy efficient computing that reduces heat generation by introducing a trade-off between energy and speed. The most developed approach to reversible computing is adiabatic CMOS, but its lowest energy dissipation is still limited by passive power, the energy wasted due to leakage current caused simply by applying a voltage to the circuit. A new approach, Adiabatic Capacitive Logic (ACL), implements reversible computing by using variable capacitors as pull-up and pull-down networks. ACL eliminates leakage current and therefore is not limited by passive power. We present the design and proposed nanofabrication of gap-closing voltage-controlled variable capacitors to implement ACL as a future computing approach.**

I. INTRODUCTION

As the end of Moore's law approaches, many efforts have emerged to pursue novel forms of energy efficient computing. One of the most promising approaches is reversible computing, which reduces the energy dissipation of a circuit by using reversibility and quasi-adiabatic transitions. Adiabatic CMOS is the most developed implementation of reversible computing. It uses CMOS circuits and replaces DC power supplies by ramping clocks to introduce a trade-off between energy and speed. However, the lowest energy dissipation in Adiabatic CMOS is limited by passive power, the energy dissipation caused by leakage.

Much research has been devoted to reducing passive power in CMOS circuits. Some of these efforts focus on creating steep devices [1] that reduce the leakage current in the subthreshold region of transistors. Other approaches try to eliminate the leakage current completely by using nano-electro-mechanical systems (NEMs) as relays to implement computing [2]. Unlike CMOS circuits, the NEMs relays do not have a subthreshold current since they have no electrical contact when they are off. However, the relays will wear out over time primarily due to degradation of the current-carrying contacts.

II. RESULTS AND DISCUSSION

Adiabatic Capacitive Logic (ACL) was proposed by Pillonnet and Houri [3] as a novel approach to reversible computing that eliminates leakage current. ACL implements variable capacitors as pull-up and pull-down networks instead of relays to implement reversible computing. Since capacitors do not need to be electrically connected, then the degradation of the contacts due to their mechanical operation can be avoided. The schematic symbol of a variable capacitor proposed for ACL is presented in Fig. 1(a). Logic gates can be created by using the variable capacitors as pull-up and pull-down networks, such as the inverter gate shown in Fig. 1(b) and the NAND gate shown in Fig. 1(c).

A. Simulation results

Gap closing voltage-controlled variable capacitors as seen in Fig. 2 are used to implement ACL digital gates. The gap closing structure is controlled with an input voltage, that is electrically isolated from the output capacitors with an insulator introduced in the cantilever. With no input voltage applied, the end of the cantilever is close to the top electrode, making CS1 much larger than CS2. When a voltage is applied to the input capacitor CL the cantilever moves down, increasing the value of CS2 and reducing that of CS1. Therefore, creating both pull-up and pull-down networks to implement digital gates. COMSOL Multiphysics was used to simulate the variable capacitors and ensure the gap-closing structure changes its capacitance, as presented in Fig. 3.

B. Fabrication

We propose to fabricate these voltage-controlled variable capacitors on silicon wafers covered with LPCVD silicon nitride as a base insulating layer. The conducting features of the electrodes and the cantilever are made of highly doped polysilicon, which is deposited by LPCVD, doped, and then polished by CMP, while the bottom electrodes are made of TiN. The cantilever is divided by a silicon nitride spacer and the top view of the design is presented in Fig 4.

The bottom electrodes are made of conducting TiN. TiN is used instead of a metal since metals enhance the deposition of the polysilicon used in overlying layers and create whisker defects [4] as seen in Fig. 5. Sacrificial layers of SiO_2 are deposited in between the conducting features that comprise the variable capacitors. The moving mechanical structures are released using a wet BHF etch to remove the sacrificial SiO_2 layers surrounding the cantilever.

The devices are currently under fabrication, as seen in Fig. 6, and present the first experimental implementation of ACL.

ACKNOWLEDGMENTS

This work was supported by the National Science Foundation. (NSF ECCS-1914061).

REFERENCES

[1] H. Kam, T.J.K. Liu, and E. Alon, "Design requirements for steeply switching logic devices", *IEEE transactions on electron devices*, 59(2), pp.326-334, 2011.

[2] A. Peschot, C. Qian, and T.J.K. Liu, "Nanoelectromechanical switches for low-power digital computing", *Micromachines*, 6(8), pp.1046-1065, 2015.

[3] G. Pillonnet, H. Fanet, and S. Houri, "Adiabatic capacitive logic: a paradigm for low-power logic", In 2017 IEEE International Symposium on Circuits and Systems (ISCAS) (pp. 1-4). May 2017.

[4] G. Choi, "The Metallic Contamination Related Whisker Defect During Deposition and Its Removal" *ECS transactions* 11, no. 2. 133, 2007.

P3-2

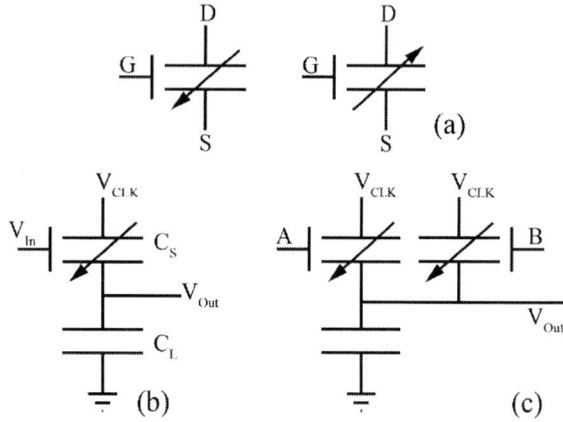

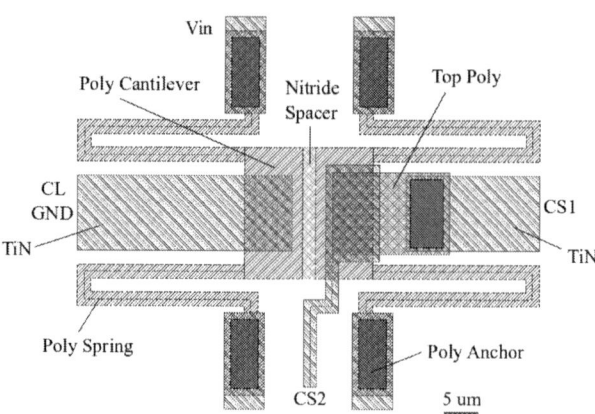

Fig. 1. Adiabatic capacitive logic. (a) Variable capacitor schematic symbols. (b) Inverter. (c) NAND gate.

Fig. 4. Top view of gap-closing voltage-controlled variable capacitor for ACL.

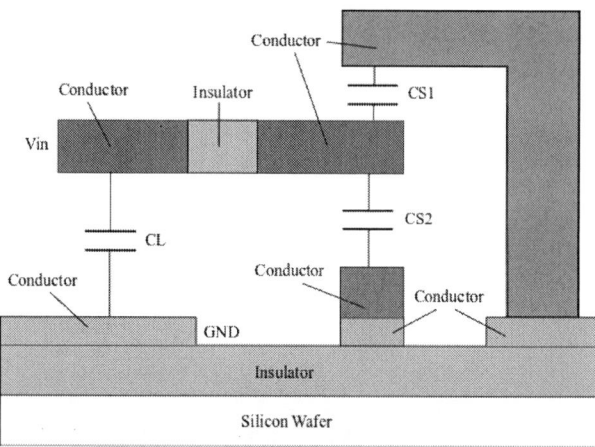

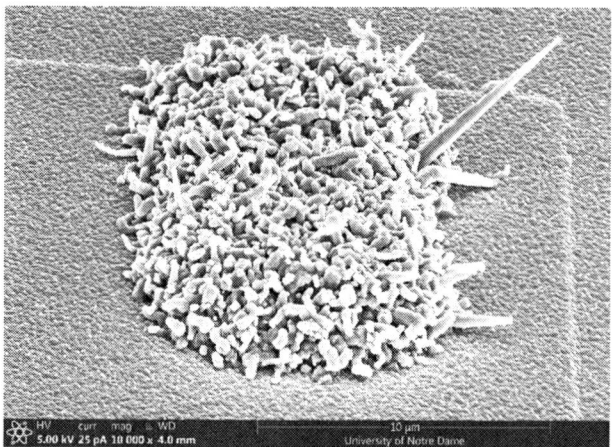

Fig. 2. Voltage-controlled variable capacitor gap-closing structure.

Fig. 5. SEM micrograph of LPCVD polysilicon deposited on Nickel showing whisker defects created where the metal enhances the polysilicon deposition.

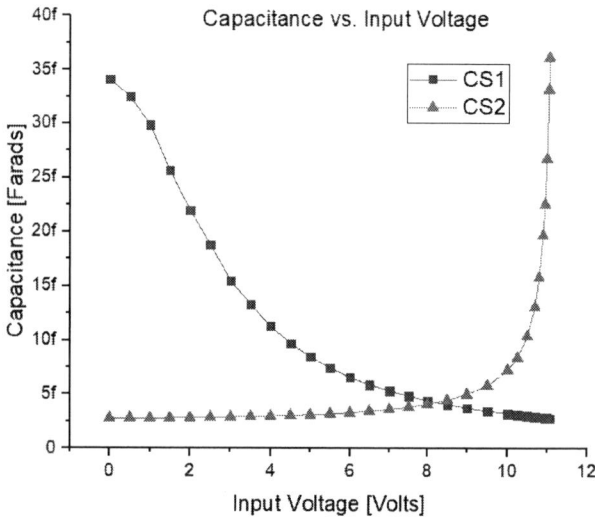

Fig. 3. Simulated capacitance of gap-closing ACL devices using COMSOL Multiphysics. The polysilicon cantilever is 300 nm thick and the distance between the cantilever and the top/bottom electrodes changes between 20 nm and 500 nm.

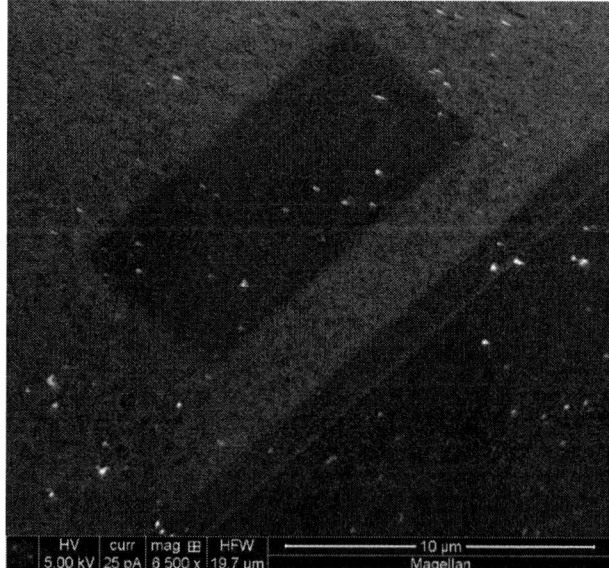

Fig. 6. SEM micrograph of LPCVD polysilicon deposited on conducting Titanium Nitride, after Chemical Mechanical Polishing.

2020 IEEE Silicon Nanoelectronics Workshop

978-1-7281-9736-4/20 $31.00 © 2020 IEEE

Resistive Approach for Extraction of Bias-Dependent Parasitic Resistance, Mobility and Virtual Gate Length in GaN HEMT

Pragyey Kumar Kaushik[1], Sankalp Kumar Singh[2], Ankur Gupta[1], Ananjan Basu[1], Edward Yi Chang[2]

[1]Centre of Applied Research in Electronics, Indian Institute of Technology Delhi, India

[2]National Chiao Tung University, Hsinchu Taiwan

Email: pragyeykumarkaushik@gmail.com

Abstract — **In this work, we introduce a resistive approach to extract various device performance parameters such as contact resistance, mobility and effective gate length. Without passivation of the un-gated region, HEMT device leads to forming of an extra gate length either side of gate called virtual gate length (δ_{LG}). Carrier mobility (μ_C) of 2DEG and δ_{LG} also depends upon V_{GS} (gate bias). The depleted channel also reduces the device current causing a significant increase in source/drain resistance (R_S/R_D).**

Keywords: Virtual gate, two-dimensional electron gas, parameter extraction, high electron mobility transistor.

I. INTRODUCTION

III-V nitride devices suffer from current collapse phenomena. This current collapse occurs due to electron trapping in AlGaN [1-4] and the virtual gate formation [2]. Available DC parameter extraction methods [5-6] provide change in mobility with gate bias but provide limited information to extract R_S and R_D. Parameter extraction by current through the channel by approximating the linear current is done in [7]. In this work, we introduce a relatively simple resistive method to extract δ_{LG}, μ_C, R_S and R_D from I_D-V_{GS} data and by biasing drain such, that device operates in linear region (Fig.1-3).

II. PARAMETER EXTRACTION

R_P is contact resistance between probe and source/drain. R_C is contact resistance between 2DEG to source/drain. We consider $R_P + R_C$ equal to R_{CP}, like [7]. The total probe to probe resistance between source and drain can be written as

$$R_T = R_D + R_S + R_{GCH} \quad (1)$$

Where R_T denotes total resistance, R_{GCH} denotes the channel resistance beneath the gate length L_G. Drain R_D and source R_S resistance can be written as

$$R_D = \left[\rho_{s0} \left(L_{DG} - \delta_{LG} \right) + \frac{\delta_{LG}}{(qn_s\mu_C)} \right] \frac{1}{W_G} + R_{CP} \quad (2)$$

$$R_S = \left[\rho_{s0} \left(L_{SG} - \delta_{LG} \right) + \frac{\delta_{LG}}{(qn_s\mu_C)} \right] \frac{1}{W_G} + R_{CP} \quad (3)$$

$$R_{GCH} = \left[\frac{L_G}{(qn_s\mu_C)} \right] \frac{1}{W_G} \quad (4)$$

Putting the value of R_D, R_S and R_{GCH} from (2), (3) and (4) respectively into (1), gives

$$R_T = \left[\rho_{s0} \left(L_{DG} + L_{SG} - 2\delta_{LG} \right) + \frac{2\delta_{LG} + L_G}{(qn_s\mu_C)} \right] \frac{1}{W_G} + 2R_{CP} \quad (5)$$

Total resistance for two different devices can be expressed as:

$$R_{TS} = \rho_{s0} P_S + \frac{2\delta_{LG}}{W_G} \left(\frac{Q}{\mu_C} - \rho_{s0} \right) + \frac{L_{GS}}{W_G} \left(\frac{Q}{\mu_C} \right) + 2R_{CP} \quad (6)$$

$$R_{TL} = \rho_{s0} P_L + \frac{2\delta_{LG}}{W_G} \left(\frac{Q}{\mu_C} - \rho_{s0} \right) + \frac{L_{GL}}{W_G} \left(\frac{Q}{\mu_C} \right) + 2R_{CP} \quad (7)$$

Solve equation 6 and 7 to get the expression for μ_C and δ_{LG}. After solving, we will get the expression equation 8 and 9 as,

$$\mu_C = \frac{\left(L_{GL} - L_{GS} \right) Q}{W_G \left[\left(R_{TL} - R_{TS} \right) - \rho_{s0} \left(P_L - P_S \right) \right]} \quad (8)$$

$$\delta_{LG} = \frac{1}{2} \left[\frac{\left(R_{TS} L_{GL} - R_{TS} L_{GS} \right) + \left(P_L L_{GS} - P_S L_{GL} \right) \rho_{s0} - 2R_{CP} \left(L_{GL} - L_{GS} \right)}{\left(R_{TL} - R_{TS} \right) - \left(P_L - P_S \right) \rho_{s0} - \rho_{s0} \left(L_{GL} - L_{GS} \right) / W_G} \right] \quad (9)$$

III. DEVICE STRUCTURE

Reference data for GaN HEMT device is taken from [7] and listed in table 1. V_T is taken as -5.43 V, from the long device at zero current voltage value because the long device is more linear than the short device (fig. 4(a)). ρ_{s0} calculated in [7] with TLM method and found the value of 336 Ω/mm. R_{CP} calculated by the fact that for $V_{GS} = 0V$, $\delta_{LG} = 0$, and comes 2.86 Ω.

Aluminium fraction taken from [7] is 28%. So 30nm barrier layer of $Al_{0.28}Ga_{0.72}N$ has considered over GaN relative permittivity of the $Al_{0.28}Ga_{0.72}N$ barrier layer is 9.36.

IV. RESULTS AND DISCUSSION

Extracted total resistance R_{TS} and R_{TL} with respect to V_{GS} in the linear region is shown in fig 4(b). Channel resistance increases when we increase V_{GS} because channel depletes towards pinch-off. Mobility can be calculate in the linear region by putting R_{TS} and R_{TL} values in (8). This provides similar behaviour as provided in [7] (Fig.4(c)).

Virtual gate phenomenon arises due to electron injection from gate being trapped in high-density surface state at gate periphery. Fig. 4(d) shows the comparison of calculated virtual gate length δ_{LG} and the data reported in [7]. It shows good agreement between validated values and the modelled equation (9). Calculated R_D and R_S are compared with [7] in Fig. 4(e) and shows good agreement. Calculated current is shown in fig.4(f) in blue as compared to measured current (red) which again shows good agreement between our modelled expression and measured data from [7].

V. CONCLUSION

In this paper, we introduce yet simple resistive approach to calculate drain-source parasitic resistance, mobility and virtual gate length on the basis of I_D (V_{GS}) graph of short and long HEMT, which also follow the results given in [7].

2020 IEEE Silicon Nanoelectronics Workshop

978-1-7281-9736-4/20 $31.00 © 2020 IEEE

P3-3

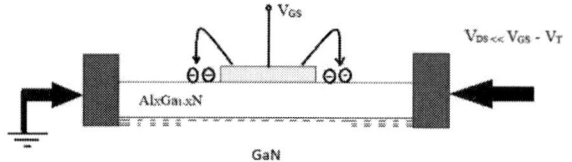

Fig.1 Cross-section view of surface quantum state occupied by an electron supplied by gate bias.

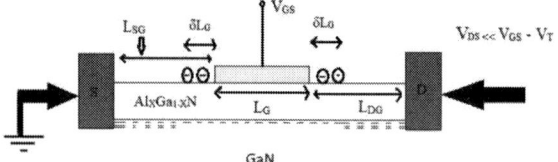

Fig.2 Dimensions ofrespective region of general HEMT.

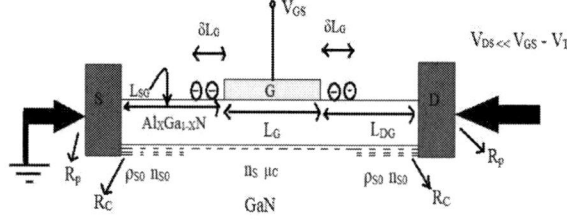

Fig.3 Resistance, charge concentration, mobility in different region of HEMT.

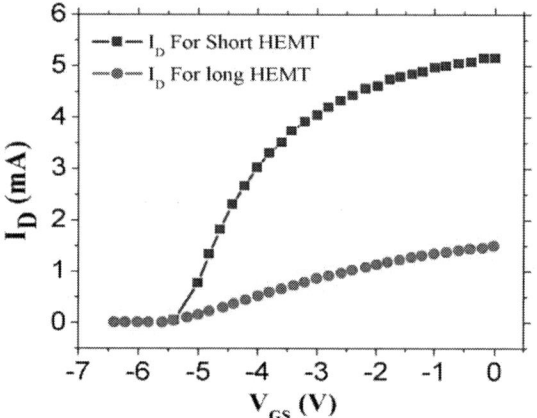

Fig.4 (a) Measured I_D– V_{GS} curve against V_{GS} [7].

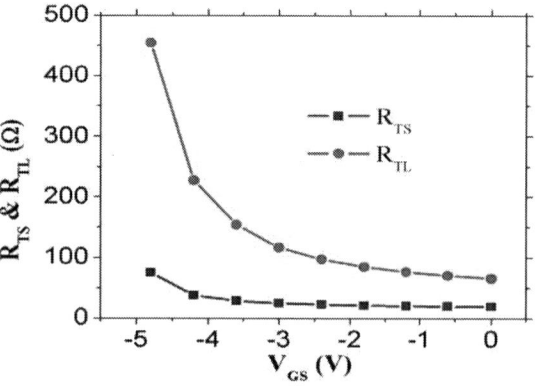

Fig.4 (b) Total resistance variation against V_{GS} of short and long HEMTs.

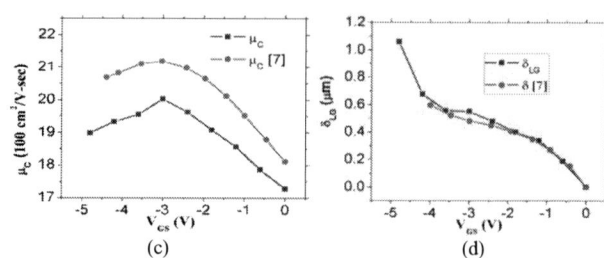

Fig. 4(c) Calculated mobility (blue) and reported mobility (red) against V_{GS} in [7]. (d) Calculated virtual gate length (blue) and reported gate length (red) against V_{GS} in [7].

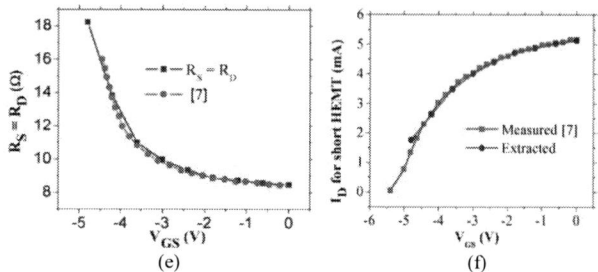

Fig. 4(e) Extracted and reported [7] source and drain resistance. (f) Comparison of measured [7] and extracted drain current for short gate HEMT.

Table. 1 Corresponding dimensions of short and long HEMT.

	W_G (µm)	L_{SG} (µm)	L_{Gs} (µm)	L_{DG} (µm)	d (nm)
Short HEMT	150	2.5	1	2.5	30
Long HEMT	150	2	20	2	30

REFERENCE

[1]. M. A. Khan, Shur, Chen, and Kuznia,"Current/voltage characteristic collapse in AlGaN/GaN hetero structure insulated gate field effect transistors at high drain bias," Electron. Lett., vol. 30, no. 25, pp. 2175–2176, Dec. 1994.

[2]. Vetury, Zhang, Keller, Mishra ,"The impact of surfacestates on the DC and RF characteristics of AlGaN/GaN HFETs." IEEE T Electron Dev 2001;48:560–6.

[3]. Binari, Ikossi, Roussos, Kruppa, Park, Dietrich, et al.,"Trapping effects and microwave power performance inAlGaN/GaN HEMTs. IEEE T Electron Dev 2001;48:465–71.

[4]. Klein, Binari, Ikossi nastasiou, Wickend, KoleskeDD, HenryRL, et al. "Investigation of traps producing current collapse in AlGaN/GaN high electron mobility transistors". Electronics Lett 2001;37:661–2.

[5]. A. M. Wells, M. J. Uren, R. S. Balmer, K. P. Hilton, T. Martin, and M. Missous, "Direct demonstration of the virtual gate mechanism for current collapse in AlGaN/GaN HFETs," Solid-State Electron., vol. 49, no. 2, pp. 279–282, Feb. 2005

[6]. R. Menozzi, G. A. U. Membreno, B. D. Nener, G. Parish, G. Sozzi, L. Faraone, and U. K. Mishra, "Temperature- dependent characteriza-tion of AlGaN/GaN HEMTs: Thermal and source/drain resistances," IEEE Trans. Device Mater. Rel., vol. 8, no. 2, pp. 255–264, Jun. 2008.

[7]. D. Pradeep, Amit and S. Karmalkar, "DC Extraction of Gate Bias Dependent Parasitic Resistances and Channel Mobility in a HEMT", IEEE Electron Device Lett., vol. 37, no. 11, pp. 1403-1406, Nov. 2016.

2020 IEEE Silicon Nanoelectronics Workshop

Understanding Transient Responses of Silicon Nanowire Photoconductors

Yaping Dan[1,2*], JiajingHe[1], Chulin Huang[2],

[1]University of Michigan-Shanghai Jiao Tong University Joint Institute, Shanghai Jiao Tong University, China, [2]School of Microelectronics, Xi'an Jiao Tong University, China
*Email: yaping.dan@sjtu.edu.cn

Abstract — **Understanding the transient responses of nanowire photoconductors helps to design the bandwidth of photoconductors. Here, we derived the theoretical transient responses of silicon nanowire photoconductors. The theoretical transient responses consist of two exponentials and fit well with the experimental results. The minority carrier lifetime and emission time of the surface trap states can be extracted from the fitting.**

Keywords: Photoconductors, transient process, surface states

I. INTRODUCTION

It is widely reported that nanoscale photoconductors often have extraordinarily high photogain in quantum efficiency. The classical gain theory[1] predicts that the high gain originates from the majority photocarrier circulation in the circuit after minority photocarriers are captured by surface trap states. The long trap time inevitably results in a long response time. However, this classical gain theory is not only implicit but also problematic as we previously found that the classical gain theory was derived on two misplaced assumptions.[2] Recently, we managed to derive explicit gain equations for single-crystalline photoconductors.[3] The explicit gain equations clearly show that the gain comes from the photocarrier separation by surface depletion region instead of photocarrier circulation.

In this work, we aim to find the physics behind the long response time in photoconductors. The results show that the fall transient response consists of two time-dependent exponentials with one associated with the depletion region capacitance and the other with the emission process of surface trap states.

II. RESULTS AND DISCUSSION

Silicon nanowire photoconductors (shown in Fig. 1) were fabricated by patterning the p-type device layer of a silicon-on-insulator wafer. These nanowires all have a surface depletion region due to the net charges on the nanowire surfaces. Photo Hall measurements indicate that the high gain in the nanowire photoconductors comes from the depletion region width modulation by light illumination.[3] As a result, a high gain photoconductor can be modelled as a resistor with floating Schottky junctions on its surfaces. Under light illumination, the photogenerated electron-hole pairs are separated by the electric field in the depletion region with electrons moving onto surfaces and holes into the channel, creating a photocurrent (J_L) across the depletion region. Since the Schottky junction is floating, a forward photovoltage (V_{ph}) and forward current (J_F) will be established, as a result of which the depletion region narrows down. When the light illumination is cut off, the forward current persists and the photogenerated electrons captured by surface trap states will emit, as a result of which the surface depletion region width will expand. At any time of t, we can establish a differential equation as eq.(1) which however is difficult, if not impossible, to analytically solve. If the electron emission from the surface trap state is very slow, the associate current can be ignored as shown in the case (a). In this case, eq.(1) can be simplified as eq.(2) and the solution can be found as in eq.(3). After the photovoltage vanishes, the depletion region variation is dominated by the electron emission from the surface states. In this case, eq.(1) can be simplified as eq.(4), the solution of which is given in eq.(5). In the end the photocurrent can be approximately written as eq.(6).

Fig.3 exhibits the transient photocurrent when the illumination laser is cut on/off periodically. One of the fall transient traces is plotted in Fig.4 which can be fitted well with eq.(6). From the fitting, we can find that the lifetime of the minority carriers is 96.74ns and the electron emission from trap state is 5.28s as shown in Table 1.

III. CONCLUSIONS

In this work, we found the transient photocurrent equation for photoconductors which can fit well with the experimental data. The fall photocurrent has two exponentials associated with depletion region capacitance and emission process of the surface trap states. The recombination lifetime of the minority carriers and the emission time of the surface trap states can be extracted from the fitting.

ACKNOWLEDGMENTS

This work was supported by the special-key project of the "Innovative Research Plan", Shanghai Municipality Bureau of Education (2019-01-07-00-02-E00075) and National Science Foundation of China (NSFC) (No. 61874072).

REFERENCES

[1] Petritz, R, "Theory of photoconductivity in semiconductor films," *Phys. Rev.* 104, pp 1508-1516, 1956.
[2] Yaping Dan, et al, "A photoconductor intrinsically has no gain," *ACS Photonics*, vol. 5, no. 10, pp. 4111-4116, September 2018.
[3] Jiajing He, et al, "Explicit gain equations for single crystalline photoconductors," *ACS Nano,* vol. 13, no. 3, pp.3405-3413, March 2020.
[4] Neamen, D. A. Semiconductor physics and devices: basic principles, 4th ed.; Mc-Graw-Hill Education: New York, 2011.

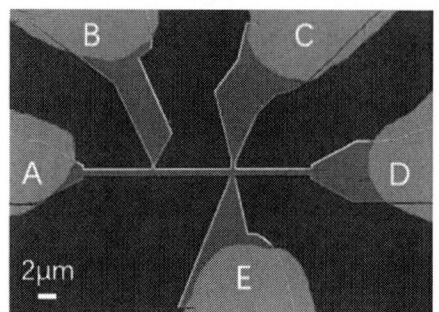

Figure 1. Optical microscopy image of the silicon nanowire photoconductor.

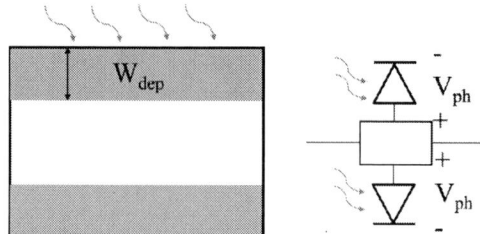

Figure 2. Schematic of equivalent device model for photoconductors.

Theoretical Derivation:

$$\frac{d(V_{bi}C_{Si})}{dt} - \frac{dQ_s}{dt} - J_F = 0 \qquad (1)$$

, in which $V_{bi} = \frac{qN_AW_{dep}^2}{2\varepsilon_s}$, $C_{Si} = \frac{\varepsilon_s}{W_{dep}}$, $Q_s = Q_{fixed} + Q_{s0}[1 - \exp(-\frac{t}{\tau_{em}})]$, $J_F = J_s[\exp\left(\frac{qV_{bi0}-qV_{bi}}{nk_BT}\right) - 1]$

Suppose $\tau_{em} \gg \tau$ where τ is the decay time when there are no surface states.

(a) In the first times of τ, $\frac{dQ_s}{dt}$ is negligible. Eq.(1) can be simplified as eq.(2).

$$\frac{1}{2}qN_A\frac{dW_{dep}}{dt} = J_s[\exp\left(\frac{qV_{bi0}-qV_{bi}}{nk_BT}\right) - 1] \qquad (2)$$

The solution is

$$W_{dep} = W_{dep0} + \frac{W_{dep0}}{\varphi}\ln\{1 - \exp[-\frac{2J_s}{qN_A}\frac{\varphi}{W_{dep0}}(t-t_0)]\} \quad (3)$$

, in which $\varphi = \frac{2qV_{bi0}}{nk_BT}$, $t_0 = \frac{qN_AW_{dep0}}{2J_s\varphi}\ln[1 - \exp(-\frac{2qV_{bi0}I_{ph0}}{nk_BTI_{ph}^S})]$

(b) After the first times of τ, the photovoltage will be nearly zero, i. e. $J_F \approx 0$. As a result, eq.(1) will become

$$\frac{d(V_{bi}C_{bi})}{dt} - \frac{dQ_s}{dt} = 0 \qquad (4)$$

, the solution of which is

$$W_{dep} = W_{dep0} + \frac{Q_{s0}}{N_A}[1 - \exp(-\frac{t}{\tau_{em}})] \qquad (5)$$

with τ_{em} being the emission time of the surface trap states.

The photocurrent can be derived as following by adding up the variation of the depletion region width in case (a) and (b).

$$I_{ph} = -\frac{nkT}{2qV_{bio}}I_{ph}^S\ln\left[1 - Aexp\left(-\frac{t}{\tau}\right)\right] + 2\frac{I_{ph}^SQ_{s0}}{qN_AW_{dep0}}\exp\left(-\frac{t}{\tau_{em}}\right)$$

$$(6)$$

, in which $A = 1 - \exp\left(-\frac{2qV_{bio}I_{pho}}{nk_BT\,I_{ph}^S}\right)$, $I_{ph}^S = \frac{\hbar\omega J_sH}{\alpha qW_{dep}}$, Js is the leakage current density of the Schottky junction, $\hbar\omega$ is the photon energy, α is the nanowire photo absorption ratio, H is the nanowire thickness, W_{dep} is the surface depletion width, Q_{s0} is the surface charge concentration in darkness, q is the unit charge, V_{bi0} is the built-in voltage of surface depletion, I_{ph0} is the photocurrent at t = 0. The physical meaning of other constants can be found in Ref [3].

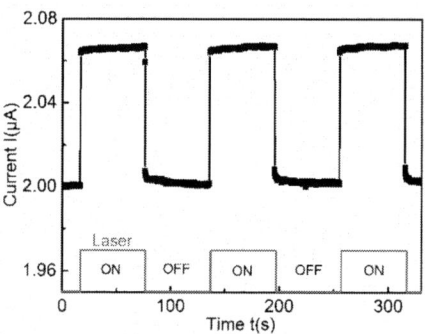

Figure 3. Transient photocurrent when the light illumination is switched ON/OFF periodically.

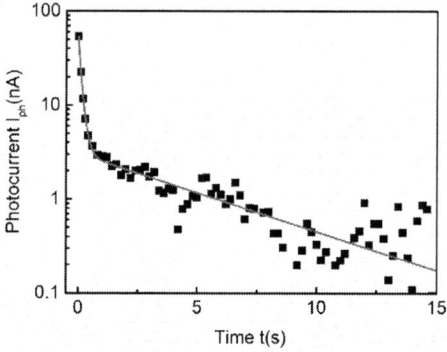

Figure 4. Experimental photocurrent fitting for one period.

Table 1. Extracted parameter values from Fig. 4.

	Value	Standard Error
Js	43.42 nA/cm²	1.19 nA/cm²
τ_{em}	5.28 s	0.36 s
τ	96.74 ns	2.65 ns

2020 IEEE Silicon Nanoelectronics Workshop

Optimal Approach to Scaling of the NEMS for Low Stand-*by* CMOS Applications

Sumit Saha[*,1,4], Sanjog Joshi[1,4], Tejas Naik[1], [2]Mayank Goel, V. Ramgopal Rao[1,3], and Maryam Shojaei Baghini[1]

[1]*Indian Institute of Technology Bombay, Powai, Mumbai, Maharashtra 400076, India,* [2]*Intel Corporation, Portland, Oregon Area, USA,*[3]*Indian Institute of Technology Delhi, Hauz Khas, New Delhi 110016, India.*
[4]*These authors contributed equally to this work *Email:* <u>sumit.saha@ee.iitb.ac.in</u>

Abstract: We report here for the first time, a simple novel scaling approach is proposed to achieve low pull-in voltage (V_{pi}), delay (t_{delay}), energy (U) and mechanical stress (σ) in the NEMS analogous to MOSFETs dimensional scaling. The study provides an efficient design methodology to achieve user specified percentage improvement of a specifically targeted parameter (V_{pi}, t_{delay}, U or σ) with the improvement in other target parameters. The approach is validated with reported experimental data and simulations.

Introduction: V_{pi}, t_{delay}, U and σ are the crucial parameters for the electrical performance, reliability and mechanical strength of the NEMS device. The proposed scaling method combines the first order models that evaluate the dependence of the geometry (length L (technology node), width W, thickness t and air gap g) on technological parameter (V_{pi}), performance parameters (t_{delay} & U) and reliability parameter (σ). It improves overall performance for all four critical parameters, thus ensuring improved electrical performance and mechanical stability, while scaling down. Scaling rules have been reported in [1-4]. However, they have considered only the electrical parameters. In [1], they start with scaling the geometrical parameters and eventually improve the target parameters. In contrast, we start itself with the user specified improvement in the target performance parameters as desired and correspondingly scale down the physical dimensions of the NEMS. This provides an effective way of implementation from the user end. In our study, we have taken into account the mechanical stress (σ), which indicates the device failure due to yielding or fatigue owing to the increase in stresses, due to miniaturization. Our approach with appropriate normalization also incorporates material properties along with the structural mechanics. We successfully validate our proposed scaling method with CoventorWare simulations and reported experimental data for fabricated NEMS device.

Scaling method and the design rule: Fixed-fixed beam structures [3] are taken into consideration for the demonstration of our analysis. The important physical parameters of the NEMS are L, W, t and g. We scale these four parameters while scaling down the devices. Fig. 1 shows our scaling model. It first describes the generalized equations for the desired target parameters. L and W have been scaled by a constant scaling parameter S (<1). Thickness t and gap g are scaled by factors S^a and S^b, respectively (where a, b are the scaling exponents). Substitution of the scaled geometry parameters into the generalized equations gives updated parameters. To ensure the reduced values, we set up the linear inequality problem. The solution set guarantees the lower values of these target

parameters for the scaled devices. Fig. 2 shows the lines along which the mentioned target parameter is invariant for a given value of S. From the convex solution set, we choose a, b away from these boundaries which ensures larger reduction for all target parameters (e.g. a=1.5 and b=1, however, any value can be chosen within the solution set) as well as it takes care of non-linearity and leads to the performance improvement. Fig. 3 gives the simple design rules based on the above discussion. It first shows the scaling factors for each targeted parameter. These are calculated by substituting a, b (a=1.5, b=1) values into eq. (III) as given in Fig. 1. Let us consider, the user wants to scale a particular target parameter (e.g. V_{pi}) by fraction X (0.8). From this X, we obtain S and calculate the new device dimensions using eq. (II) in Fig. 1 (with a=1.5, b=1). The target parameter is certainly scaled X times, and the other target parameters are also scaled by the scaling factors as mentioned in the eq. (III) in Fig.1. S is less than 1 and a, b is greater than zero to ensure the scaling down and performance improvement for all target quantities.

Validation experiments and Results: For experimental validation, three devices are selected from Fig. 5 & 6 in [4], one assigned as a reference and other two as scaled structures. In Fig 2, the marked points (A&B) which show the scaled devices lie on the appropriate side of the pull in invariant line, i.e. the one with smaller pull in value satisfies the inequalities and vice versa. Hence the model is validated using experimental data. We have also validated the theoretical results with CoventorWare simulations [6]. Five different device geometries varying the value of S as 1 (reference model [5]), 0.9, 0.8, 0.7 and 0.6 (a=1.5, b=1) are considered. Fig. 4(a) and 4(b) show the variation of the V_{pi} and σ with 1/S, and the variation of the U and t_{delay} with 1/S, respectively using the proposed scaling model. The experimental value of the V_{pi} and t_{delay} [5] is well matched with the theoretical and the simulation value (S=1). The detailed simulation results for the reference structure are shown in Fig. 5 and 6. The same structure is fabricated as shown using SEM imaging of NEMS device in Fig 7. OFF-state capacitance and the operation voltage is used for the calculation of U. The simulated values of the target parameters for the three devices (S=1, 0.8, 0.6) are listed in Table I. The values match well with the theoretical values.

Conclusion: An electro-mechanical scaling model is proposed and validated with reported experimental data as well as simulation, which match with our in-house fabrication. The proposed scaling model could be useful for a variety of electrostatic NEMS applications ranging from switches in low stand-*by* CMOS circuits to sensor nodes.

P3-5

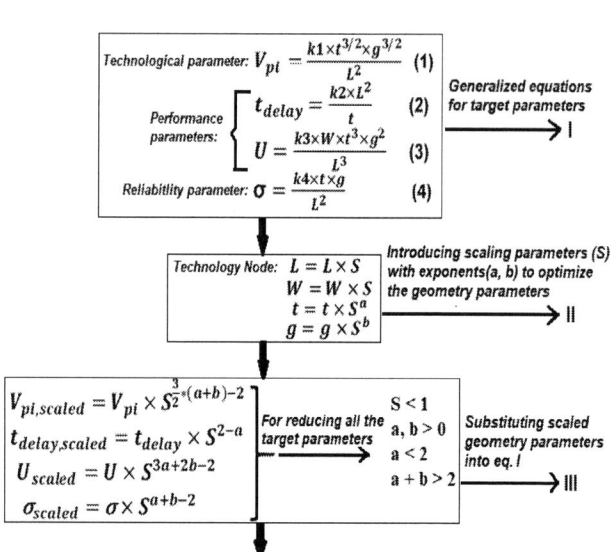

Technological parameter: $V_{pi} = \dfrac{k1 \times t^{3/2} \times g^{3/2}}{L^2}$ (1)

Performance parameters:
$\begin{cases} t_{delay} = \dfrac{k2 \times L^2}{t} & (2) \\[2mm] U = \dfrac{k3 \times W \times t^3 \times g^2}{L^3} & (3) \end{cases}$

Reliability parameter: $\sigma = \dfrac{k4 \times t \times g}{L^2}$ (4)

→ Generalized equations for target parameters → I

Technology Node:
$L = L \times S$
$W = W \times S$
$t = t \times S^a$
$g = g \times S^b$

Introducing scaling parameters (S) with exponents(a, b) to optimize the geometry parameters → II

$V_{pi,scaled} = V_{pi} \times S^{\frac{3}{2}*(a+b)-2}$
$t_{delay,scaled} = t_{delay} \times S^{2-a}$
$U_{scaled} = U \times S^{3a+2b-2}$
$\sigma_{scaled} = \sigma \times S^{a+b-2}$

For reducing all the target parameters →
$S < 1$
$a, b > 0$
$a < 2$
$a + b > 2$
→ Substituting scaled geometry parameters into eq. I → III

Choose appropriate S, a & b to scale the geometry parameters for the required improvement in target parameters and proceed for the simulation/ device fabrication

Fig.1 Flowchart of a simple NEMS optimization procedure.

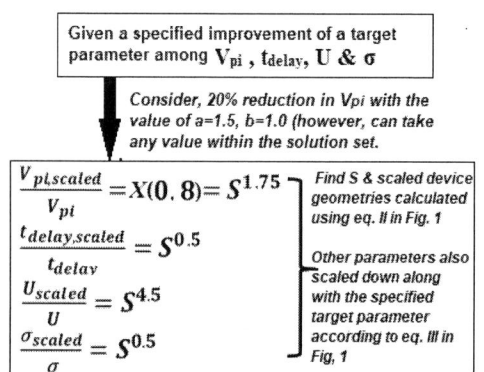

Given a specified improvement of a target parameter among V_{pi}, t_{delay}, U & σ

Consider, 20% reduction in Vpi with the value of a=1.5, b=1.0 (however, can take any value within the solution set.

$\dfrac{V_{pi,scaled}}{V_{pi}} = X(0.8) = S^{1.75}$

$\dfrac{t_{delay,scaled}}{t_{delay}} = S^{0.5}$

$\dfrac{U_{scaled}}{U} = S^{4.5}$

$\dfrac{\sigma_{scaled}}{\sigma} = S^{0.5}$

Find S & scaled device geometries calculated using eq. II in Fig. 1

Other parameters also scaled down along with the specified target parameter according to eq. III in Fig, 1

Fig.3 Flowchart of an example using our simple scaling model.

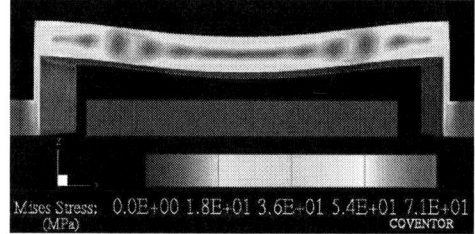

Fig.6 Mises stress distribution using simulation

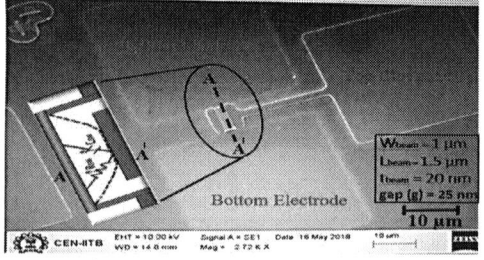

Fig.7 SEM image of the fixed-fixed NEMS fabricated at IIT Bombay.

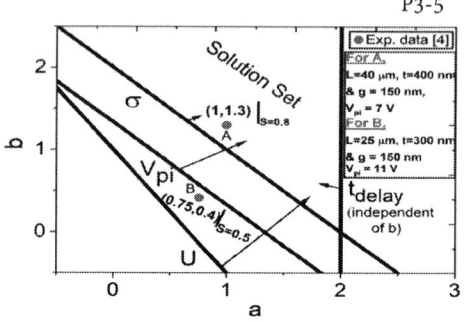

Fig.2 Model validation with measured data [4]. Assigned reference: L = 50 µm, t = 500 nm, g = 200 nm and V_{pi} = 9 V [4]. For Scaled devices, V_{pi} (A) = 7 V, V_{pi} (B) = 11 V from which S, a & b obtained as marked in the graph.

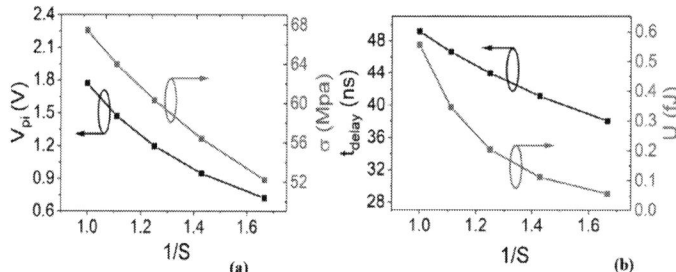

Fig.4 (a) Variation of V_{pi} and σ with 1/S, (b) Variation of t_{delay} and U with 1/S using the proposed scaling model.

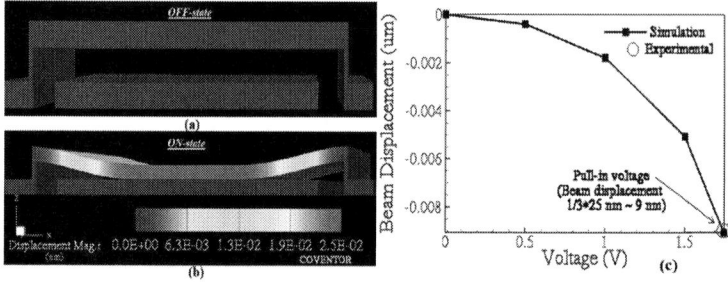

Fig.5 CoventorWare simulation (a) OFF-state (b) ON-state, (c) Beam displacement with the applied voltage at pull-in condition using simulation.

TABLE I: SIMULATED VALUE OF THE TARGET PARAMETERS

S	L (µm)	W (µm)	t (nm)	g (nm)	V_{pi} (V) sim.	V_{pi} (V) exp. [5]	t_{delay} (ns) sim.	t_{delay} (ns) exp. [5]	U (fJ) sim.	σ (Mpa) sim.
1 (ref.)	1.5	1	20	25	1.72	1.8	49.64	50	0.52	67.21
0.8	1.2	0.8	14.3	20	1.09		44.76		0.19	60.19
0.6	0.9	0.6	9.29	15	0.58		38.22		0.04	52.01

Reference:
[1] H. Kam et al., IEEE Transaction on Electron Device, Vol. 58, No. 1, 2011.
[2] J. Oen Lee et al., Nature nanotechnology, Vol. 8, 2013.
[3] C. Pawashe et al., IEEE Transaction on Electron Device, Vol. 60, No. 9, 2013.
[4] Y. Qian et al., Transducer, Anchoroga, 2015.
[5] N. Gilda et al., IEEE Journal of Microelectromechanical System, Vol. 26, No.3, 2017.
[6] www.coventor.com/coventorware.

2020 IEEE Silicon Nanoelectronics Workshop

Electroluminescence of Er:O-doped nano pn diode in silicon-on-insulator[P3-6]
and its current-voltage characteristics at room temperature

Takafumi Fujimoto[1], Keinan Gi[1], Stefano Bigoni[2], Michele Celebrano[2], Marco Finazzi[2], Giorgio Ferrari[2],
Takahiro Shinada[3], Enrico Prati[4], Takashi Tanii[1]

[1]School of Science and Engineering, Waseda University, Japan
[2]Politecnico di Milano, Italy
[3]Center for Innovative Integrated Electronic Systems, Tohoku University, Japan
[4]Istituto di Fotonica e Nanotecnologie, Consiglio Nazionale delle Ricerche, Italy
Email: fujimoto@tanii.nano.waseda.ac.jp

Abstract — **Electroluminescence from erbium-doped nanoscale pn diodes was achieved. Decreasing the number of erbium ions in scaling light-emitting silicon devices decreases the emission intensity. This trend opens new possibility of single-photon emission — key function of quantum communication. Furthermore, doping by ion implantation takes advantage of controlling the number and position of erbium ions in the device. According to this trend, we fabricated pn-diodes with dimensions of telecom wavelength and observed the electroluminescence from the erbium doped region at the forward bias of 1.2 V. We discuss the photoemissivity and the current-voltage characteristics of the device, toward the single-photon emission.**

Keywords: electroluminescence, erbium, nanoscale pn diode

I. INTRODUCTION

The use of inner-shell transition of Er^{3+}: $^4I_{13/2} \rightarrow {}^4I_{15/2}$, which is enhanced by the formation of $Er:O_x$ complexes in silicon, is promising for single-photon sources in quantum communication at telecom wavelength [1]. Low-energy ion implantation and subsequent rapid thermal annealing take advantages of controlling the number and position of erbium and oxygen atoms as well as the compatibility of CMOS processes. Yet, the electrically controllable practical single-photon source has not been developed.

Single-ion implantation is a powerful tool to implant ions one by one in a target material. Thus far, threshold voltage trimming [2] and Anderson-Mott transition [3] in arrays of a few dopant atoms in a silicon transistor have been demonstrated using single-ion implantation techniques. Also, single-photon emission from a colour centre in diamond has been achieved by controlling the number and position of impurities implanted in the diamond [4,5]. These demonstrations suggest that such deterministic ion implantation has potential to open the possibility of fabricating CMOS-compatible single-photon emitters by combining photoemissive impurities or defects with silicon devices.

Here, we report on the electroluminescence of Er:O-doped silicon pn diode fabricated on silicon-on-insulator (SOI) substrates. Er:O complexes are formed by co-implantation of erbium and oxygen to a specially designed cuboid with the dimension of 440 nm or 880 nm width (single or double wavelength in silicon) and 220 nm thickness (half wavelength) located at the pn junction. The cuboid emits photons at the wavelength around 1.5 µm when the forward bias is applied. Comparison between diodes with and without

Er:O doping reveals the difference in the current-voltage (I-V) characteristics and in photoemissivity. We discuss such differences with results from device simulation, toward the single-photon emission.

II. RESULTS AND DISCUSSION

Figure 1 shows the schematics of the pn diode. The diode was fabricated in a n-type SOI layer with the initial phosphorus concentration of $\sim 10^{15}$ cm^{-3} using electron beam lithography and dry etching. The diode has the boron-doped (P$^+$) diffusion layer and the phosphorus-doped (N$^-$) extension on both sides and a nanoscale cuboid in-between. The cuboid was doped by co-implanting erbium and oxygen at 20 keV with the fluence of 10^{14} cm^{-2}. The activation of dopants and the formation of Er:O complexes were performed simultaneously by thermal annealing at 900 °C for 30 min. The formation of Er:O complexes by thermal annealing has already been confirmed by atom probe tomography [6].

Fig. 2 shows the representative confocal microscopy image of the electroluminescence from the diodes with and without Er:O doping at room temperature. Both the diodes are biased at 1.2 V in the forward direction, and the images were taken with a 1.3 µm long pass filter. The confocal images clearly show the photoemission from the Er:O-doped cuboid, as compared with the slight emission from the non-doped one. The number of photons detected by the avalanche photodiode was proportional to the current (data not shown), showing the emissivity of $\sim 10^6$ photons per second at the forward current of 10 µA.

Fig.3 shows the electroluminescence and the current trend versus the bias voltage. Spectroscopy confirmed photon emission peaked around 1.5 µm, which can be explained by both active erbium ions as from photoluminescence at this concentration [7] and possible D1 dislocation [8] introduced by the doping process. We conjecture that the background emissions can be discriminated by improving the dry etching and ion implantation condition.

Fig. 4 shows the I-V characteristics of the pn diodes at room temperature. Intriguingly, the Er:O-doped diodes exhibit a supressed forward current below 0.4 V, whereas the diode without Er:O doping exhibits normal exponential characteristics. The I-V characteristics calculated by using 2-dimensional device simulator PISCES-II [9] are shown in Fig. 5. We found that the erbium ions implanted in the cuboids, which is known to act as donor, [10] increases the potential barrier for electrons diffusing across the P$^+$/N$^-$ junction, supressing the forward current. We confirmed that all the Er:O-doped diodes showing the supressed forward current

2020 IEEE Silicon Nanoelectronics Workshop

exhibit photoemission at the forward bias of 1.2 V, indicating the characteristics is originated from the erbium doping. The potential barrier height decreases with increasing forward bias and becomes negligible at the forward bias of 1.2 V where ohmic current appears in the I-V characteristics. The experimental results that the nanoscale structures exhibit distinguishable electroluminescence indicate the possibility of exploring erbium-doped nanodiodes towards single-photon emission [11].

ACKNOWLEDGMENTS

This work was supported by JSPS KAKENHI (18H03766, 17H0251), and partly by the Waseda University Grant for Special Research Projects (2018K-214) and QUASIX Grant from Italian Space Agency.

REFERENCES

[1] M. Celebrano *et al*, *Nanomaterials* **9**, 416, 2019.
[2] T. Shinada *et al*, *Nature* **437**, 1128, 2005.
[3] E. Prati *et al*, *Nat. Nanotechnol.* **7**, 443, 2012.
[4] S. Tamura *et al*, *Appl. Phys. Express* **7**, 115201, 2014.
[5] R. Fukuda *et al*, *New J. Phys.* **20**, 083029, 2018.
[6] Y. Shimizu *et al*, Silicon Nanoelectronics Workshop 2017.
[7] M. Celebrano *et al*, *Opt. Lett.* **48**, 3311, 2017.
[8] B. Zheng *et al.*, *Appl. Phys. Lett.* **64**, 2842, 1994.
[9] PISCES, Advanced 1D and 2D Device Simulation for Silicon, 1994, The Board of Trustees of the Leland Stanford Junior University.
[10] F. Priolo *et al*, *Phyt.Rev.* B **57**, 4443, 1998.
[11] F. Cavaliere *et al*. *Quantum Reports* **2**, 80-106, 2020.

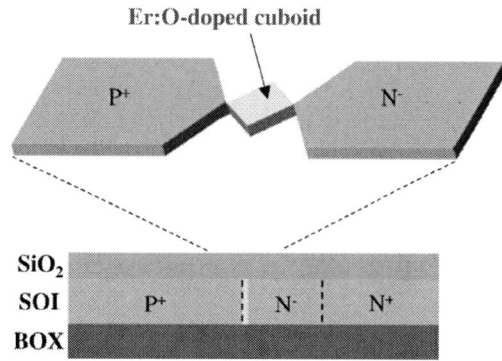

Fig.1: Schematics of the pn diode.

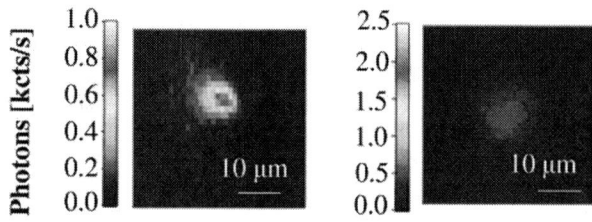

Fig:2 Confocal microscopy images of the emission from (left) Er:O-doped diode and (right) no-doped one. The region around the cuboids were observed at room temperature.

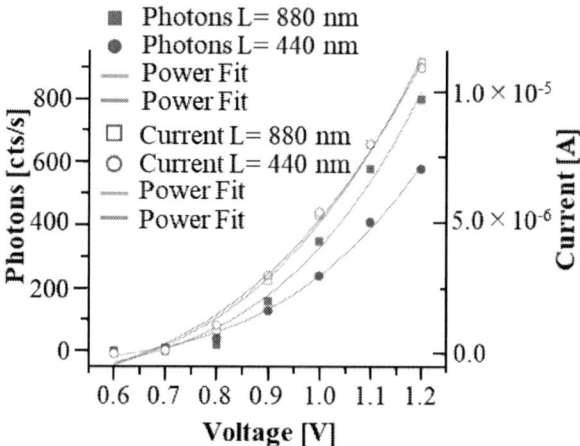

Fig:3 Relationship between photoemission and applied forward bias voltage.

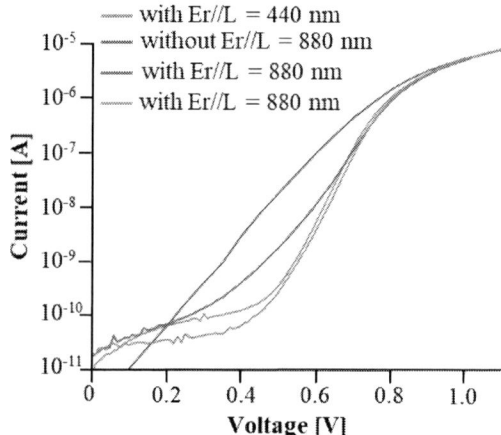

Fig.4: I-V characteristics of the pn diodes at room temperature.

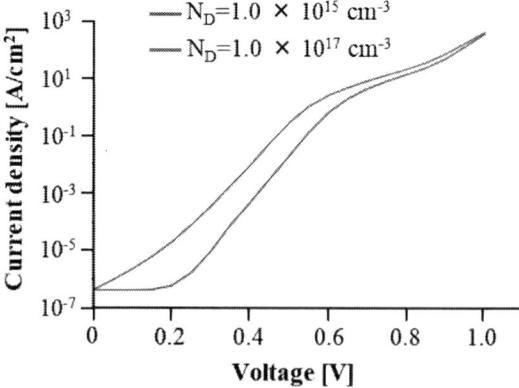

Fig:5 Simulated I-V characteristics of Er-doped ($N_D = 10^{17}$ cm^{-3}) and non-doped one ($N_D = 10^{15}$ cm^{-3}). We assume that implanted erbium ions act as donor.

2020 IEEE Silicon Nanoelectronics Workshop

Performance Investigation of Universal Gates and Ring Oscillator using Doping-free Bipolar Junction Transistor

Abhishek Sahu, Abhishek Kumar, and Shree Prakash Tiwari[#]

Department of Electrical Engineering, Indian Institute of Technology Jodhpur, Karwar, Jodhpur 342037, India

[#]Email: sptiwari@iitj.ac.in

Abstract — **Performance of symmetric lateral doping-free bipolar junction transistor (BJT) on silicon on insulator (SOI) in universal gates and ring oscillator were investigated. Charge carriers in SOI at emitter and collector regions are induced with two unique approaches, i.e., the charge plasma (CP) and polarity control (PC). Four types of devices (CP-NPN, CP-PNP, PC-NPN, and PC-PNP) was used for bipolar CMOS type NAND and NOR gates. Excellent transient response with rise and fall time less than 5 ns and propagation delay less than 2.4 ns were obtained.**

Keywords: Charge plasma (CP), doping free BJT, polarity control (PC), TCAD simulation reconfigurable gates,

I. INTRODUCTION

Symmetric lateral bipolar junction transistor (BJT) is a promising candidate for low power mixed-signal applications because of its CMOS process compatibility and higher cut off frequency [1] – [3]. However, with the continuous scaling, these structures require complicated fabrication process and high thermal budget for doping and annealing processes. Meanwhile, devices with low fabrication costs and simplified processes are reported [4], but with unrealistic gain of 4000-10000 and a reduced frequency response, i.e., a low cut-off frequency (range of 3.6-10 GHz). Recently, a device with good frequency response and decent gain with no doping requirement was demonstrated [5]. However, reports on circuit performance with these devices are scarce.

Here, we present the performance of universal gates with doping-free symmetric lateral BJT. The charges are induced with two approaches viz. charge plasma (CP) and polarity control (PC) techniques. PC based devices have additional capability of reconfigurability, where the same circuit can work as NAND/NOR by changing the voltage polarity [5]. NAND and NOR gates are designed with standard four transistors CMOS-like complementary bipolar logic, and mixed-mode simulations are performed. 1-D transport equations were solved using 2D simulations in Sentaurus TCAD O-2018.06 for analysis of these devices and circuits.

II. DEVICE STRUCTURE, RESULTS, AND DISCUSSION

A. Device Structure

Fig. 1 shows cross-sectional schematic of both types of devices, where L_E is length of emitter and collector regions, L_B is length of base, and L_{Si} represents length of lightly doped (order of 10^{15} cm^{-3}) silicon (Si) film where emitter, base and collector regions are formed by both CP and PC approaches. For device and circuit simulations, $L_B = 65$ nm, $T_B = 15$ nm, $T_{Si} = 20$ nm, low input = 0V, high input = 1V, and power supply voltage of 1V are taken. Induced carrier concentration are extracted along the device length, as shown in Fig. 2.

B. Transient Analysis of Logic Gates

The balanced device design [5], is used to design complimentary NAND and NOR gates. The Voltage-Time Characteristics (VTCs) for both the gates support their logical expression (Fig 3). Extracted parameters are summarised in Table II. The peak and average switching power consumed by individual devices are shown in Fig 4, indicating that the average switching power is almost constant for all device combinations, while peak power for PC based devices is less than CP based devices. On the other hand, PC devices are slower than CP based devices. Moreover, it was found that that NAND is more power-efficient than NOR gate. The input noise margin is calculated by drawing the butterfly curves for both the inputs V_a & V_b for all the device configurations. The butterfly curves shown in Fig. 5 for NAND gates show a worst-case noise margin of 0.41 V and 0.42 V for CP and PC based devices, respectively. For NAND, worst case input is V_b, while for NOR, it is V_a because the transistors corresponding to these inputs face a series impedance of next transistor.

C. Ring Oscillator Performance Analysis

Device design parameters optimized for the best inverter characteristics were used to create 3-stage ring oscillator. Promising ring oscillator characteristics are achieved with both the CP and PC based devices (Fig. 5). Obtained oscillating frequency with CP and PC based devices are is 1.11 GHz and 0.70 GHz respectively.

II. SUMMARY

We have demonstrated universal gates and 3-stage ring oscillators based on the doping-free BJT devices. These devices can be utilized in designing cost-effective circuits, as these devices knock-down the complex doping process while offering comparable performance matrices. These performance matrices can be further improved by selecting appropriate sizing. This study indicates that these devices have potential to be used for more complex complementary bipolar logic circuits.

REFERENCES

[1] T. H. Ning et al., IEEE JEDS, vol. 1, no. 1, pp. 21-27, Jan. 2013.

[2] G. G. Shahidi et al., IEDM 1991 pp. 663-666.

[3] Parke et al., IEDM 1992, pp. 453-456.

[4] M. J. Kumar et al., IEEE TED, vol. 59, no. 4, pp. 962-967, April 2012.

[5] A. Sahu et al., IEEE TED, vol. 63, no. 7, pp. 2684-2690, July 2016.

P3-7

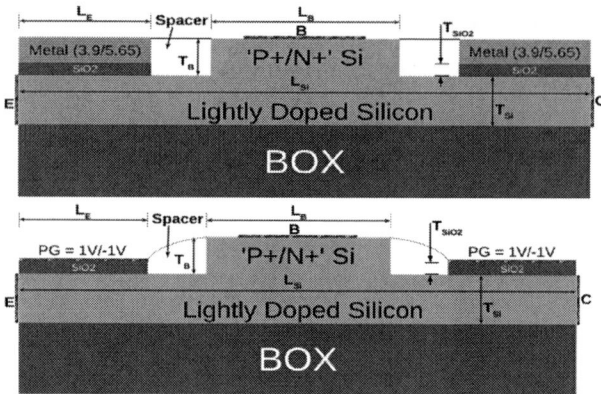

Fig. 1. Cross-sectional views of NPN/PNP lateral symmetrical bipolar junction transistors based on CP (top), and PC (bottom) approaches.

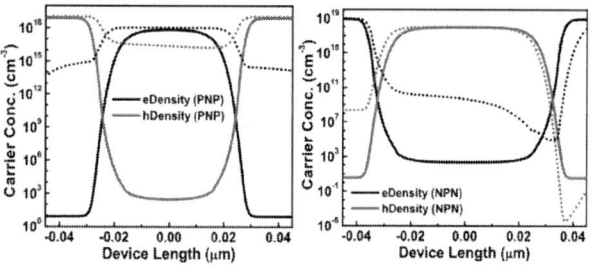

Fig. 2. Induced carrier concentration profile below 2 nm below the CP metal contact under cut-off region (solid lines) and in saturation region (dotted lines) for PNP and NPN transistors

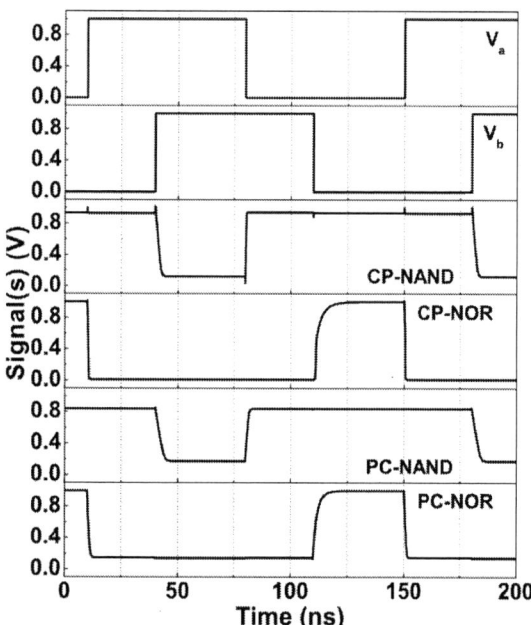

Fig. 3. Voltage-time characteristics for NAND and NOR gates based on CP and PC based devices

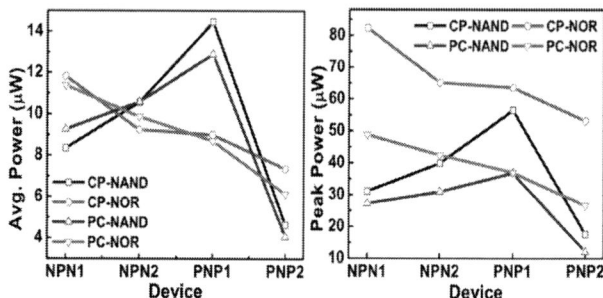

Fig. 4. Average and Peak power consumed by each device for all configuration

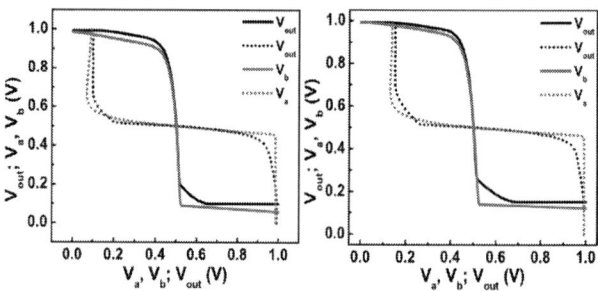

Fig. 4. Butterfly curves for NAND gates based on CP (left) and PC (right) approaches.

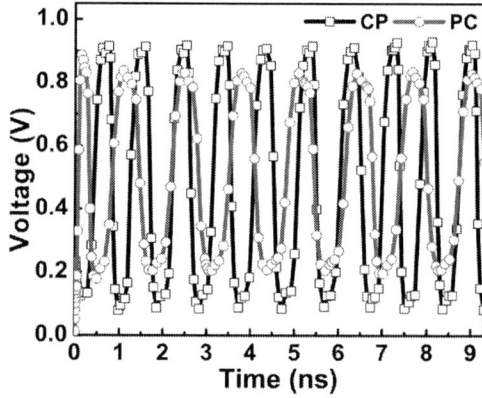

Fig. 5. Characteristics of 3-stage ring oscillators with CP and PC based devices

TABLE I
PARAMETER OBTAINED FROM TRANSIENT ANALYSIS

Doping Approach – Gate	Rise Time (ns)	Fall Time (ns)	Peak Power (μW)	Avg. Switching Power (μW)
CP-NAND	0.50	2.98	56.50	9.64
CP-NOR	4.80	0.83	83.10	9.38
PC-NAND	1.60	3.59	36.90	9.20
PC-NOR	4.00	0.75	48.80	9.09

2020 IEEE Silicon Nanoelectronics Workshop

978-1-7281-9736-4/20 $31.00 © 2020 IEEE

Towards a Chip-Scale Millimeter-Wave Spectrum/Signal Analyzer Using Spin-Wave Diffraction and Interference

H. Aquino[1], D. Connelly[1], A. Papp[2], A. Orlov[1], J. Chisum[1], G. H. Bernstein[1] and W. Porod[1]

[1]Department of Electrical Engineering, University of Notre Dame, Notre Dame, IN, USA
[2]Pazmany Peter Catholic University, Budapest, Hungary
Email: haquino@nd.edu

Abstract — **A new class of device has been proposed that converts millimeter electrical signals into spin waves with micrometer wavelengths, with which computing can be done in a small footprint. The spin waves are then converted back into electrical signals. Essential for this is the design of proper transducers between the two domains. We have simulated an electrical-to-spin-wave transducer that shows improved bandwidth. We are developing a numerical simulation design tool that crosses between these domains.**

Keywords: Magnonics, Microwave magnetics

I. INTRODUCTION

In the presence of an RF small-signal magnetic field perpendicular to a DC magnetic field, the magnetic moments of a magnetic film will precess. This precession propagates with a periodic spatial phase shift and is called a spin wave. This has been demonstrated in magnetic materials deposited on silicon substrates [1], allowing for the possibility of integrating spin-wave devices with CMOS.

One direction for using spin waves is to convert millimeter or microwave electrical signals to spin waves having micrometer wavelengths. All signal processing is then done by the interference of spin waves traveling through a magnetic thin film. These waves are then converted back into electrical signals [2]. Our goal is to demonstrate the feasibility of this new device type by building a chip-scale, real-time spectrum analyzer [2].

II. CPW SPIN-WAVE TRANSDUCER AND SIMULATION TOOL

Coplanar waveguides (CPW) placed on top of a magnetic film can be used to convert electrical signals into spin waves [3] through the coupling of the current-induced magnetic fields. However, because of the interference between the waves, the spin-wave excitation strength from the CPW exhibits a comb-like frequency response, making it unsuitable for the device described above.

We performed simulations using Ansys HFSS [4] and Mumax3 [5] that show that launching high-amplitude spin waves in an out-of-plane bias field launches waves with long wavelengths, which reduces the destructive interference between simultaneously launched spin waves but introduces additional unwanted structure in the wave. When this is followed by a gap in the magnetic material (Fig. 1), both the amplitude and wavelength of the spin wave are decreased, and the unwanted structure is removed. This produces a more uniform frequency response, as shown in Fig. 2. Simulation results of a null are shown in Fig. 3.

Essential to the design of this new class of device is a tool that links the micromagnetic and electromagnetic domains. The flow chart of our tool is shown in Fig. 4 [6]. The tool makes use of Ansys HFSS [4] to simulate the electromagnetics of the transducer. The results of this first simulation are then fed to OOMMF [7] where micromagnetic simulations are performed. Steps 1 and 2 have been validated with measurement (Fig. 5). Steps 3-5 will solve for the effect of the spin waves on the waveguide, and is currently being pursued.

III. SUMMARY

We have taken steps towards building a chip-scale, real-time spectrum analyzer using the diffraction and interference of spin waves. We found that launching large amplitude spin waves followed by a gap in the magnetic material produces a spin-wave launcher with a more uniform response and reaches higher frequencies than using the CPW under normal conditions.

A numerical simulation design tool that connects between electromagnetic and micromagnetic domains is currently being worked on. The first half of the tool has been verified with experimental data.

ACKNOWLEDGMENTS

This work was supported by the National Science Foundation (NSF) through the Spectrum Efficiency, Energy Efficiency, and Security (SpecEES) program.

REFERENCES

[1] V. Vlaminck and M. Bailleul, "Spin-wave transduction at the submicrometer scale: experiment and modeling," *Physical Review B*, vol. 81, no. 1, p. 014425, 2010.

[2] Á. Papp, W. Porod, Á. Csurgay and G. Csaba, "Nanoscale spectrum analyzer based on spin-wave interference," *Scientific Reports*, vol. 7, no. 1, 2017.

[3] S. Maendl, I. Stasinopoulos and D. Grundler, "Spin waves with large decay length and few 100 nm wavelengths in thin yttrium iron garnet grown at the wafer scale," *Applied Physics Letters*, vol. 111, no. 1, p. 012403, 2017.

[4] "ANSYS HFSS: high frequency electromagnetic field simulation software," Ansys.com, 2020. [Online]. https://www.ansys.com/products/electronics/ansys-hfss. [Accessed: 19- Apr- 2020].

[5] A. Vansteenkiste, J. Leliaert, M. Dvornik, M. Helsen, F. Garcia-Sanchez and B. Van Waeyenberge, "The design and verification of MuMax3," *AIP Advances*, vol. 4, no. 10, p. 107133, 2014.

[6] D. Connelly, G. Csaba, G.H. Bernstein, A. Orlov, J. Chisum and W. Porod, "Towards a simulation framework for coupled microwave and micromagnetic structures." *International Workshop on Computational Nanotechnology*, May 2019.

[7] M.J. Donahue, D. G. Porter, "OOMMF User's Guide, Version 1.0," *National Institute of Standards and Technology*, NISTIR 6376, 1999.

P3-8

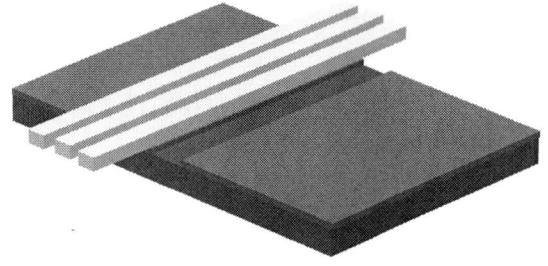

Fig. 1. Schematic of a CPW placed on top of a magnetic film with a gap in the film.

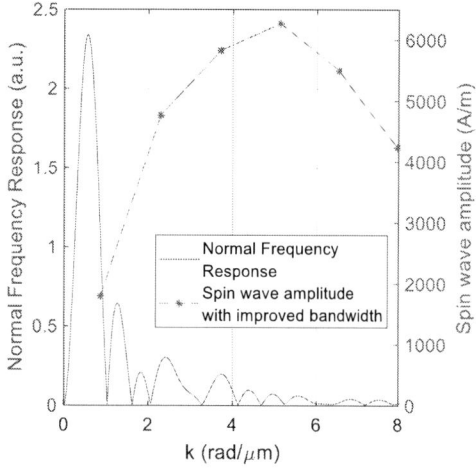

Fig. 2. The relative spin wave intensity as a function of wavenumber when launched from a CPW under the condition of no gap in the magnetic film. The plot also shows the spin wave amplitude from simulations of a CPW launching high-amplitude spin-waves with a gap in the film.

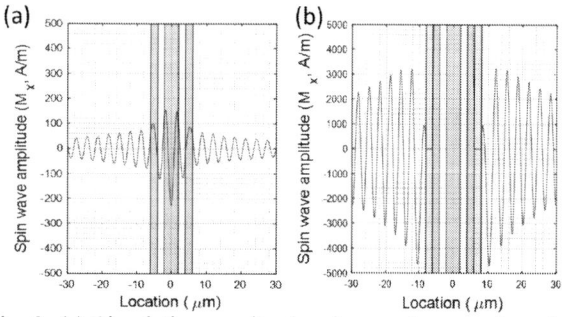

Fig. 3. (a) Simulation results showing a spin wave launched from a CPW with no in the film. The wavenumber used (k = 2 μm⁻¹) is a null in the response of the CPW. (b) Simulation results for the same wavenumber when launching high-amplitude, long-wavelength spin-waves which are then decreased in amplitude and wavelength by a gap in the magnetic film.

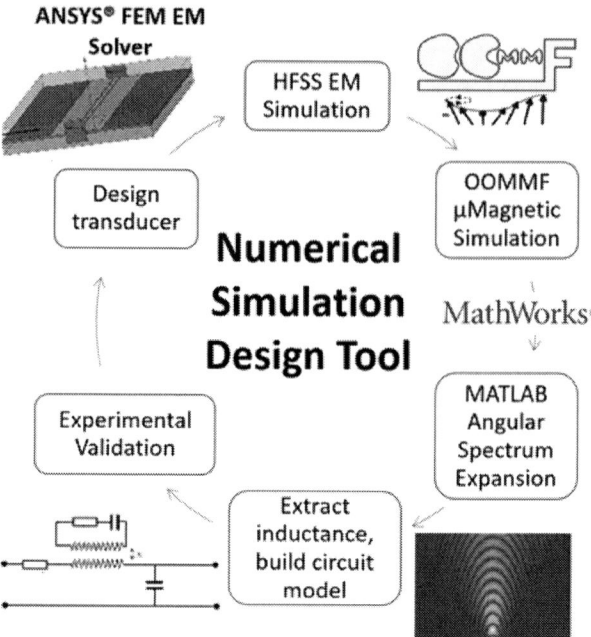

Fig. 4. Flow chart of the numerical simulation design tool that links the electromagnetic and micromagnetic domains. [6]

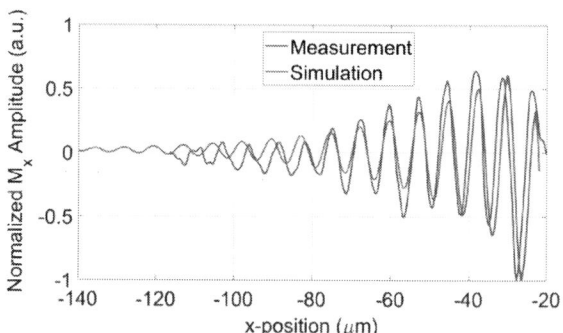

Fig. 5. Comparison of experimental data showing a spin wave launched from a CPW and comparison with simulation with ND's simulation tool.

2020 IEEE Silicon Nanoelectronics Workshop

AUTHOR INDEX

Ahn, Min-Ju 51, 55
Aihara, Hiroki 43, 47
Amano, Ikuma 37
Ang, Jia M. 79, 87, 91
Ansari, M. H. R. 25
Aquino, H. 127
Arita, Masashi 37
Asai, Yuki 37
Asselberghs, I. 67
Baghini, Maryam S. 121
Basu, Ananjan 117
Bernstein, G. H. 113, 127
Bigoni, Stefano 123
Bolshakov, P. 73
Borders, William A. 21
Brems, S. 67
Byun, Beommo 37
Campbell, Jason P. 7
Camsari, Kerem Y. 21
Cao, Rui 41, 45
Carabasse, C. 5
Castriotta, Michele 89
Celebrano, Michele 123
Celis-Cordova, Rene 115
Chakrabarti, Somsubhra 79, 87, 91
Chang, Edward Y. 117
Chang, Shu-Tong 85, 107
Chang, You-Tai 57
Chao, Tien-Sheng 53
Chee, Mun Y. 79, 87, 91
Chen, J. Z. 29, 39
Chen, Jiezhi 27, 41, 45, 103
Chen, Kuan-Ting 85, 107
Chen, Wei-Yen 53
Cheng, C. H. 23
Cheung, Kin P. 7
Chisum, J. 127
Cho, Seongjae 25
Cho, Tzuting 101
Chow, Samuel C. W. 91
Chung, Chun-Chih 53
Chung, Steve S. 23
Chung, Wonil 15
Chung, Yun-Fang 85, 107
Coignus, J. 5
Connelly, D. 127
Cott, D. 67
Dan, Yaping 119

Dananjaya, Putu A. 79, 87, 91
Datta, Supriyo 21
De Marneffe, J.-F. 67
Debnath, A. 35, 111
Devriendt, K. 67
Ding, Xiangxiang 75, 83
Dong, Yulong 9
Dupuy, E. 67
Ee, Yong C. 79, 87, 91
Feng, Y. 29, 39
Feng, Yang 27
Feng, Yulin 75, 83
Ferrari, Giorgio 89, 123
Finazzi, Marco 123
Francois, T. 5
Fujimoto, Takafumi 123
Fukami, Shunsuke 21
Gao, B. 29, 39
Gaur, A. 67
Gi, Keinan 123
Goel, Mayank 121
Gong, Xiao 59
Grenouillet, L. 5
Groven, B. 67
Gupta, Ankur 117
Gyakushi, Takayuki 37
Han, Kaizhen 59
Han, R. Z. 29, 39
Hara, Shusuke 63
Hasan, M. 35
He, Jiajing 119
Hiramoto, Toshiro 33, 51, 55
Hoffmann, Michael 1
Hou, Kunqi 79, 87, 91
Hsieh, E. R. 23
Huang, Chulin 119
Huang, P. 29, 39, 75
Huang, Yu-En 53
Hur, Jae 11
Hwang, Sungmin 81
Inokawa, Hiroshi 63, 65
Itoh, Kohei M. 3
Jia, Menghua 45
Joshi, Sanjog 121
Jun, Desmond L. J. 79, 87
Jupalli, T. T. 35, 111
Kang, J. 29, 39, 75
Kang, Won-Mook 77

Kansal, Harshit	105
Kao, Kuo-Hsing	13
Kaushik, Pragyey K.	117
Khan, Asif I.	11
Kim, Jangsaeng	77
Kim, Sihyun	95, 99
Kinzel, Edward C.	113
Kobayashi, Masaharu	33, 51, 55
Kodera, Tetsuo	31
Krishnaraja, Abinaya	17
Kumar, Abhishek	125
Kumar, P. J.	35
Lee, Jae Y.	25
Lee, Jong-Ho	77, 81
Lee, Kitae	95, 99
Lee, Soochang	77
Lee, Yao-Jen	13
Lei, Dian	59
Lew, Wen S.	79, 87, 91
Li, Pei-Wen	57
Li, Ruiyi	75, 83
Li, W.	67
Li, Xiuyan	9
Li, Yuanpeng	45
Li, Yuan	103
Liang, Gengchiau	71
Liang, Renrong	101
Lim, Gerard J.	91
Lin, D.	67
Lin, Horng-Chih	57
Liu, Fayong	69
Liu, Jingquan	9
Liu, L.	29, 39, 75, 83
Liu, X.	29, 39, 75
Lomenzo, Patrick	1
Loy, Desmond J. J.	91
Luo, Sheng	71
Luo, Yuan-Chun	11
Lyu, Xiao	7
Ma, Xiaolei	41
Ma, Yuechi	83
Maikap, Siddheswar	93
Manivannan, Revathi	65
Matsui, Chihiro	49
Maurice, T.	67
Medury, Aditya S.	105
Mikolajick, Thomas	1
Mimura, H.	111
Min, Kyung K.	81
Mizushina, Keita	43, 47
Mizuta, Hiroshi	69
Mizutani, Tomoko	33
Moraru, D.	35, 109, 111
Morin, P.	67
Morita, Yukinori	69
Mulaosmanovic, Halid	1
Muruganathan, Manoharan	69
Nagarajan, Anitharaj	63
Naik, Tejas	121
Neo, Y.	111
Nowak, E.	5
Ogawa, Shinichi	69
Ohno, Hideo	21
Orlov, A.	113, 115, 127
Panchanathan, Aruna P.	63
Pandy, C.	111
Papp, A.	127
Park, Byung-Gook	25, 77, 81, 95, 99
Peng, Kang-Ping	57
Pervaiz, Ahmed Z.	21
Phommahaxay, A.	67
Porod, W.	113, 127
Prabhudesai, G.	35, 109, 111
Prati, Enrico	89, 123
Quevedo-Lopez, M.	73
Radisic, D.	67
Radu, I. P.	67
Ramakrishnan, V. N.	111
Rao, V. R.	121
Richter, C.	5
Rodder, Mark	15
Rodriguez-Davila, R. A.	73
Saha, Sumit	121
Sahu, Abhishek	125
Sang, Pengpeng	103
Saraya, Takuya	33, 51, 55
Satoh, Hiroaki	63, 65
Schmidt, Marek	69
Schram, T.	67
Schroeder, U.	1, 5
Senapati, Asim	93
Shen, Chuan-Hui	53
Shen, Wensheng	75
Shinada, Takahiro	123
Shrestha, Pragya R.	7
Si, Mengwei	7
Singh, Sankalp K.	117
Slesazeck, Stefan	1
Smets, Q.	67
Snider, Gregory L	115
Su, Chun-Jung	13, 57
Sung, Po-Jung	13
Suzuki, Shun	43, 47, 49
Svensson, Johannes	17

Szakmany, Gergo P. 113
Tabe, M. 35, 109
Takahashi, Yasuo 37
Takeuchi, Ken 43, 47, 49
Takeuchi, Kiyoshi 33
Tan, Kuan H. 91
Tanii, Takashi 123
Tasneem, Nujhat 11
Thiam, A. 67
Thong, Jia R. 79, 87, 91
Tiwari, Shree P. 125
Toh, Eng H. 91
Triozon, F. 5
Tsurumaki-Fukuchi, Atsushi 37
Vaxelaire, N. 5
Wai, Samuel C. C. 87
Wang, Fei 41
Wang, Panni 11
Wang, Qianwen 45, 103
Wang, Wei-E 15
Wang, Xiangyu 75
Wang, Zehao 83
Wernersson, Lars-Erik 17
Wu, Ruei-Jen 57
Wu, Wen-Fa 13
Wu, Ying 59
Xiang, Y. C. 29, 39
Xu, Jun 101
Xu, Shengqiang 59
Yamaguchi, K. 109, 111
Yamaguchi, Shin 43
Yang, Cheng-Hsien 85, 107
Yang, H. Z. 29, 39
Yang, W. Y. 23
Ye, Peide D. 7, 15
Yeh, Wen-Kuan 13
Young, C. D. 73
Yu, Ao 83
Yu, Guofang 101
Yu, Shimeng 11
Zhan, Xuepeng 27, 41, 45
Zhang, Xiaoyi 71
Zheng, Dongqi 15
Zhou, Zheng 75